147
Advances in Polymer Science

Editorial Board:
A. Abe · A.-C. Albertsson · H.-J. Cantow · K. Dušek
S. Edwards · H. Höcker · J. F. Joanny · H.-H. Kausch
T. Kobayashi · K.-S. Lee · J. E. McGrath
L. Monnerie · S. I. Stupp · U. W. Suter
E. L. Thomas · G. Wegner · R. J. Young

Springer

*Berlin
Heidelberg
New York
Barcelona
Hong Kong
London
Milan
Paris
Singapore
Tokyo*

Macromolecular Architectures

Volume Editor: J. G. Hilborn

With contributions by
P. Dubois, C. J. Hawker, J. L. Hedrick,
J. G. Hilborn, R. Jérôme, J. Kiefer,
J. W. Labadie, D. Mecerreyes, W. Volksen

 Springer

Chemistry Library

This series presents critical reviews of the present and future trends in polymer and biopolymer science including chemistry, physical chemistry, physics and materials science. It is addressed to all scientists at universities and in industry who wish to keep abreast of advances in the topics covered.

As a rule, contributions are specially commissioned. The editors and publishers will, however, always be pleased to receive suggestions and supplementary information. Papers are accepted for „Advances in Polymer Science" in English.

In references Advances in Polymer Science is abbreviated Adv. Polym. Sci. and is cited as a journal.

Springer WWW home page: http://www.springer.de

ISSN 0065-3195
ISBN 3-540-65576-X
Springer-Verlag Berlin Heidelberg New York

Library of Congress Catalog Card Number 61642

This work is subject to copyright. All rights are reserved, whether the whole or part of the material is concerned, specifically the rights of translation, reprinting, re-use of illustrations, recitation, broadcasting, reproduction on microfilms or in other ways, and storage in data banks. Duplication of this publication or parts thereof is only permitted under the provisions of the German Copyright Law of September 9, 1965, in its current version, and permission for use must always be obtained from Springer-Verlag. Violations are liable for prosecution under the German Copyright Law.

© Springer-Verlag Berlin Heidelberg 1999
Printed in Germany

The use of registered names, trademarks, etc. in this publication does not imply, even in the absence of a specific statement, that such names are exempt from the relevant protective laws and regulations and therefore free for general use.

Typesetting: Data conversion by MEDIO, Berlin
Cover: E. Kirchner, Heidelberg
SPIN: 10691447   02/3020 - 5 4 3 2 1 0 - Printed on acid-free paper

Volume Editor

Prof. Jöns G. Hilborn
Départment des Matériaux
Laboratoire de Polymères
MX-C Ecublens
CH-1015 Lausanne
Switzerland
E-mail: joens.hilborn@epfl.ch

Editorial Board

Prof. Akihiro Abe
Department of Industrial Chemistry
Tokyo Institute of Polytechnics
1583 Iiyama, Atsugi-shi 243-02, Japan
E-mail: aabe@chem.t-kougei.ac.jp

Prof. Ann-Christine Albertsson
Department of Polymer Technology
The Royal Institute of Technolgy
S-10044 Stockholm, Sweden
E-mail: aila@polymer.kth.se

Prof. Hans-Joachim Cantow
Freiburger Materialforschungszentrum
Stefan Meier-Str. 21
D-79104 Freiburg i. Br., FRG
E-mail: cantow@fmf.uni-freiburg.de

Prof. Karel Dušek
Institute of Macromolecular Chemistry, Czech
Academy of Sciences of the Czech Republic
Heyrovský Sq. 2
16206 Prague 6, Czech Republic
E-mail: office@imc.cas.cz

Prof. Sam Edwards
Department of Physics
Cavendish Laboratory
University of Cambridge
Madingley Road
Cambridge CB3 OHE, UK
E-mail: sfe11@phy.cam.ac.uk

Prof. Hartwig Höcker
Lehrstuhl für Textilchemie
und Makromolekulare Chemie
RWTH Aachen
Veltmanplatz 8
D-52062 Aachen, FRG
E-mail: 100732.1557@compuserve.com

Prof. Jean-François Joanny
Institute Charles Sadron
6, rue Boussingault
F-67083 Strasbourg Cedex, France
E-mail: joanny@europe.u-strasbg.fr

Prof. Hans-Henning Kausch
Laboratoire de Polymères
École Polytechnique Fédérale
de Lausanne, MX-D Ecublens
CH-1015 Lausanne, Switzerland
E-mail: hans-henning.kausch@epfl.ch

Prof. Takashi Kobayashi
Institute for Chemical Research
Kyoto University
Uji, Kyoto 611, Japan
E-mail: kobayash@eels.kuicr.kyoto-u.ac.jp

Prof. Kwang-Sup Lee
Department of Macromolecular Science
Hannam University
Teajon 300-791, Korea
E-mail: kslee@eve.hannam.ac.kr

Prof. James E. McGrath
Polymer Materials and Interfaces Laboratories
Virginia Polytechnic and State University
2111 Hahn Hall
Blacksbourg
Virginia 24061-0344, USA
E-mail: jmcgrath@chemserver.chem.vt.edu

Prof. Lucien Monnerie
École Supérieure de Physique et de Chimie
Industrielles
Laboratoire de Physico-Chimie
Structurale et Macromoléculaire
10, rue Vauquelin
75231 Paris Cedex 05, France
E-mail: lucien.monnerie@espci.fr

Prof. Samuel I. Stupp
Department of Materials Science
and Engineering
University of Illinois at Urbana-Champaign
1304 West Green Street
Urbana, IL 61801, USA
E-mail: s-stupp@uiuc.edu

Prof. Ulrich W. Suter
Department of Materials
Institute of Polymers
ETZ, CNB E92
CH-8092 Zürich, Switzerland
E-mail: suter@ifp.mat.ethz.ch

Prof. Edwin L. Thomas
Room 13-5094
Materials Science and Engineering
Massachusetts Institute of Technology
Cambridge, MA 02139, USA
E-mail. thomas@uzi.mit.edu

Prof. Gerhard Wegner
Max-Planck-Institut für Polymerforschung
Ackermannweg 10
Postfach 3148
D-55128 Mainz, FRG
E-mail: wegner@mpip-mainz.mpg.de

Prof. Robert J. Young
Manchester Materials Science Centre
University of Manchester and UMIST
Grosvenor Street
Manchester M1 7HS, UK
E-mail: robert.young@umist.ac.uk

Preface

Thanks to recent advances in the chemistry of preparing polymers, an increasing number of tools are at our disposal for the design of polymer materials. The design level ranges form monomer synthesis, controlled stepwise or chainwise polymerization, block copolymer synthesis, over branching to crosslinking reactions. Depending on the structure of the individual polymer chains formed these will be organized in the bulk to give specific properties. Hence, this gives us two architectural levels: *The structure of invidual macromolecules and the microstructure of the material produced.* While both of these organization levels may contribute to the design of materials properties we would ultimately like to be able to tailor our material to suit desired applications in which surface properties, mechanical or thermal behavior, processability, optical or electrical characeristics etc. are crucial. The next decades should see an enormous advance in nanoscopic and supramolecular chemistry leading to novel predetermined properties. Molecular manipulation of nano and microstructures paves the way to organic polymer materials by design. Such architectures comprise both the synthesis and the kinetic and thermodynamics of macromolecular organization and is the theme of this volume.

The book consists of four articles reviewing the literature based on the authors own experiences over the last decade in this field. It does not claim to be exhaustive nor to provide complete coverage of the very extensive literature in this field. Instead, it focuses on the currently intense areas of research namely living polymerization, block copolymer synthesis, synthesis of dendrimers and finally macroporous thermosets. Hopefully, this volume will not only serve as a book on the design of macromolecular architectures but also as a source of inspiration to produce polymers combining several functional properties.

In the first chapter by P. Dubois and D. Mecerreyes, living polymerization to produce precisely defined linear polyesters is outlined and also compared to other living polymerization techniques. In chapter two, C. Hawker describes the synthesis of polymeric dendrimers which are organic globular-like nanoscopic entities of exact molecular mass and functionality synthesized either by the convergent or divergent approach. How block copolymers are produced to define micromorphology in high performance polymers and thereby tailoring their thermal, chemical, mechanical and dielectrical properties is the content of chapter three by J. Hedrick. The book concludes with a fourth chapter by J. Kiefer on the importance of kinetic and thermodynamics for microstructural organization in thermosets.

The editor would also like to acknowledge the valuable input from Professor Stanislaw Penczek, Professor Bernard Sillion, Professor Anders Hult, and Professor Vipin Kumar who served as referees for the above contributions.

Lausanne,
November 1998

Jöns G. Hilborn

Contents

Novel Macromolecular Architectures Based on Aliphatic Polyesters: Relevance of the "Coordination-Insertion" Ring-Opening Polymerization
D. Mecerreyes, R. Jérôme, P. Dubois 1

Nanoscopically Engineered Polyimides
J. L. Hedrick, J. W. Labadie, W. Volksen, J.G. Hilborn 61

Dendritic and Hyperbranched Macromolecules – Precisely Controlled Macromolecular Architectures
C. J. Hawker 113

Macroporous Thermosets by Chemically Induced Phase Separation
J. Kiefer, J. L. Hedrick, J. G. Hilborn 161

Author Index Volumes 101–147 249

Subject Index 261

Novel Macromolecular Architectures Based on Aliphatic Polyesters: Relevance of the "Coordination-Insertion" Ring-Opening Polymerization

David Mecerreyes, Robert Jérôme, Philippe Dubois*

Center for Education and Research on Macromolecules (CERM), University of Liège, Sart-Tilman, B6, B-4000 Liège, Belgium
*Laboratory of Polymeric and Composite Materials, University of Mons-Hainaut, Place du Parc 20, B-7000 Mons, Belgium
E-mail: Philippe.dubois@umh.ac.be

Recent developments in the macromolecular engineering of aliphatic polyesters have been overviewed. First, aluminum alkoxides mediated living ring opening polymerization (ROP) of cyclic (di)esters, i.e., lactones, lactides, glycolide, is introduced. An insight into this so-called "coordination-insertion" mechanism and the ability of this living polymerization process to prepare well-defined homopolymers, telechelic polymers, random and block copolymers is then discussed. In the second part, the combination of the living ROP of (di)lactones with other well-controlled polymerization mechanisms such as anionic, cationic, free radical, and metathesis polyadditions of unsaturated comonomers, as well as polycondensations, is reported with special emphasis on the design of new and well-tailored macromolecular architectures. As a result of the above synthetic breakthrough, a variety of novel materials have been developed with versatile applications in very different fields such as biomedical and microelectronics.

Keywords. Lactones, Lactides, Aliphatic polyesters, Ring opening polymerization, Living polymerization, Macromolecular engineering

List of Symbols and Abbreviations		2
1	Introduction	3
2	Aluminum Alkoxides Mediated Ring Opening Polymerization of Lactones and Lactides	6
2.1	Homopolymerization of Cyclic (Di)esters as Initiated by Al(OiPr)$_3$	8
2.2	Random and Block Copolymerization	10
2.3	Selective End-Functionalization	13
2.4	Kinetic Aspects of the "Coordination-Insertion" ROP	16
2.5	Synthesis and (Co)polymerization of Functional Cyclic Ester Monomers	19

3	Synthesis of Block and Graft Copolymers by Combination of (Di)lactones Ring Opening Polymerization with Other Living/Controlled Polymerization Processes	21
3.1	Ring Opening Polymerization	22
3.2	Anionic Polymerization	27
3.3	Cationic Polymerization	30
3.4	Radical Polymerization	32
3.5	Ring Opening Metathesis Polymerization	36
3.6	Polycondensation	39
3.7	Dendritic Construction	41
3.8	Coordination Polymerization	44
4	**Aliphatic Polyesters as Building Blocks for New Materials**	45
4.1	Biodegradable and Biocompatible Thermoplastic Elastomers	45
4.2	Polyimide Nanofoams	47
4.3	Organic-Inorganic Nanocomposites	50
4.4	Biodegradable Amphiphilic Networks	52
4.5	Nano- and Microspheres for Biomedical Applications	54
5	**Conclusions**	55
References		56

List of Symbols and Abbreviations

ATRP	atom transfer radical polymerization
BD	butadiene
DM(T)A	dynamic mechanical (thermo)analysis
DMAP	dimethylaminopyridine
DMSO	dimethyl sulfoxide
DSC	differential scanning calorimetry
EA	ethyl acrylate
MA	methyl acrylate
MMA	methyl methacrylate
MWD	molecular weight distribution
NMR	nuclear magnetic resonance
PCEVE	poly(chloro vinyl ether)
PCL	poly(ε-caprolactone)
PCS	photon correlation spectroscopy
PDI	polydispersity index
PEO	poly(ethylene oxide)
PLA	polylactide, including (D,L) and P(L)LA
PMCP	poly(methylene-1,3-cyclopentane)

PNB	polynorbornene
PS	poly(styrene)
ROP	ring opening polymerization
ROMP	ring opening metathesis polymerization
RT	room temperature
SAXS	small angle X-ray scattering
TEM	transmission electron microscopy
TEOS	tetraethoxysilane
THF	tetrahydrofuran
T_g	glass transition temperature
TGA	thermal gravimetric analysis
T_m	melting temperature
TMC	trimethylene carbonate
TMEDA	tetramethylethylenediamine
UV	ultraviolet
VP	vinyl pyrrolidone

1
Introduction

Biodegradable polymers have attracted widespread attention during the last few years [1]. This important research effort has been driven by the need for specific single-use materials in the biomedical field and by the search for biodegradable substitutes of conventional commodity thermoplastics, in answer to the increasing discarded plastic waste in landfills. Among the various families of biodegradable polymers, aliphatic polyesters have a leading position since hydrolytic and/or enzymatic chain cleavage yields ω-hydroxyacids which in most cases are ultimately metabolized. As will be discussed later, aliphatic polyesters can be prepared from a large variety of starting (natural) materials and synthetic routes. By a judicious choice of the repetitive ester unit(s), one can play at will with the material properties such as crystallinity, glass transition temperature, toughness, stiffness, adhesion, permeability, degradability, etc.

Polyesters are currently synthesized by a step-growth process, i.e., a polycondensation, from a mixture of a diol and a diacid (or a diacid derivative), or from a hydroxy-acid when available. Ring opening polymerization (ROP) of cyclic esters and related compounds is an alternative method for the synthesis of aliphatic polyesters. Comparison of these two mechanisms is clearly in favor of the polyaddition process [2]. Molecular weight of the polycondensates is usually limited to a few tens of thousands (M_n<30,000), and the only way to control it in this limited range of chain length is the use of terminating (monofunctional) agents. Even though conversion of the hydroxyl and acid groups is close to completion, any departure from the reaction stoichiometry has a very detrimental effect on the chain length. Furthermore, polycondensation of ω-hydroxy acids leads to the formation of side-reaction by-products, and it requires long reaction

Table 1. Monomer structures and polymer melting point and glass transition temperatures of the most common aliphatic polyesters obtained by ROP [2, 7]

Monomer	Polymer	T_g (°C)	T_m (°C)
(lactone ring structure with R)	Polylactone Poly(ω-hydroxy acid)		
R=-(CH$_2$)$_2$-βPL, β-propiolactone	PβPL	-24	93
R=-(CH$_2$)$_3$-γBL, γ-butyrolactone	PγBL	-59	65
R=-(CH$_2$)$_4$-δVL, δ-valerolactone	PδVL	-63	60
R=-(CH$_2$)$_5$-εCL, ε-caprolactone	PεCL	-60	65
R=-(CH$_2$)$_2$-O-(CH$_2$)$_2$-DXO, 1,5-dioxepan-2-one	PDXO	-36	-
R=-(CH$_2$-CH(CH$_3$))-βBL, β-butyrolactone	PβBLisotactic[a] PβBL atactic	5 -2	180 -
R=-(C(CH$_3$)$_2$-CH$_2$)-PVL, pivalolactone	PPVL	-10	245
(dilactone ring structure with R$_1$, R$_2$, R$_3$, R$_4$)	Polydilactone Poly(α-hydroxy acid)		
R$_1$=R$_2$=R$_3$=R$_4$=H GA, glycolide	PGA	34	225
R$_1$=R$_4$=CH$_3$, R$_2$=R$_3$=H L-LA, L-lactide	PLLA	55–60	170
R$_1$=R$_4$=H, R$_2$=R$_3$=CH$_3$ D-LA, D-lactide	PDLA	55–60	170
R$_1$=R$_3$=CH$_3$, R$_2$=R$_4$=H meso-LA, meso-lactide	PmesoLA	45–55	-
D-LA/L-LA (50–50) D,L-LA, (D,L) racemic lactide	PDLLA	45–55	-

[a] Also known as poly(3-hydroxybutyrate) [3]

times together with high temperatures. In contrast, ROP is usually free of these limitations. Under rather mild conditions, high molecular-weight aliphatic polyesters can be prepared in short periods of time. Table 1 presents the monomer structures, the related aliphatic polyesters as obtained by ROP, and their abbreviations. The thermal characteristic features, i.e., the glass transition and melting temperatures, are also reported. It is worth noting that in addition to the chemical methods, many bacteria synthesize, accumulate, and deposit in the cells aliphatic polyesters which are generally known as poly(hydroxy alkanoic acids) (PHA). The high stereoselectivity of the enzymatic synthesis produces as

Table 2. Different chemistries involved in the synthesis of aliphatic polyesters by ROP [2]

Mechanism	Initiator and/or catalyst
Cationic	Protonic acids: HCl, HBr, RCOOH, RSO$_3$H
	Lewis acids: AlCl$_3$, BF$_3$, FeCl$_2$, ZnCl$_2$
	Alkylating agents: CF$_3$SO$_3$CH$_3$, Et$_3$O$^+$ $^-$BF$_4$, (CH$_3$)$_2$I$^+$ $^-$SbF$_6$
	Acylating agents: CH$_3$C(O)$^+$ $^-$OCl$_4$
Anionic	Alkoxides: RO^{-+}M (M=alkali metal, complexed or not by crown ether)
	Carboxylates: RCOO$^-$ $^+$M (M=alkali metal)
	Alkali metal: naphthalenides
	Alkali metal supramolecular complexes
	Grafitides: KC$_{24}$
Free Radical	Peroxides (monomers: cyclic ketene acetals)
Via Active Hydrogen	Amines and alcohols
Zwitterionic	Tertiary amines and phosphines
Coordination	Alkoxides: ROM (M=metal with free p, d, or f orbitals of a favorable energy)
	Carboxylates: RCOOM (M=metal with free p, d, or f orbitals of a favorable energy)
	Metal oxides and halogenides (mainly of Sn and transition metals)
Enzymatic	Lipase

a rule polyesters with high crystallinity which have attracted a great deal of attention during the last few years [3].

The first attempts at ROP have been mainly based on anionic and cationic processes [4, 5]. In most cases, polyesters of low molecular weight were recovered and no control on the polymerization course was reported due to the occurrence of side intra- and intermolecular transesterification reactions responsible for a mixture of linear and cyclic molecules. In addition, aliphatic polyesters have been prepared by free radical, active hydrogen, zwitterionic, and coordination polymerization as summarized in Table 2. The mechanistic considerations of the above-mentioned processes are outside the scope of this work and have been extensively discussed in a recent review by some of us [2]. In addition, the enzyme-catalyzed ROP of (di)lactones in organic media has recently been reported; however, even though this new polymerization procedure appears very promising, no real control of the polyesters chains, or rather oligomers, has been observed so far [6].

Above all, the discovery that some organometallic compounds are effective in the synthesis of high molecular weight PCL [7] promoted a renewed interest in the ROP of lactones, particularly with alkyl metals, metal halides, oxides, carboxylates, and alkoxides. These metal compounds were first classified as anionic or cationic initiators [8]. Nevertheless, various studies have shown that most metal derivatives initiate the chain reaction through active covalent bonds [9]. Accordingly some authors classified those ROP as pseudoionic processes, which commonly involved coordination active species. Although this pseudoionic ROP allows the synthesis of polyesters of a high molecular weight, control of the polymerization is very difficult to achieve and is rather an exception. Actually depending on the structure of the organometallic derivatives, they can act either as catalyst, e.g., metal oxides, halides, and carboxylates, or as initiators, which is the case for metal alkoxides, the metal of which contains free p-, d-, or f- orbitals of a favorable energy (see next section). In the former case, the Lewis acid-type catalysts would not be chemically bonded to the growing chains, so that they can activate more than one chain. As a result, the average degree of polymerization is not directly controlled by the monomer-to-catalyst molar ratio. Moreover, transesterification side-reactions also perturb chain propagation which makes the molecular weight distribution broader (PDI~2). On the other hand, the "active covalent" bonds of some of the above metal alkoxides display a good compromise of reactivity so that an acceptable control for the lactones ROP could be achieved. Among them, aluminum alkoxides have proved to promote a ROP with a restricted occurrence of termination, transfer, and transesterification side-reactions, showing a high degree of livingness and an unequal versatility in the preparation of high molecular weight polyesters and novel macromolecular architectures [10].

The purpose of this review is to report on the recent developments in the macromolecular engineering of aliphatic polyesters. First, the possibilities offered by the living (co)polymerization of (di)lactones will be reviewed. The second part is devoted to the synthesis of block and graft copolymers, combining the living coordination ROP of (di)lactones with other living/controlled polymerization mechanisms of other cyclic and unsaturated comonomers. Finally, several examples of novel types of materials prepared by this macromolecular engineering will be presented.

2
Aluminum Alkoxides Mediated Ring Opening Polymerization of Lactones and Lactides

Two different mechanisms have been proposed for the ROP of (di)lactones depending on the nature of the organometallic derivatives. Metal halides, oxides, and carboxylates would act as Lewis acid catalysts in an ROP actually initiated with a hydroxyl-containing compound, such as water, alcohol, or ω-hydroxy acid; the later would result more likely from the "in-situ" hydrolysis of the (di)lactone [11]. Polymerization is assumed to proceed through an insertion mechanism, the details of which depends on the metal compound (Scheme 1a). The most frequently encountered Lewis acid catalyst is undoubtedly the stannous 2-ethylhexanoate, currently referred to as stannous octoate ($Sn(Oct)_2$). On the other hand, when metal alkoxides containing free *p*-, *d*-, or *f*- orbitals of a favo-

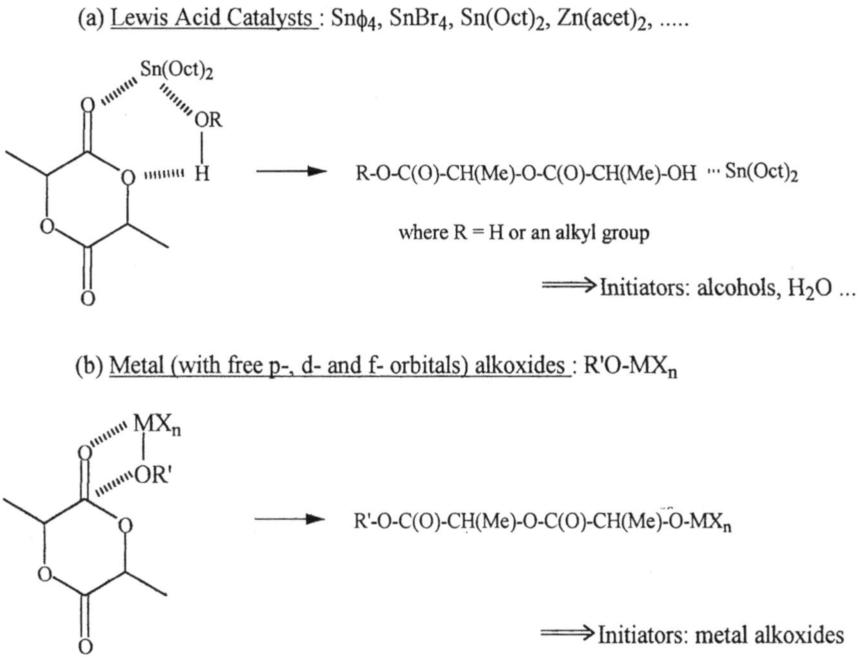

(a) <u>Lewis Acid Catalysts</u>: $Sn\phi_4$, $SnBr_4$, $Sn(Oct)_2$, $Zn(acet)_2$, ...

R-O-C(O)-CH(Me)-O-C(O)-CH(Me)-OH ··· $Sn(Oct)_2$

where R = H or an alkyl group

⟹ Initiators: alcohols, H_2O ...

(b) <u>Metal (with free p-, d- and f- orbitals) alkoxides</u>: $R'O-MX_n$

R'-O-C(O)-CH(Me)-O-C(O)-CH(Me)-O-MX_n

⟹ Initiators: metal alkoxides

Scheme 1. Currently proposed insertion mechanisms in ROP of (di)lactones (schematized here for lactide monomers)

rable energy (Mg-, Sn-, Ti-, Zr-, Fe-, Al-, Y-, Sm-, Zn-alkoxides) are used as initiators, a two-step "coordination-insertion" mechanism would prevail, which consists of the lactone complexation onto the propagating species, i.e., the growing metal alkoxide, followed by a rearrangement of covalent bonds leading to the cleavage of the metal-oxygen bond of the propagating species and the acyl-oxygen bond of the cyclic monomer (Scheme 1b) [2, 12].

Although some organometallic compounds can allow for the synthesis of polyesters of a high molecular weight, control of the ROP process usually remains a problem. As an example, in the case of Lewis acid catalyst, molecular weights are difficult to predict and the molecular weight distribution is broad. M_w/M_n is close to 2 as the result of the occurrence of side transesterification reactions. On the other hand, Kricheldorf has studied different metal alkoxides and he has reported that "active covalent" bonds of the investigated metal alkoxides were reactive enough to generate intramolecular transesterification reactions – also known as "back-biting" reactions – yielding cyclic oligomers as by-products. Within the limits of the studied initiators, the reactivity would be $Al(OiPr)_3 < Zn(OnPr)_2 < Ti(OnBu)_4 < Bu_3SnOMe < Bu_2Sn(OMe)_2$. In agreement with these observations, some of us, and more recently Inoue and Penczek, have reported on the living polymerization of ε-CL, as initiated by aluminum alkoxides species such as bimetallic (Zn, Al) μ-oxo alkoxides and aluminum triisopropoxide [13–15], tetraphenylporphinato aluminum alkoxides [16], and diethyl aluminum methoxide [17], respectively. Confirming the observed living character of the polymerization promoted by aluminum alkoxides, the extent of side reactions, mainly the "back-biting" ones, has been studied by Penczek et al., who have quantitatively described the degree of livingness by the so-called selectivity parameter (β). β is defined as the ratio of the rate constant of propagation to the rate constant of side reactions ($\beta=k_p/k_s$) and shows the highest value in ROP initiated by Al(OiPr)$_3$ [18]. Recent works have reported on the living polymerization of εCL and LA as initiated by lanthanide alkoxides. Although these initiators look very promising because of the very high polymerization rate, their selectivity appears to be much less than in the case of aluminum trialkoxides [19–21].

This section aims at reviewing recent advances in ROP of cyclic (di)esters initiated by aluminum alkoxides. The controlled synthesis of high molecular weight polymers with narrow molecular weight distribution obtained by ROP initiated by Al(OiPr)$_3$ will be discussed first. Next, the preparation of random and diblock copolymers, α- and α,ω-functional (co)polyesters, will be presented. Then the different factors which affect the polymerization kinetics will be analyzed. Finally, our attention will be devoted to the synthesis and subsequent polymerization of functional monomers, the main objective being the straightforward preparation of aliphatic polyesters bearing functional pendant groups all along the main backbone.

2.1
Homopolymerization of Cyclic (Di)esters as Initiated by Al(OiPr)$_3$

Aluminum isopropoxide has proved to be a very effective initiator for the polymerization of lactones: βPL, δVL, εCL, DXO, and βBL and dilactones: D,L- and L-LA, and GA. It is worth noting that cyclic carbonates, e.g., 2,2-dimethyltrimethylene carbonate (DTC) and cyclic anhydrides such as adipic anhydride (AA) have also been polymerized by aluminum trialkoxides with the unique possibility to control the molecular parameters of the polycarbonate (PDTC) and polyanhydrides (PAA). This is illustrated in Scheme 2 and the preferred reaction conditions are given in Table 3.

The ring-opening proceeds through a "coordination-insertion" mechanism which involves the insertion of the monomer into the "Al-O" bond of the initiator (Scheme 3). The acyl-oxygen bond of the cyclic monomer is cleaved in a way which maintains the growing chain attached to aluminum through an alkoxide bond. Polymerization is currently stopped by hydrolysis of the active aluminum alkoxide bond which leads to the formation of a hydroxyl end-group. Rather as an exception, the active growing species in polymerization of adipic anhydride is not an aluminum alkoxide, but an aluminum carboxylate, which leads to a carboxylic acid end-group after ultimate hydrolytic deactivation. The second chain extremity is capped with an ester carrying the isopropoxy radical of the initiator. Note that as far as the ROP of DTC is concerned, this second end-group turns out to be an isopropoxy carbonate function.

Table 3. Preferred reaction conditions for the ROP of different monomers as initiated by Al(OiPr)$_3$

Monomer	Solvent	Temperature °C	Reference
βPL	Toluene THF	40	[22]
δVL	Toluene THF	0–25	[22]
εCL	Toluene THF	0–25	[22–24]
βBL	Toluene	75	[25]
DXO	Toluene	25	[26]
GA[a]	Toluene	40	[27]
D,L-LA	Toluene	70	[28]
L,L-LA	Bulk	110–180	[29–31]
AAα	Toluene CH$_2$Cl$_2$	25	[32, 33]
TMC	Toluene	25	[34]

[a] Due to very limited solubility of PGA and PAA chains in toluene, it is advised to initiate the polymerization by living PCL oligomers, [iPrO-PCL-O]$_3$Al, as prepared by εCL ROP promoted by Al(OiPr)$_3$

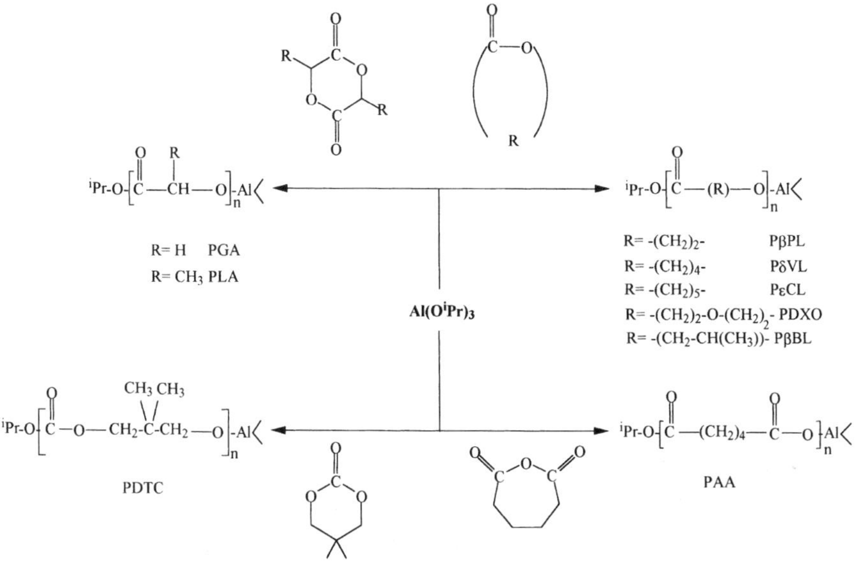

Scheme 2

Whatever previously listed monomer is considered, the "living" character of the polymerization has been confirmed by kinetic studies, where polymerization is first order in both monomer and initiator. Moreover, the molecular weight can be predicted by the monomer-to-initiator molar ratio, the molecular weight distributions are narrow (PDI=1.05–1.20), and the monomer resumption

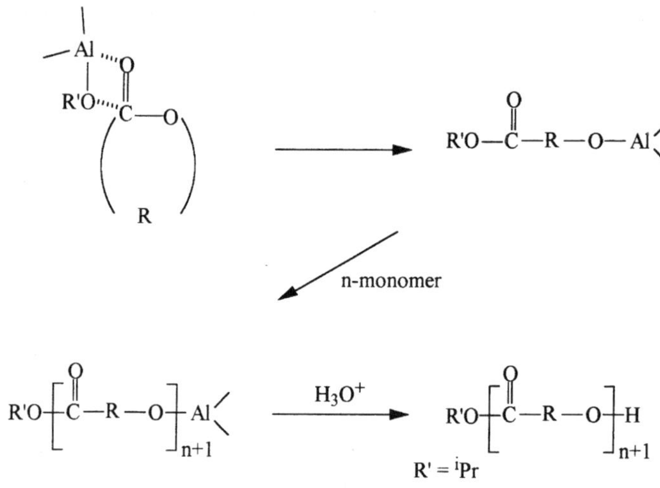

Scheme 3

experiments are quantitative. Worthy of particular mention is the "living" mechanism reported for the bulk polymerization of lactide in absence of any solvent [29]. Actually, P(L- and D,L-)LA chains of M_n higher than 100,000 can be synthesized in bulk at a temperature ranging from 110 to 180 °C. Again, the degree of polymerization is in close agreement with the starting feed monomer-to-Al(OiPr)$_3$ ratio. However, some limitations have been observed in ROP of βBL. At βBL monomer-to-initiator ratios higher than 150, a competition between propagation and (inter- and intra-molecular) transesterifications, and elimination reactions, i.e., proton abstraction with the concomitant formation of crotonic ester, takes place and the control over the PβBL molecular structure is lost definitively [25].

2.2
Random and Block Copolymerization

Generally, when two or more monomers of similar reactivity polymerize according to the same "living" mechanism, their simultaneous polymerization leads to random copolymers and their sequential polymerization to block copolymers. Accordingly, copolymerization of εCL with δVL [35], TOSUO [36], and γBL [37] initiated with Al(OiPr)$_3$ in toluene at 25 °C leads to highly statistical copolymers as confirmed by triad analysis by ^{13}C NMR spectroscopy. At the same time, the control over the molecular parameters and the narrow molecular weight distributions already observed in the respective homopolymerizations are maintained. Of particular interest is the case of γBL, the five-membered cyclic ester, usually classified as a lactone unable to be ring-opening polymerized under conventional experimental conditions. However, in the presence of fresh-

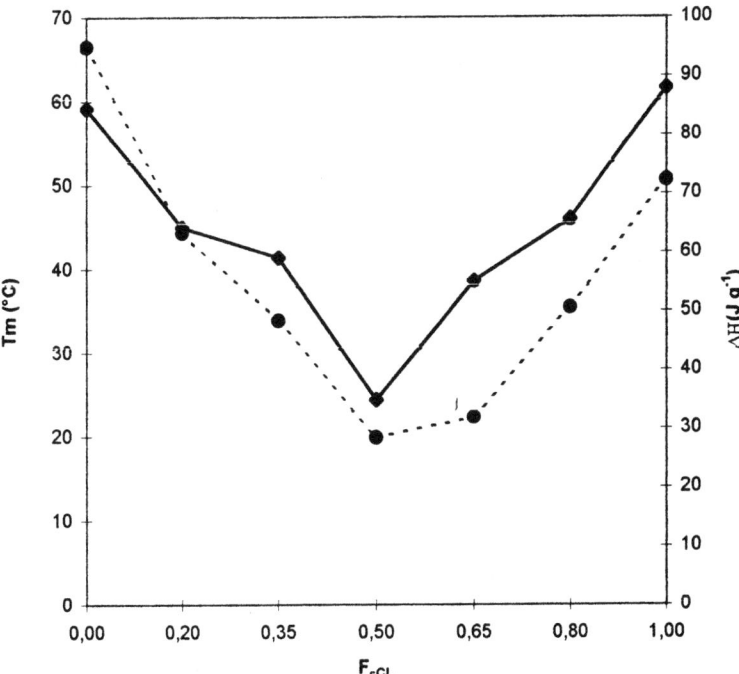

Fig. 1. Melting temperature (—,T_m) and enthalpy (--,ΔH) of poly(εCL-co-δVL) random copolymers at different compositions, as synthesized by "coordination-insertion" ROP initiated with Al(OiPr)$_3$ in toluene at 0 °C. $F_{εCL}$ is the molar fraction of εCL in the copolyester. (T_m and ΔH were determined by DSC at a heating rate of 10 °C/min)

ly distilled Al(OiPr)$_3$, i.e., predominantly aluminum isopropoxide associated in trimer known for being the most active species in ROP, γBL has proved to oligomerize up to a DP of ca. 10, in toluene at RT. Furthermore, γBL copolymerizes with εCL up to about 50 mol% of γ-oxybutyryl units content, with high and controlled molecular weights and narrow MWD [37].

Interestingly, the lactones copolymerization is responsible for a decrease in both T_m and degree of crystallinity of the copolyesters when compared to the parent homopolymers. This behavior is illustrated in Fig. 1 in the case of poly(εCL-co-δVL) random copolymers [35].

On the other hand, when εCL is copolymerized with dilactones such as GA [38] and (D- and D,L-)LA [39], tapered or pseudoblock copolymers are obtained with a reactivity ratio much in favor of the dilactone. As an example, the reactivity ratios in the copolymerization of εCL and D,L-LA in toluene at 70 °C are r_1= 0.92 (ε-CL) and r_2=26.5 (D,L-LA). Very similar reactivity ratios were calculated for copolymerization between εCL and L-LA, other experimental conditions being kept unchanged. However the control over the polymerization is lost due to transesterification side reactions perturbing the propagation step. Such a behavior is characterized by the broadening of molecular weight distributions (PDI~2), the formation of cyclic oligomers, and the randomization of the comonomer units along the polyester chain, as the monomer conversion increases and approaches completion.

Block copolymers have been prepared by sequential addition of the comonomers as well. So far several combinations have been attempted by copolymerizing successively different lactones or mixtures of them, e.g., (poly(εCL-b-δVL), poly[εCL-b-(εCL-co-δVL)] [35] and poly(εCL-b-DXO) [26]. In every one of the above-mentioned cases, perfectly well-tailored diblock copolyesters are prepared whatever the nature of the comonomer that is polymerized first. It is worth recalling that sequential copolymerization involving 1,5-dioxepan-2-one (DXO, see Table 1) forms a block copolymer in which one block is nothing but an alternating poly(ester-alt-ether). As will be discussed in Sect. 2.5, diblock copolyesters with one block selectively functionalized by ethylene ketal groups spread all along the backbone can be readily prepared through the sequential copolymerization of εCL with 1,4,8-trioxa(4,6)spiro-9-undecanone (TOSUO) [40]. The resulting poly(εCL-b-TOSUO) copolyesters offer the advantage of being easily converted in block copolymers with one sequence bearing either ketone or alcohol pendant functions, by successive and highly selective ketal deprotection and ketone reduction reactions, respectively. More detail is made available in the "synthesis and (co)polymerization of functional cyclic ester monomers" section. As far as the sequential copolymerization of lactones and dilactones is concerned, block copolymers (poly[εCL-b-(D,L-LA)] and poly(εCL-b-GA)) have been recovered with the exclusive condition that the lactone was added first, followed by the addition of the dilactone after complete consumption of εCL [27, 41]. The same methodology has led successfully to the synthesis of block copolymers between εCL and adipic anhydride AA, that is poly(εCL-b-AA). The polymerization of εCL is initiated first by Al(OiPr)$_3$ and then AA is added to the living PCL chains. Reverse addition is forbidden since the ROP of cyclic anhydrides proceeds through aluminum carboxylate growing species, which are known to be totally inactive in promoting the polymerization of εCL. Furthermore, PAA chains are poorly soluble in apolar solvent so that above an M_n of ca. 5000 the polyanhydride as directly initiated by Al(OiPr)$_3$ would precipitate quite rapidly. Starting the polymerization with εCL is a key solution to overcome this drawback, the living PCL preformed chains responsible for keeping the whole P(εCL-b-AA) growing copolymer soluble in toluene [33].

The exclusive formation of the block copolymer has been confirmed by selective fractionation, NMR spectroscopy, and SEC analysis. For instance, the copolymerization of εCL and δVL has been followed by SEC. Figure 2 compares the SEC chromatograms of the first PCL block and the final poly(εCL-b-δVL) diblock copolymer. The molecular weight of the macroinitiator is shifted towards higher values in close agreement with the theoretical value expected from the comonomer-to-Al(OiPr)$_3$ molar ratio, and the MWD remains very narrow during the copolymerization process (PDI=1.10).

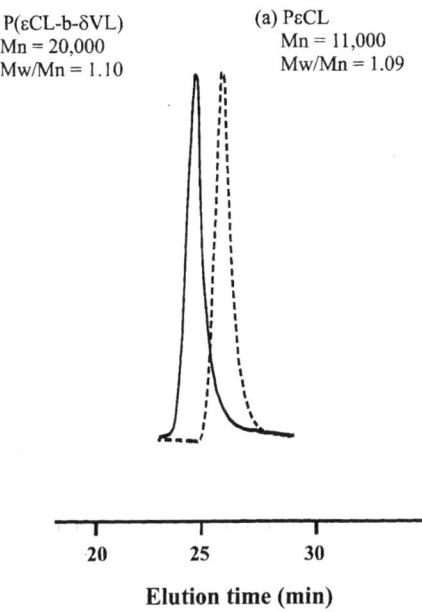

Fig. 2. SEC traces of PCL macroinitiator (*curve a*) and poly(εCL-*b*-δVL) block copolymer (*curve b*). Polymerization conditions: [monomer]$_0$=1 mol/l; ([εCL]$_0$+([δVL]$_0$)/[Al(O*i*Pr)$_3$]$_0$= 200; toluene at 0 °C

Interestingly, the ROP of GA by living PCL macroinitiators as initiated by aluminum alkoxides has led to PGA chains of unprecedented molecular weights (M_n as high as 70,000). Note also that symmetrical triblock copolyesters which present interesting elastomeric properties (see Sect. 4.1), have also been prepared by sequential addition of the εCL and DXO comonomers as will be discussed in Sect. 4.1.

2.3
Selective End-Functionalization

The synthesis of end-reactive polymers (telechelic macromolecules) of a precisely controlled molecular weight is an important step for any macromolecular engineering. In this regard, Al derivatives containing p (p=1,3) functional alkoxide groups associated with 3-p inactive alkyl groups have proved to be highly effective. In agreement with the "coordination-insertion" mechanism discussed above, the functional group (OCH$_2$X) of the aluminum alkoxides is selectively attached to the α-chain end via an ester linkage, whereas a hydroxyl end group is systematically present in the ω- position, as the result of the hydrolytic deactivation of the propagating aluminum alkoxide species (Scheme 4).

Functional diethyl aluminum alkoxides have been prepared by reaction of AlEt$_3$ with an equimolar amount of the corresponding alcohol (XCH$_2$OH). The

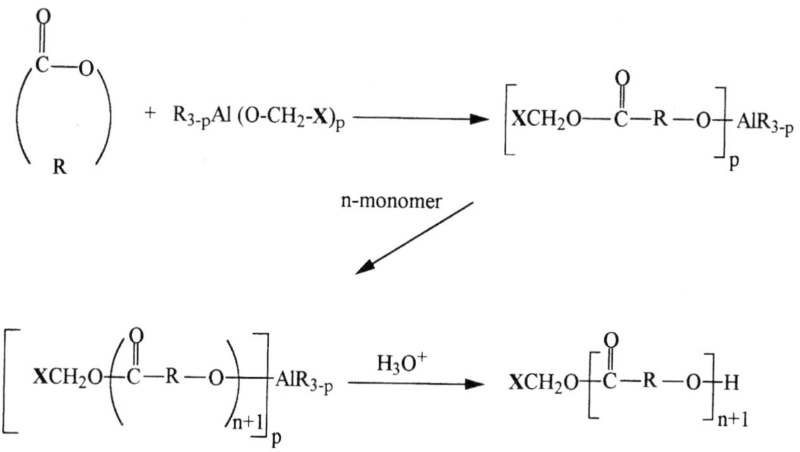

Scheme 4

$$Et_3Al + HO\text{-}CH_2\text{-}X \xrightarrow{toluene} Et_2\text{-}Al\text{-}O\text{-}CH_2\text{-}X + C_2H_6 \uparrow$$

$$(^iPrO)_3Al + 3\ HO\text{-}CH_2\text{-}X \xrightleftharpoons{toluene} (X\text{-}CH_2\text{-}O)_3Al + 3\ ^iPrOH$$

where X =	Ref.
CH_2Br	[42,43]
$(CH_2)_2\text{-}CH=CH_2$	[42,43]
$CH_2\text{-}N\text{-}(CH_2CH_3)_2$	[42]
$CH_2\text{-}OCO\text{-}CH(CH_3)=CH_2$	[45,46]
$CH_2\text{-}CBr_3$	[47]
Ph	[48]
$CH_2\text{-}Ph\text{-}NO_2$	[49]
 norbornyl	[50]

Scheme 5

reaction equilibrium is favorably displaced by formation and elimination of ethane [42, 43]. On the other hand, aluminum trialkoxides have been synthesized by substituting the three alkoxy groups of $Al(OiPr)_3$ by the selected functional alcohol. Distillation of the toluene/2-propanol azeotrope allows the substitution to be complete (Scheme 5) [44].

Polymerization of ε-CL and (L- or D,L-)LA is perfectly "living" when initiated with any of the aforementioned functional aluminum alkoxides in toluene at 25 and 70 °C, respectively. This is supported by the close agreement between the mean degree of polymerization (DP) at total monomer conversion and the monomer-to-initiator molar ratio, the narrow molecular weight distributions (PDI=

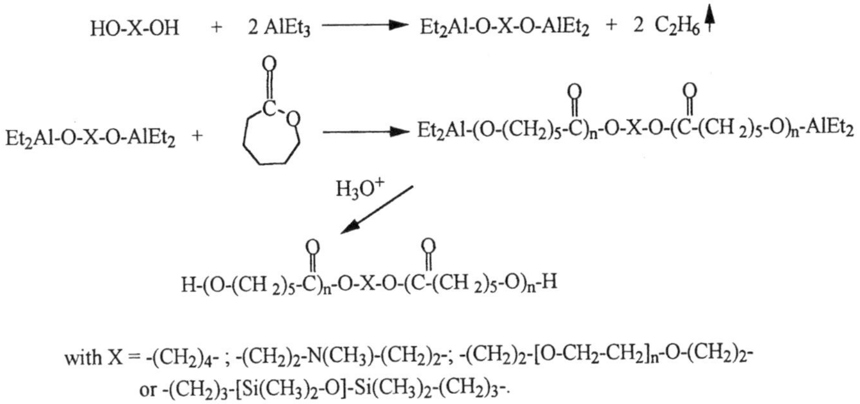

$$HO\text{-}X\text{-}OH \ + \ 2\ AlEt_3 \longrightarrow Et_2Al\text{-}O\text{-}X\text{-}O\text{-}AlEt_2 \ + \ 2\ C_2H_6 \uparrow$$

$$Et_2Al\text{-}O\text{-}X\text{-}O\text{-}AlEt_2 \ + \ \underset{\text{(caprolactone)}}{\bigcirc} \longrightarrow Et_2Al\text{-}(O\text{-}(CH_2)_5\text{-}\underset{\|}{\overset{O}{C}})_n\text{-}O\text{-}X\text{-}O\text{-}(\underset{\|}{\overset{O}{C}}\text{-}(CH_2)_5\text{-}O)_n\text{-}AlEt_2$$

$$\downarrow H_3O^+$$

$$H\text{-}(O\text{-}(CH_2)_5\text{-}\underset{\|}{\overset{O}{C}})_n\text{-}O\text{-}X\text{-}O\text{-}(\underset{\|}{\overset{O}{C}}\text{-}(CH_2)_5\text{-}O)_n\text{-}H$$

with X = -(CH$_2$)$_4$- ; -(CH$_2$)$_2$-N(CH$_3$)-(CH$_2$)$_2$-; -(CH$_2$)$_2$-[O-CH$_2$-CH$_2$]$_n$-O-(CH$_2$)$_2$- or -(CH$_2$)$_3$-[Si(CH$_3$)$_2$-O]-Si(CH$_3$)$_2$-(CH$_2$)$_3$-.

Scheme 6

1.1–1.3), and the quantitative end-group functionalization as detected by NMR spectroscopy. Clearly a great variety of asymmetric telechelic α-functional ω-hydroxy PCL and PLA have been recovered owing to the great versatility of the initiators that can be used. Similarly, the preparation of symmetric telechelic α,ω-hydroxy polyesters has been accomplished by using alkylaluminum di-alkoxides as difunctional initiators, e.g., Et$_2$Al-O-X-O-AlEt$_2$ where X is either an aliphatic group or a polymer chain (Scheme 6) [51]. A hydroxyl group is selectively attached at both chain ends as demonstrated by FTIR and ^{1}H NMR spectroscopies, and confirmed by titration of the derived α,ω-dicarboxylic acid polyesters. Diethylaluminum dithiolates have also been investigated as initiator for the ROP of εCL. Dithiols have been reacted with a two-fold molar excess of AlEt$_3$, making Al dithiolates available for the controlled synthesis of hydroxy telechelic PCL [51]. Synthesis and use of H$_2$N-(CH$_2$)$_3$-OAlEt$_2$ as an initiator also deserve interest, since the nucleophilic addition of the amino group onto the carbonyl of the monomer is nothing but an initiation process in addition to the aluminum alkoxide itself. This is another way to prepare α,ω-hydroxy PCL in a living manner [51].

Another general approach to end-functional polyesters relies upon the proper control of the termination step instead of the initiation one. In this regard, the aluminum alkoxide propagating species or the resulting ω-hydroxy end group after hydrolytic deactivation can be readily modified by reacting with carboxylic acid derivatives such as acid chlorides, acid anhydrides, or isocyanates. By this method, different ω-functional polyesters have been prepared [10, 43, 51]. The main advantage of this method is being able to avoid the preparation of the functional aluminum alkoxide. However, the functionalization may not be quantitative, specially when molecular weight is relatively high (Scheme 7).

A more general strategy to approach symmetrical telechelic polyesters consists of the control of both initiation and termination steps. Indeed, combination of a functional initiator for the (di)lactones ROP with an effective coupling

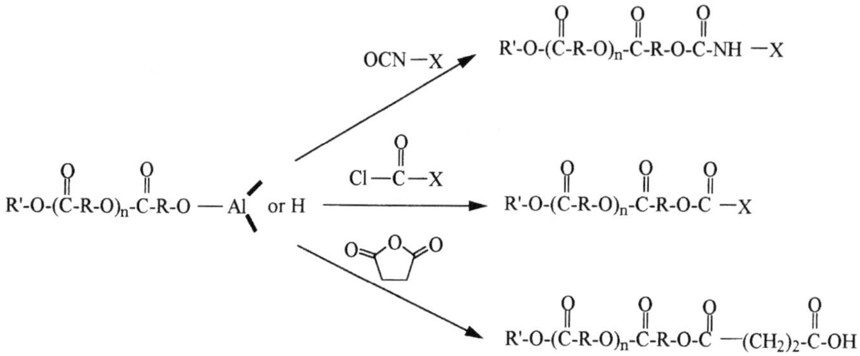

Scheme 7

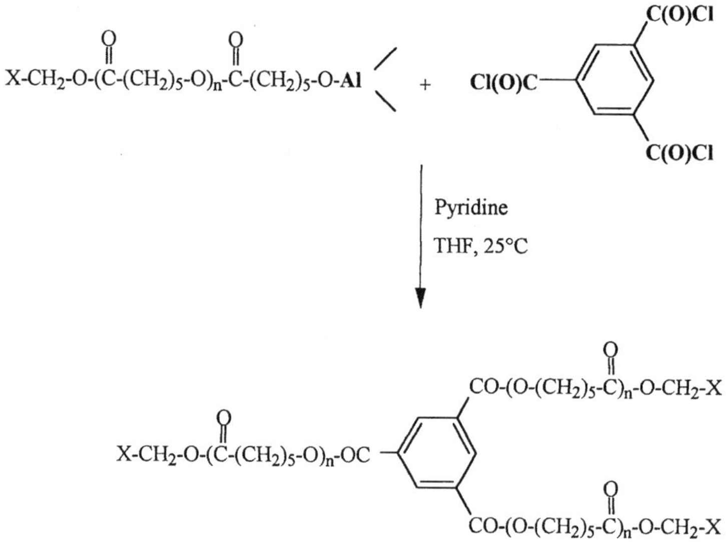

Scheme 8

agent, such as terephthalic acid chloride, quantitatively yields an α,ω-functional polyester of a two fold molecular weight [52]. Similarly, end-functional tri-arm star-shaped PCLs have been prepared using a trifunctional coupling agent (Scheme 8) [53].

2.4
Kinetic Aspects of the "Coordination-Insertion" ROP

Significant advances in the understanding of the "coordination-insertion" ROP mechanism have been made owing to the kinetic studies by Duda and Penczek.

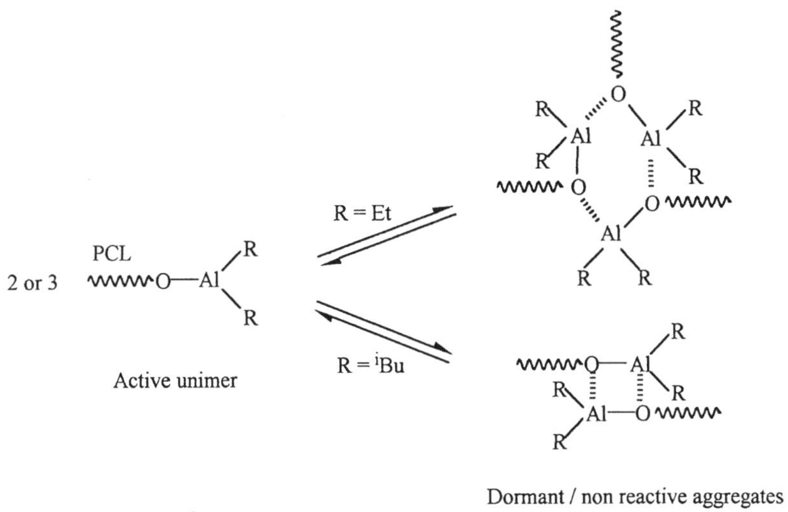

Scheme 9

As mentioned above, the major advantage of aluminum alkoxide mediated ROP when compared with most anionic and cationic processes is its living character. Transesterification side reactions or termination reactions are fully depressed at least until complete monomer conversion, and polyesters with predictable molecular weight and narrow MWD, are obtained. However, aluminum alkoxides, like other metal alkoxides, are known to be aggregated in solution. This aggregation plays an important role in the polymerization kinetics. For instance, polymerizations are characterized by an induction period of time before the initiation step which appears to correspond to a monomer/initiator coordinative rearrangement [10]. This fact can also determine the initiation efficiency. For example, Al(OiPr)$_3$ in toluene leads to a coordinative aggregation equilibrium between tetramers (A4) and trimers (A3). Depending on the monomer and reaction conditions both species (A3+A4) can act as initiators for the polymerization (case LA in toluene at 70 °C) or only the more reactive A3 can be involved (case εCL) [54]. On the other hand, the aggregation of some aluminum alkoxides was also observed during the propagation step. For instance, in the polymerization of εCL initiated in apolar solvent with dialkyl aluminum monoalkoxides, R$_2$AlOR', the covalent "Al-O" growing species proved to associate and form inactive aggregates. The type and size of the aggregates depend on the solvent polarity, the nature of the alkyl R substituents, and the presence of coordinative ligands such as amines and alcohols. Actually, only the small fraction of disassociated aluminum alkoxides (in equilibrium with the aggregates) are active in ROP of (di)lactones. This polymerization behaves as a "living-dormant" polymerization, often known as "pseudoliving" polymerization (Scheme 9) [55].

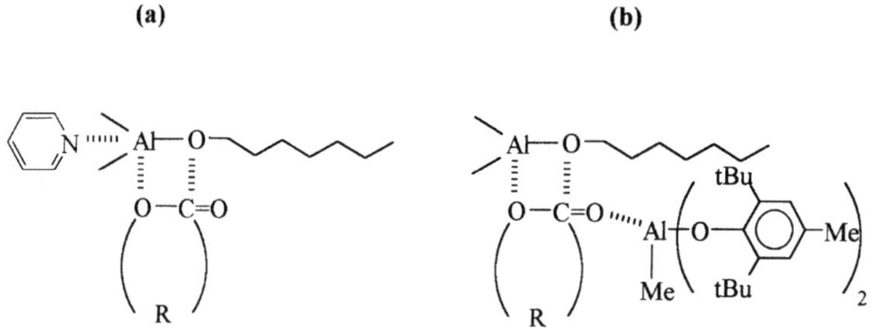

(a) (b)

Scheme 10

Apart from the alkoxides aggregation, the different factors which affect the kinetics of the (di)lactones ROP initiated by aluminum alkoxides might be summarized as follows.

a) Nature of the alkoxides: aluminum trialkoxides propagate faster than dialkyl aluminum monoalkoxides due to the higher ionicity of the Al-O bond in the former case [56].
b) Temperature: as expected, the overall polymerization rate increases with temperature. However, at high temperature the transesterification side reactions may occur, limiting the molecular weight, broadening the MWD, and yielding cyclic oligomers [56].
c) Solvent: the polymerization proceeds faster in non-polar solvents. This is explained by the fact that polar solvents compete with the monomer in coordinating the reaction sites, i.e., the aluminum atom of the growing species. In other words, the solvent prevents or at least limits the access of the monomer to the reaction site, thus decreasing the overall rate of polymerization [56, 57].
d) Additives: external additives can modify the rate of polymerization by three different mechanisms.
   1. By acting as transfer agents. It has been established by kinetic studies that low-molecular weight alcohols and diols, which have been introduced into the polymerization mixture of εCL and Al(OiPr)$_3$, operate as chain-transfer agents. Accordingly, they participate in the initiation step in such a way that the total initiation species equals three isopropoxy groups of Al(OiPr)$_3$, plus the added alcohol (diol) molecules. Moreover, they can also act either as decelerators in polymerization initiated with highly active Al(OiPr)$_3$ trimers (A3), or as accelerators in polymerization initiated with much more stable and accordingly less active Al(OiPr)$_3$ tetramers (A4). Alcohols actually modify and shift the equilibrium between the A3/A4 aggregates, facilitating the transformation of the low reactive A4 into A3 (or other more reactive initiating species) [58].
   2. By provoking the complete deaggregation of the growing chains as initiated by R$_2$AlOR'. This is the case of secondary amines, the addition of which increases the overall rate of polymerization [55].

3. By increasing the Al-O bond reactivity. This effect has been observed when Lewis bases, such as pyridine, picoline, and nicotine, were added to the reaction mixture in equimolar amount relative to the Al alkoxide active species [30–33]. This kinetic effect is triggered by a specific coordination of the Lewis base onto the Al atom of the growing site which may increase the ionicity of the Al-O bond (Scheme 10a). These additives have to be added in the right amounts. An excess generally produces a decrease in the polymerization rate. The ligand does not act as an activator any more but rather blocks the growing sites, preventing the monomer from coordinating the aluminum alkoxide. It is worth noting that Inoue et al. have recently reported on the kinetic activation of δVL polymerization initiated by tetraphenylporphinato (TPP)Al alkoxide in the presence of bulky Lewis acid compounds [59]. In that precise case, the extremely fast polymerization has been attributed to the coordinative interaction between the monomer and the bulky Lewis acid (Scheme 10b).

2.5
Synthesis and (Co)polymerization of Functional Cyclic Ester Monomers

Special attention has been paid to the preparation of biodegradable polymers bearing functional pendant groups. The availability of functional pendant groups is highly desirable for the fine tuning of properties like crystallinity, glass transition temperature, hydrophilicity, and chemical reactivity in view of, e.g., attachment of drugs, improvement of biocompatibility, and promotion of bioadhesion. Several examples of functional monomers and their (co)polymerization have been reported in the literature (Scheme 11).

Although important efforts have been devoted to the synthesis of functionalized monomers, which is very often complex and tedious, only one example of controlled/living (co)polymerization has been reported so far. This is the case of the 5-ethylene ketal ε-caprolactone, precisely 1,4,8-trioxa(4,6)spiro-9 undecanone (TOSUO), which is readily synthesized by a one-step oxidation reaction of the commercially available monoethylene ketal of the 1,4-cyclohexanedione [36]. TOSUO has proved to homopolymerize in a "living" way when initiated by Al(OiPr)$_3$ in toluene at 25 °C (Scheme 12) [66]. As a result, well-defined homo PTOSUO, but also random and block copolymers with εCL, were prepared [36, 40]. Treatment of the (co)polyester chain with $(C_6H_5)_3CBF_4$ in CH_2Cl_2 at RT allows for the quantitative deprotection of the pendant ketone functions. Furthermore the reduction of the formed ketone into hydroxyl groups has been performed selectively by $NaBH_4$ in CH_2Cl_2/EtOH solvent mixture at RT. Importantly, no chain scission has been detected by SEC and ^{1}H NMR spectroscopy, neither after the deprotection step nor after the reduction one. Thus aliphatic polyesters bearing either ketone pendant groups or hydroxyl pendant groups can be readily synthesized, which opens new application prospects. It is worthwhile pointing out that the controlled copolymerization between TOSUO and lactides has also been successfully performed. Aliphatic polyesters with hydroxyl groups highly reactive toward AlEt$_3$ provided a macroinitiator for (di)lactones polymerization, so that synthesis of biodegradable functional comb, graft, and dendrigraft aliphatic polyesters has been possible (see Sect. 3.1). The poly(εCL-*co*-TOSUO) copolyesters have proved to be easily redispersed in water. Compared to pure PCL in the same molecular weight range that gives a crude precipitate, both homo PTOSUO and copolymers with εCL provided stable dispersions in water of a mean size below 100 nm (see Sect. 4.4). On the other hand, poly(εCL-*co*-TOSUO) copolymers with at least 15 mol% TOSUO units statistically distributed along the backbone displayed a low T$_g$ of ca. −40 °C, making them valuable rub-

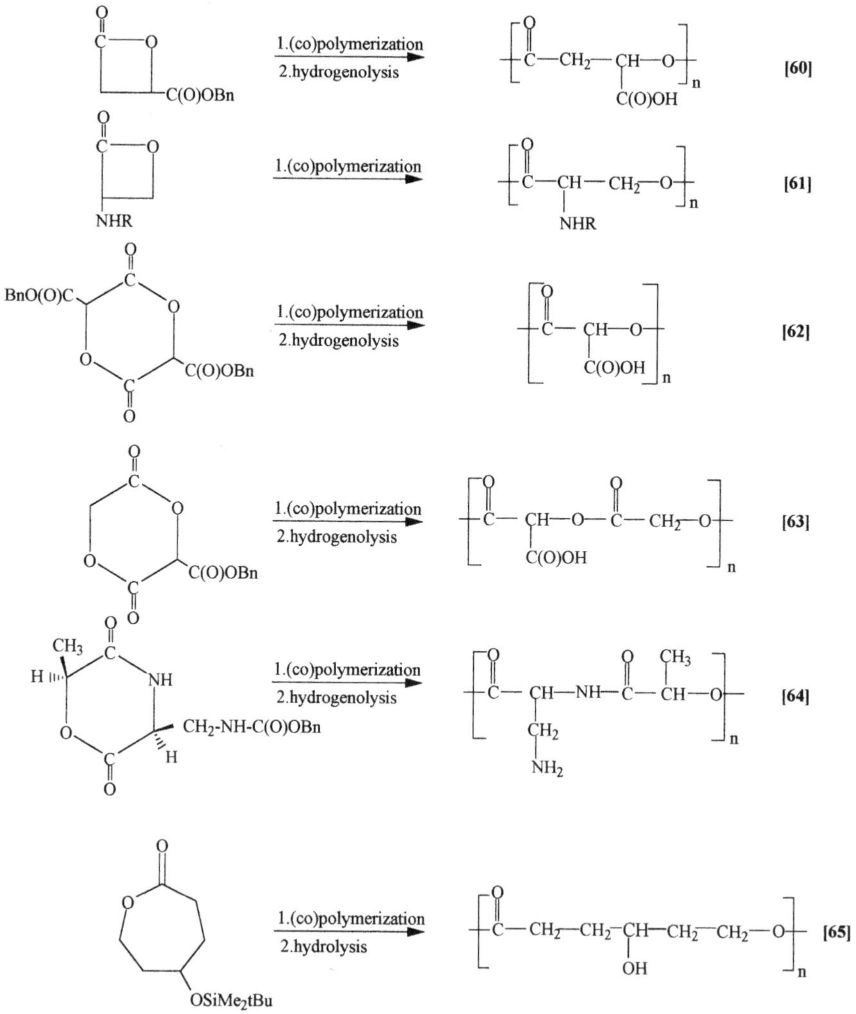

Scheme 11

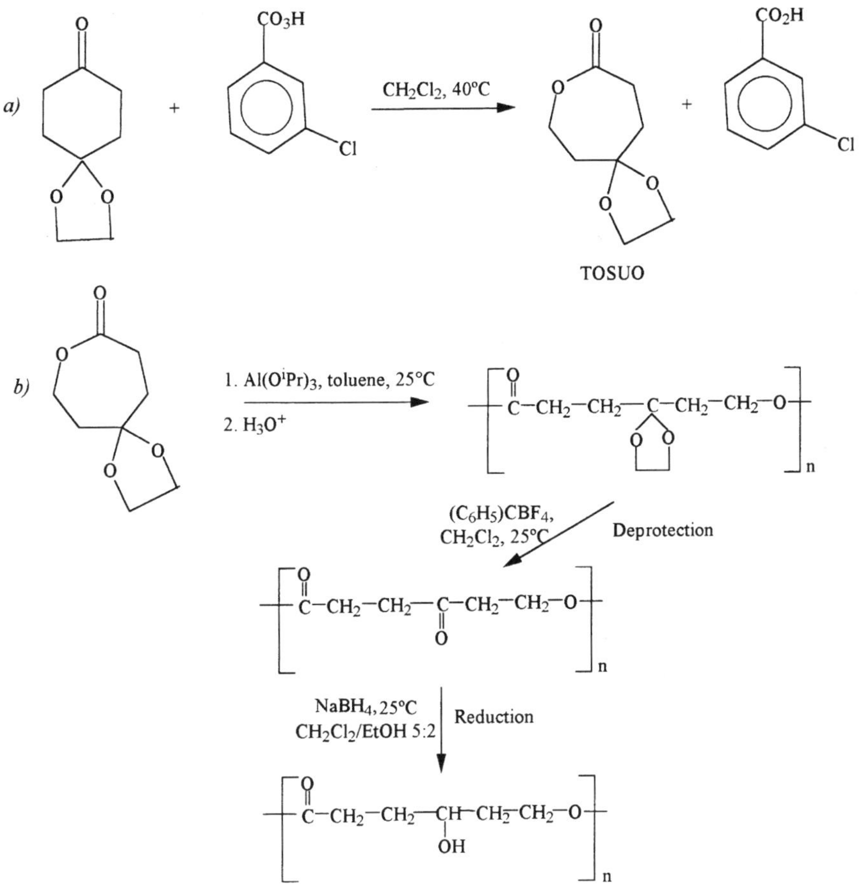

Scheme 12

bery materials to be used in combination with semicrystalline PCL. Therefore, poly[εCL-b-(εCL-co-TOSUO)-b-εCL] symmetrical triblock copolyesters represent potential biodegradable and biocompatible thermoplastic elastomers.

3 Synthesis of Block and Graft Copolymers by Combination of (Di)lactones Ring Opening Polymerization with Other Living/Controlled Polymerization Processes

Macromolecular engineering is the ultimate goal of the polymer chemist when he has a monomer or a family of monomers at his disposal. Once each step of the polymerization process is carefully controlled, every molecular parameter of the polymer is predictable: molecular weight, tacticity, molecular weight distribution, nature of the end groups, microstructure, and composition, and block length in the case of copolymers. Like pieces of a construction set, these properly tailored macromolecules can then be used to design new polymeric materials [67, 68]. Block and graft copolymers are convincing examples of well-controlled molecular architectures that comprise at least two polymeric components and lead to original materials, e.g., thermoplastic elastomers, polymeric emulsifiers, surface modifiers, etc.

With the idea of extending the scope of the macromolecular engineering of aliphatic polyesters, the "coordination-insertion" ROP of lactones and dilactones has been combined with other polymerization processes. This section aims at reviewing the new synthetic routes developed during the last few years for building up novel (co)polymer structures based on aliphatic polyesters, at least partially.

3.1
Ring Opening Polymerization

Considerable effort has been carried out by different groups in the preparation of amphiphilic block copolymers based on poly(ethylene oxide) PEO and an aliphatic polyester. A common approach relies upon the use of preformed ω-hydroxy PEO as macroinitiator precursors [51, 70]. Actually, the anionic ROP of ethylene oxide is readily initiated by alcohol molecules activated by potassium hydroxide in catalytic amounts. The equimolar reaction of the PEO hydroxy end group(s) with triethyl aluminum yields a macroinitiator that, according to the coordination-insertion mechanism previously discussed (see Sect. 2.1), is highly active in the εCL and LA polymerization. This strategy allows one to prepare di- or triblock copolymers depending on the functionality of the PEO macroinitiator (Scheme 13a,b). Diblock copolymers have also been successfully prepared by sequential addition of the cyclic ether (EO) and lactone monomers using tetraphenylporphynato aluminum alkoxides or chloride as the initiator [69].

Of special interest is the heterogeneous catalytic coordination ROP process recently proposed by Hamaide et al. [117–119]. The catalytic system was ob-

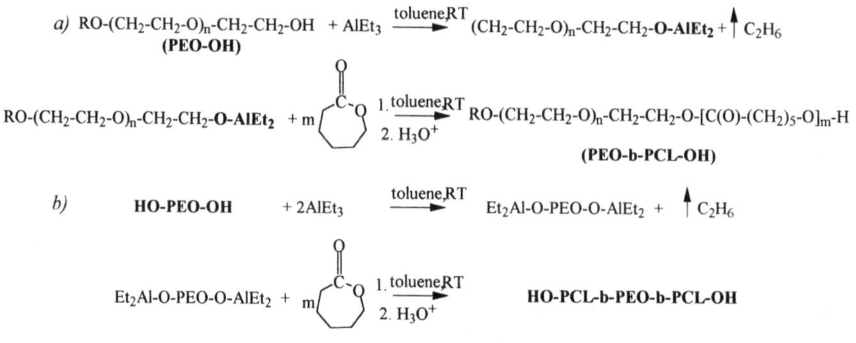

Scheme 13

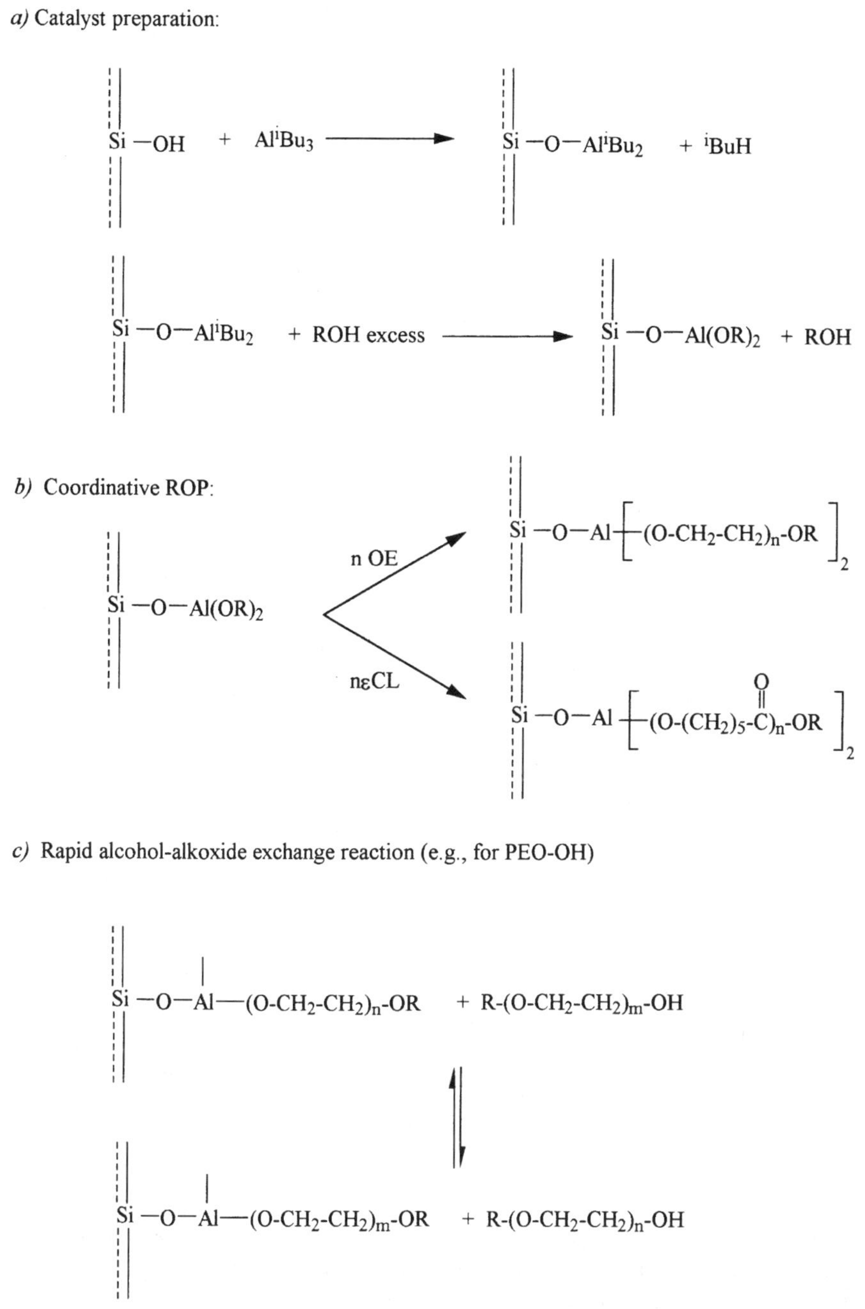

a) Catalyst preparation:

$$\text{Si-OH} + \text{Al}^i\text{Bu}_3 \longrightarrow \text{Si-O-Al}^i\text{Bu}_2 + {}^i\text{BuH}$$

$$\text{Si-O-Al}^i\text{Bu}_2 + \text{ROH excess} \longrightarrow \text{Si-O-Al(OR)}_2 + \text{ROH}$$

b) Coordinative ROP:

$$\text{Si-O-Al(OR)}_2 \xrightarrow{n\,\text{OE}} \text{Si-O-Al}\left[\text{(O-CH}_2\text{-CH}_2)_n\text{-OR}\right]_2$$

$$\text{Si-O-Al(OR)}_2 \xrightarrow{n\varepsilon\text{CL}} \text{Si-O-Al}\left[\text{(O-(CH}_2)_5\text{-}\overset{\overset{O}{\|}}{C})_n\text{-OR}\right]_2$$

c) Rapid alcohol-alkoxide exchange reaction (e.g., for PEO-OH)

$$\text{Si-O-Al-(O-CH}_2\text{-CH}_2)_n\text{-OR} + \text{R-(O-CH}_2\text{-CH}_2)_m\text{-OH}$$

$$\updownarrow$$

$$\text{Si-O-Al-(O-CH}_2\text{-CH}_2)_m\text{-OR} + \text{R-(O-CH}_2\text{-CH}_2)_n\text{-OH}$$

Scheme 14

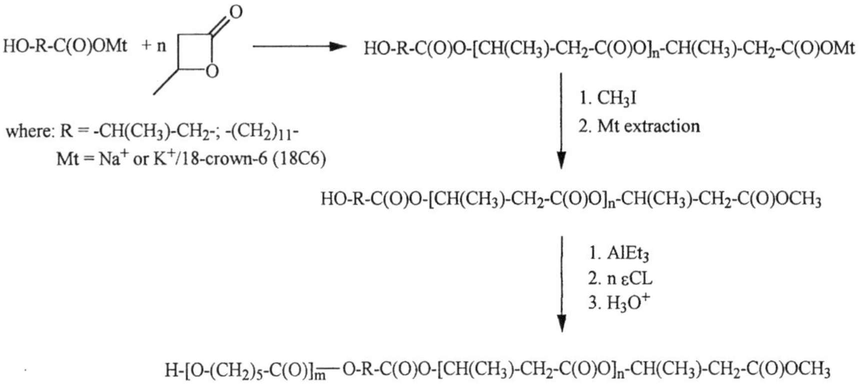

HO-R-C(O)OMt + n [β-propiolactone] ⟶ HO-R-C(O)O-[CH(CH$_3$)-CH$_2$-C(O)O]$_n$-CH(CH$_3$)-CH$_2$-C(O)OMt

where: R = -CH(CH$_3$)-CH$_2$-; -(CH$_2$)$_{11}$-
Mt = Na$^+$ or K$^+$/18-crown-6 (18C6)

1. CH$_3$I
2. Mt extraction

↓

HO-R-C(O)O-[CH(CH$_3$)-CH$_2$-C(O)O]$_n$-CH(CH$_3$)-CH$_2$-C(O)OCH$_3$

1. AlEt$_3$
2. n εCL
3. H$_3$O$^+$

↓

H-[O-(CH$_2$)$_5$-C(O)]$_{\overline{m}}$-O-R-C(O)O-[CH(CH$_3$)-CH$_2$-C(O)O]$_n$-CH(CH$_3$)-CH$_2$-C(O)OCH$_3$

Scheme 15

tained by grafting alkyl aluminum moieties onto silica. Porous silica (surface area: 320 m^2/g) was reacted with triisobutylaluminum in order to replace all the silanol groups by Si-O-AlR$_2$ bridges (Scheme 14a). Adding alcohol in excess caused the formation of aluminum alkoxides and allowed the catalytic mode, because of a rapid exchange reaction between the grafted active centers and the free alcohols present in the medium. All the polymer chains were end-capped by these alcohol molecules, which gave functionalized oligomers (Scheme 14b). The catalytic systems based on aluminum alkoxide grafted onto silica has proved to be efficient in ROP of epoxides, such as ethylene oxide (EO) and propylene oxide (PO), lactones (εCL) and lactides [118]. Despite the fact that heterogeneous catalyst generally results in a dispersion of properties, the exchange reaction between alcohols and active alkoxide species averages the structures of the chains and allows narrow MWD (Scheme 14c). Not only tailor-made ω-functionalized oligomers but also random and block (oligo)copolymers have been produced. Note that a continuous process has been set up and that the recycling of the catalyst can be performed after washing the support [119].

Block copolymers of (R,S)-β-butyrolactone and εCL have been synthesized by combining the anionic ROP of the first monomer with the coordinative ROP of the second one (Scheme 15) [71]. The first step consisted of the synthesis of hydroxy-terminated atactic PβBL by anionic polymerization initiated by the alkali-metal salt of a hydroxycarboxylic acid complexed with a crown ether. The hydroxyl end group of PβBL could then be reacted with AlEt$_3$ to form a macroinitiator for the εCL ROP.

Using the opposite strategy, preformed ω-aliphatic primary amino PCL has been shown to be an efficient macroinitiator for the ROP of γ-benzylglutamate N-carboxy anhydride (BG-NCA) leading to original biodegradable P(εCL-b-BG) diblock copolymers. Interestingly, the polybenzylglutamate segment can be selectively hydrogenated into polyglutamic acid, i.e., a hydrophilic polypeptide block, making available new biocompatible and biodegradable block copolymer surfactants (Scheme 16a) [72]. The key problem for the successful block copol-

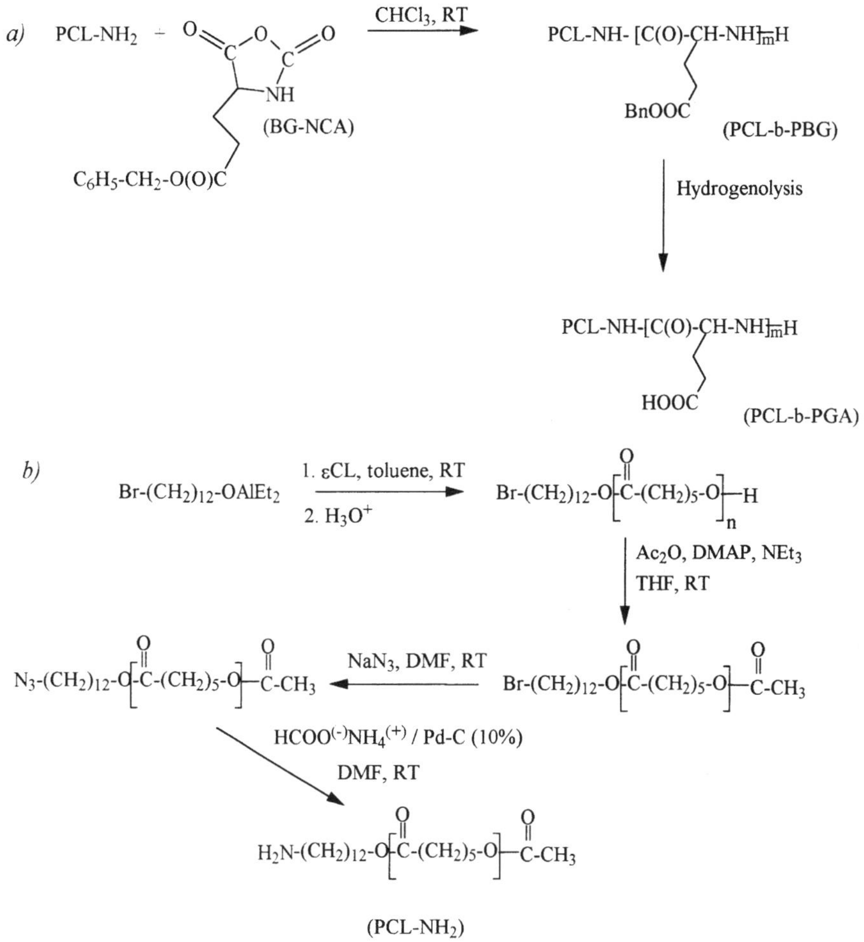

Scheme 16

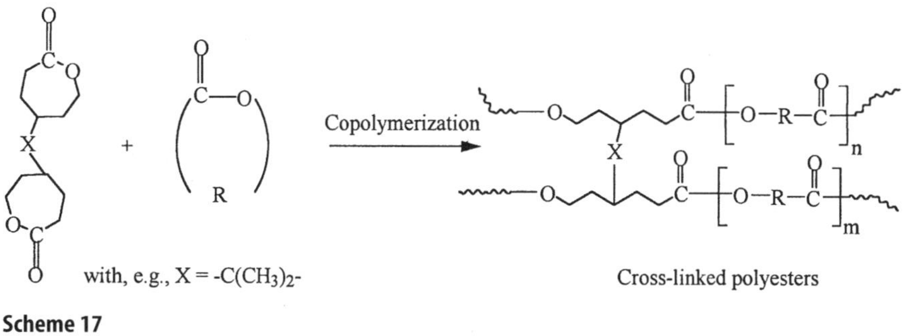

Scheme 17

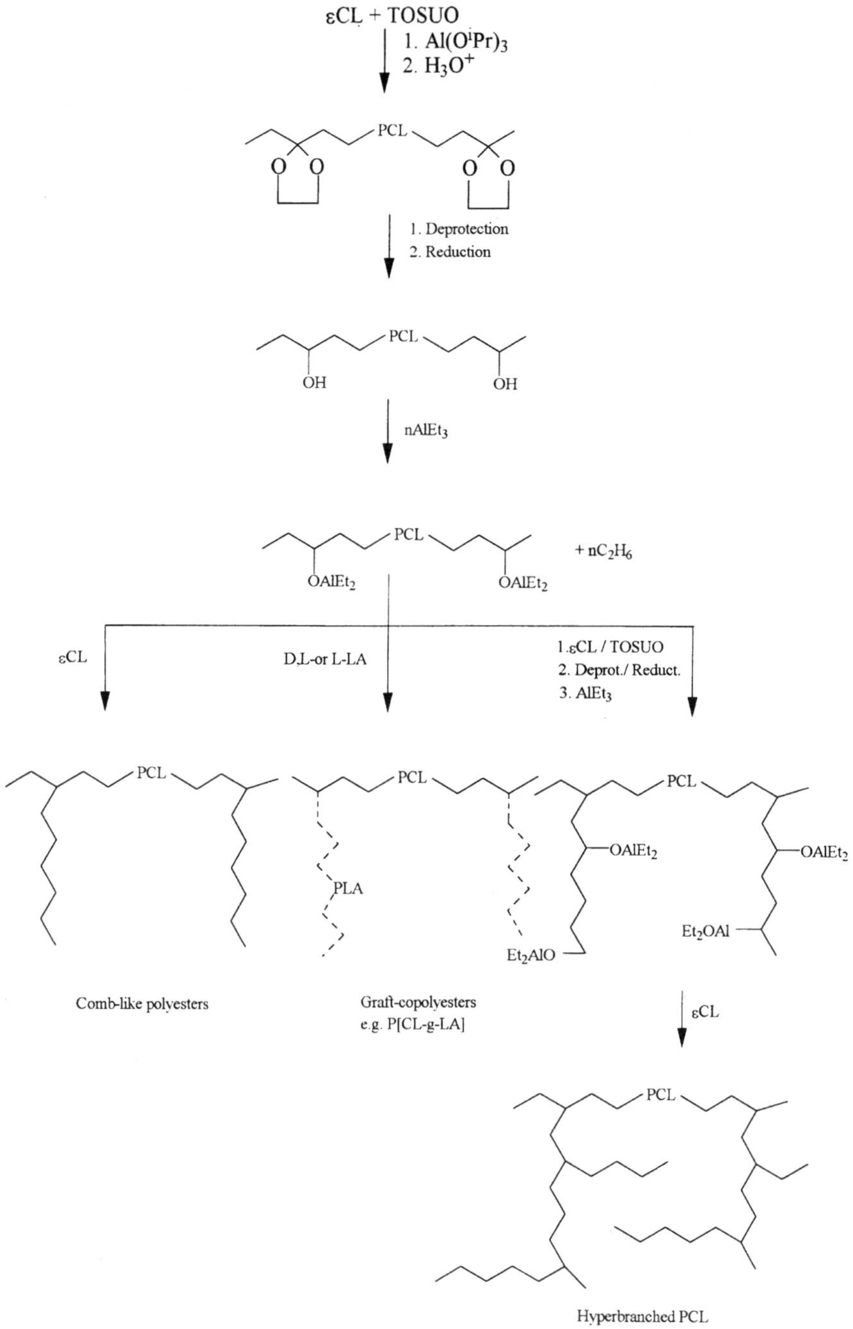
εCL + TOSUO
1. Al(O^{i}Pr)$_3$
2. H$_3$O$^+$
PCL
1. Deprotection
2. Reduction
OH
nAlEt$_3$
OAlEt$_2$
+ nC$_2$H$_6$
εCL
D,L-or L-LA
1.εCL / TOSUO
2. Deprot./ Reduct.
3. AlEt$_3$
PLA
Et$_2$AlO
Et$_2$OAl
Comb-like polyesters
Graft-copolyesters
e.g. P[CL-g-LA]
Hyperbranched PCL

Scheme 18

ymerization of εCL and amino acid-NCAs was the synthesis of ω-aliphatic primary amino PCL. As previously reported (see Sect. 2.3), initiation of εCL polymerization with an aluminum alkoxide containing a primary amino group leads to an α,ω-hydroxy PCL instead of the expected α-NH_2, ω-OH polymer. An initiator that contains a masked amine was then proposed. In this respect, a bromine-containing diethylaluminum alkoxide, $Br(CH_2)_{12}OAlEt_2$, has proved to be a valuable candidate, since it is easily converted into the expected primary amine (Scheme 16b). The polypeptide (PBG) block is thus terminated with a primary amine so that a coupling reaction would lead to the related PCL-*b*-polypeptide-*b*-PCL triblock copolymer. After selective hydrogenation, this type of amphiphilic triblock copolymers are precursors of physically cross-linked hydrogels.

Pitt et al. [65], and more recently, Albertsson et al. [73], have prepared chemically cross-linked aliphatic polyesters by ROP of the corresponding cyclic ester monomers in the presence of γ,γ'-bis(ε-caprolactone)-type comonomers (Scheme 17). The cross-linked films displayed different swelling behaviors, degradability, and elastomeric properties depending on the nature of the lactone and composition of the comonomers feed.

Recently, our laboratory reported on the first synthesis of fully biodegradable graft copolyesters by coordinative ROP. A PCL with pendant hydroxyl groups has been prepared by copolymerization with a few molar percent of TOSUO and subsequent ketone deprotection and reduction (see Scheme 12b). Reaction of the pendant hydroxyl groups with a slight excess of triethyl aluminum provides a multifunctional macroinitiator for the controlled ROP of εCL and lactides with formation of comb-like or graft copolymers, respectively (Scheme 18). If the grafts also contain TOSUO units, the release of the hydroxyl groups through the same procedure as previously, and their reaction with $AlEt_3$, allow an additional grafting reaction from being performed, leading to hyperbranched structures [74]. By judiciously controlling the lactone, e.g., εCL/TOSUO molar ratio and the initial content in Al(O*i*Pr)$_3$ initiator, one has the possibility of tailoring the grafting density, the graft length, and the grafts distribution along the polyester backbone by, e.g., sequential addition of εCL and εCL/TOSUO comonomer mixtures.

3.2
Anionic Polymerization

ω-Hydroxy terminated polymers prepared by anionic polymerization, e.g., polystyrene (PS-OH) or polybutadiene (PBD-OH), have been transformed in efficient macroinitiators leading to the synthesis of poly(S-*b*-CL) or poly(BD-*b*-CL) diblock copolymers [75]. For that purpose, soluble zinc and aluminum μ-oxo-alkoxides have been investigated as initiator precursors. One alkoxide group of the μ-oxo-alkoxides was substituted by either PS-OH or PBD-OH, all the other alkoxide groups being replaced by inactive carboxylate moieties as schematized hereafter (Scheme 19).

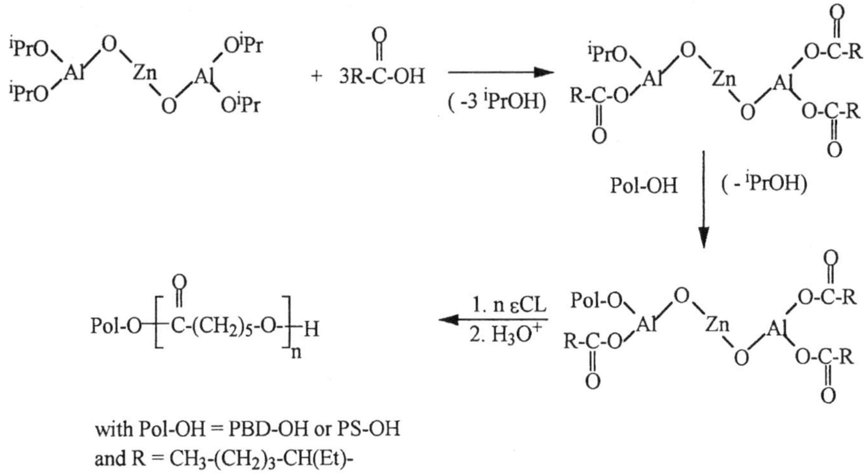

Scheme 19

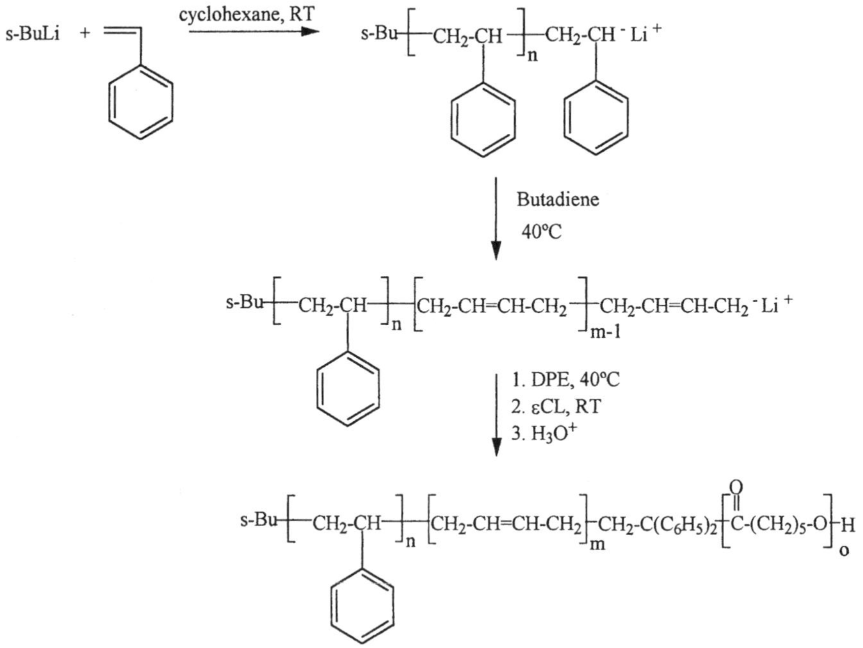

Scheme 20

Similar block copolymers, i.e., poly(S-b-CL), poly(BD-b-CL) as well as poly(S-b-BD-b-CL) ABC triblock copolymers have recently been prepared by Stadler et al. by sequential anionic polymerization (Scheme 20) [76]. Addition of

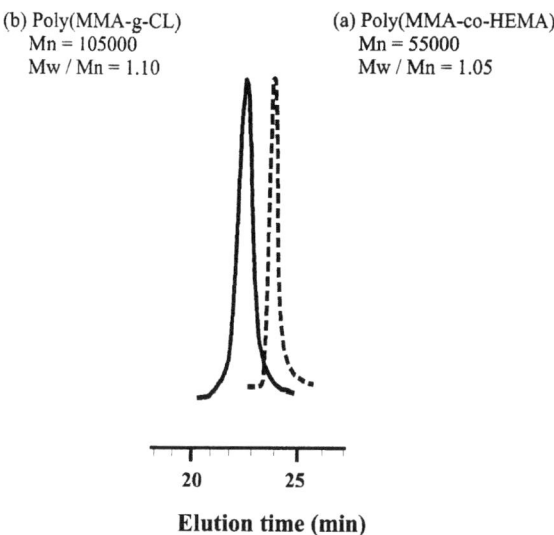

Fig. 3. Comparison of SEC traces for poly(MMA-co-HEMA) starting macroinitiator (*curve a*) and poly(MMA-g-CL) final graft copolymer (*curve b*). For synthetic conditions, see Scheme 21

1,1-diphenylethylene is required in order to decrease down the too high reactivity of the polybutadienyl anions toward the initiation of the purely anionic ROP of εCL. Furthermore, due to the inevitable inter- and intramolecular transesterification reactions in anionic ROP, especially at higher monomer conversion, the εCL polymerization time has been strictly controlled and the copolymerization was stopped by hydrolytic deactivation well before reaching quantitative εCL conversion.

Recently, our laboratory has developed a new method for the synthesis of poly(alkyl methacrylate-g-lactone) graft copolymers [77]. It consists of a set of two consecutive living polymerizations, i.e., an anionic polyaddition followed by a "coordination-insertion" ROP. In the first step, an alkyl methacrylate, e.g., methyl or *n*-butyl methacrylate, was copolymerized with a few molar percent of (trimethylsiloxy)ethyl methacrylate via a living anionic (co)polymerization as promoted by 1,1-diphenylhexyllithium (DPHLi) in the presence of lithium chloride salt (LiCl) in THF at –78 °C. Deprotection of the hydroxyl function followed by the equimolar reaction with triethyl aluminum led to a multifunctional methacrylic macroinitiator for lactone ROP, with several initiating sites along the main backbone (Scheme 21). The second step consisted of the lactone ROP. The ROP was again perfectly well controlled and yielded graft copolymers of methyl and *n*-butyl methacrylates and lactones such as εCL, δVL, and TOSUO. The graft copolymers were characterized by a polymethacrylate main backbone and polyester grafts of predetermined molecular weight, a controlled branch density, and a narrow apparent polydispersity that ranged from 1.05 to 1.20. SEC analysis

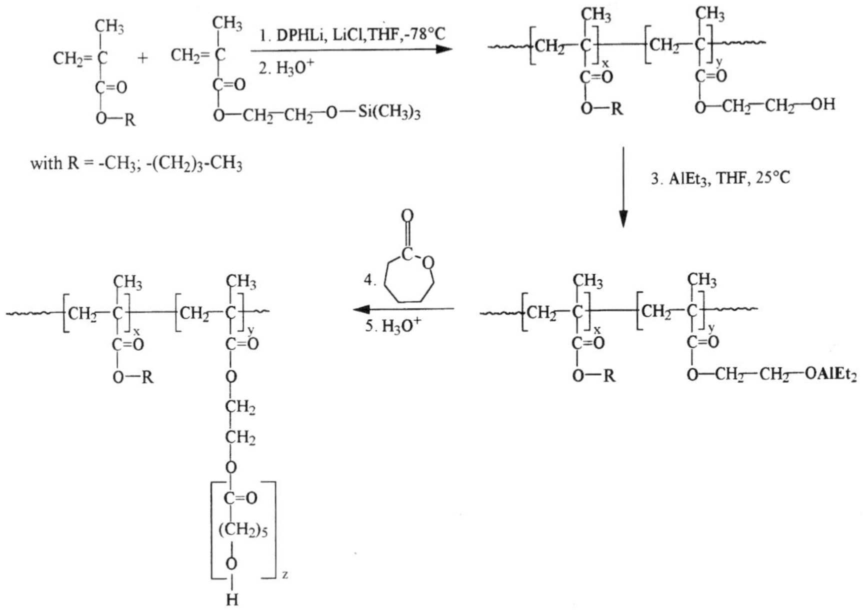

Scheme 21

of the poly(MMA-g-CL) copolymer shows the expected increase in molecular weight with no remaining trace of unreacted starting macroinitiator (Fig. 3).

3.3
Cationic Polymerization

Harris and Sharkey have converted isobutylene polymers into thermoplastic elastomers by the grafting of semi-crystalline polypivalolactone segments, the melting temperature of which is of ca. 245 °C [78]. The "grafting-from" technique was investigated on poly(isobutylene-co-p-methylstyrene) random copolymers, which were synthesized by cationic copolymerization. Active initiation species were generated by metalation of the benzylic carbons, followed by a carboxylation reaction and then formation of ammonium carboxylate pendant groups (Scheme 22). The ROP of PVL was initiated by those carboxylic acid salts positioned all along the polyisobutylene backbone. Nevertheless, the control over the graft copolymers parameters was very poor.

In a similar way, the "grafting-from" technique has been applied to the synthesis of poly(chloroethylvinylether) chains by grafted PCL segments, i.e., poly(CEVE-g-CL) graft copolymers. Purposely cationically prepared PCEVE were partially modified by the introduction of 5–10% hydroxyl groups [79]. An equimolar reaction of the pendant hydroxyl functions with HAliBu$_2$ provided diisobutyl aluminum monoalkoxides dispersed along the polyether backbone,

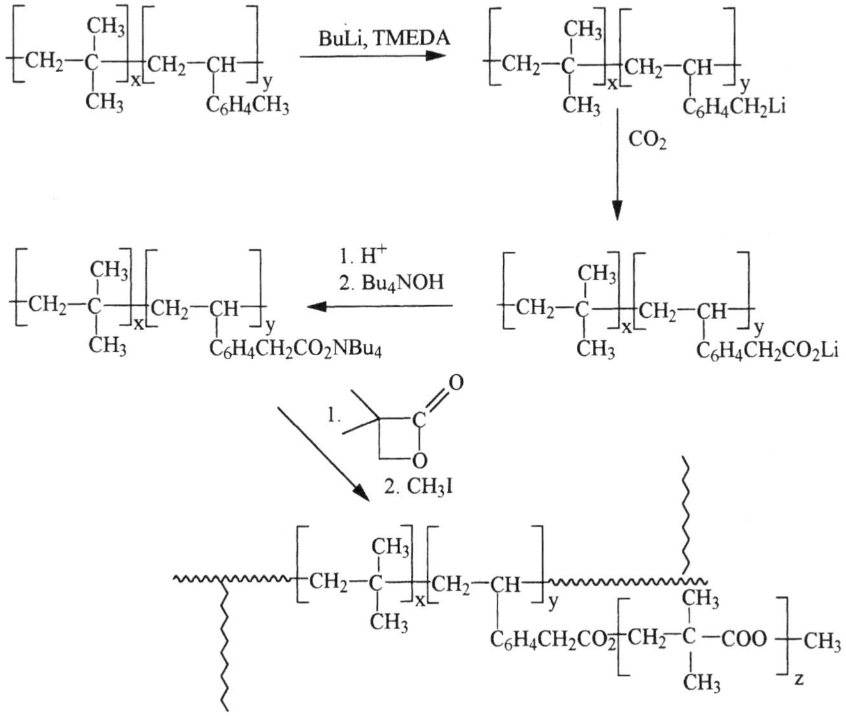

Scheme 22

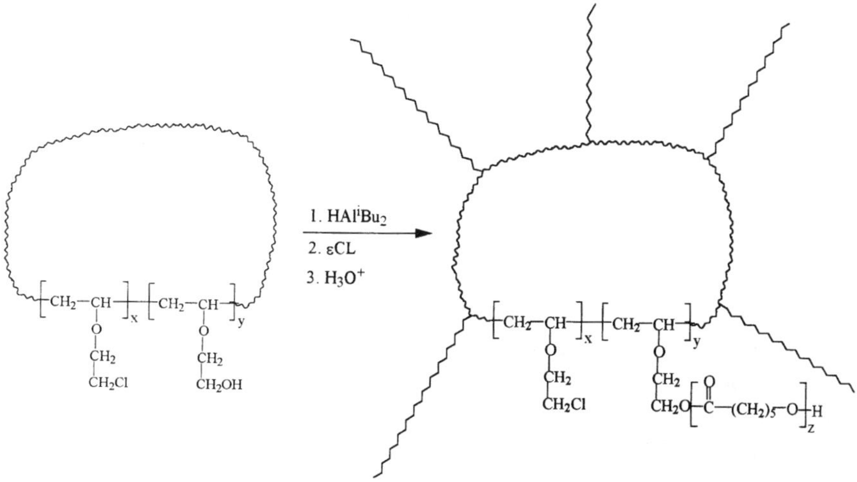

Scheme 23

which again proved to be active in coordinative ROP of εCL. Substitution of PCEVE macrocycles for the PCEVE linear chains led to novel "sun-like" graft copolymers with well-defined molecular parameters (Scheme 23) [80].

3.4
Radical Polymerization

The last decades have witnessed the emergence of new "living"/controlled polymerizations based on radical chemistry [81, 82]. Two main approaches have been investigated; the first involves mediation of the free radical process by stable nitroxyl radicals, such as TEMPO while the second relies upon a Kharash-type reaction mediated by metal complexes such as copper(I) bromide ligated with 2,2'-bipyridine. In the latter case, the polymerization is initiated by alkyl halides or arenesulfonyl halides. Nitroxide-based initiators are efficient for styrene and styrene derivatives, while the metal-mediated polymerization system, the so called ATRP (Atom Transfer Radical Polymerization) seems the most robust since it can be successfully applied to the "living"/controlled polymerization of styrenes, acrylates, methacrylates, acrylonitrile, and isobutene. Significantly, both TEMPO and metal-mediated polymerization systems allow molec-

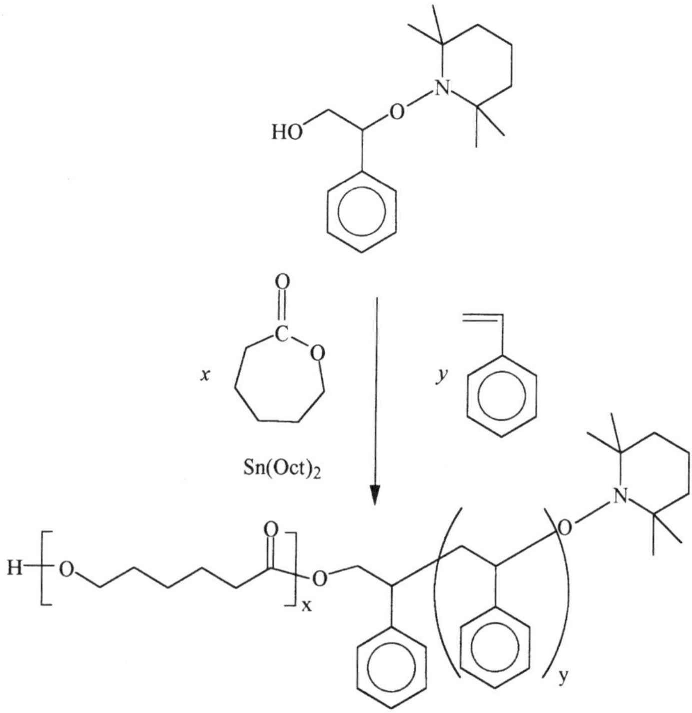

Scheme 24

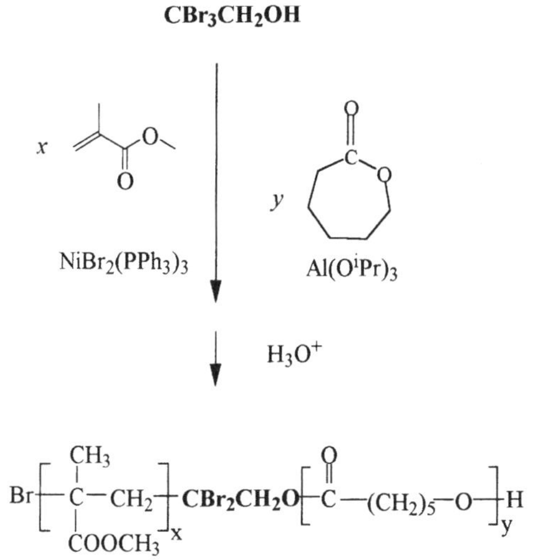

Scheme 25

ular weight and chain end functionality to be controlled accurately, while polydispersity remains narrow (PDI<1.5). In collaboration with Hawker et al., we have recently combined this new radical chemistry with aluminum alkoxide initiated ROP of lactones in order to obtain well-defined block and graft copolymers. Quite interestingly, we introduced the original concept of performing simultaneous dual living polymerizations as a novel one-step approach to block copolymers [83]. For example, from a hydroxy functionalized alkoxyamine initiator, both living ROP of εCL catalyzed by $Sn(Oct)_2$ and controlled free radical polymerization of styrene take place without the need for any intermediate steps (Scheme 24). It means that two different chains, i.e., PCL and PS chains, are growing away from the "biheaded" initiator, simultaneously and via two drastically different mechanisms.

In a very similar way, hydroxy functionalized ATRP initiators such as 2,2,2-tribromoethanol can be used for the simultaneous polymerization of εCL and MMA (Scheme 25) [83]. Purposely, the ROP of εCL is promoted by $Al(OiPr)_3$ added in catalytic amount so that the rapid alcohol-alkoxide exchange reaction (see Sect. 2.4) activates all the hydroxyl functions. In order to avoid initiation by the isopropoxy groups of $Al(OiPr)_3$. The in-situ formed *i*PrOH is removed by distillation of the *i*PrOH/toluene azeotrope. On the other hand, the ATRP of MMA is catalyzed by $NiBr_2(PPh_3)_3$. The two aforementioned one-step methods provide block copolymers with controlled composition and molecular weights, but with a slightly broad MWD (PDI=1.5–2).

In a more conventional approach, poly(S-*b*-CL) and poly(MMA-*b*-CL) block copolymers have been prepared from the same components as described previously, but in a two-step process via macromolecular initiators [47]. In a first step,

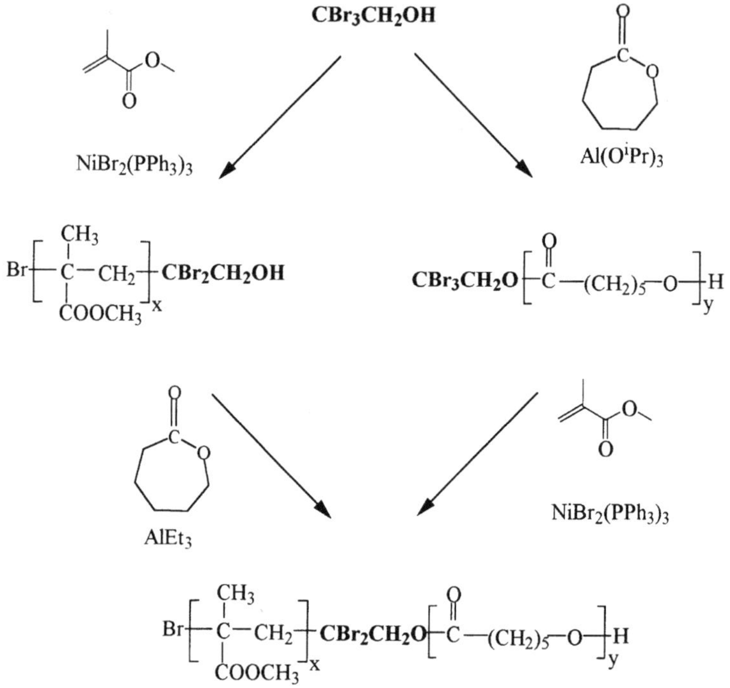

Scheme 26

α-hydroxy PS and α-hydroxy PMMA have been prepared by controlled radical polymerization. After transformation of the hydroxyl end groups into aluminum alkoxide functions, these polymeric initiators can be used to initiate the ROP of εCL (Scheme 26, shown for poly(MMA-b-CL) copolymers). On the other hand, α-TEMPO and α-CBr$_3$ PCL prepared by living coordinative ROP initiated with the hydroxy functionalized alkoxyamine and CBr$_3$CH$_2$OH have been able to initiate quantitatively the controlled polymerization of styrene and methyl methacrylate, respectively. By this two-step method, block copolymers with controllable composition and molecular weight for each blocks were recovered with narrow MWD (PDI=1.1–1.3). This strategy has been extended successfully to other monomers such as D,L-LA and L-LA and functional methacrylates, e.g., 2-hydroxyethyl methacrylate [84]. Furthermore, the isolated diblock copolymers are again potential macroinitiators for additional ROP of (di)lactones or ATRP of unsaturated monomers, paving the way to novel ABC triblock or other multiblock copolymers.

Concerning the synthesis of graft copolymers, Jedlinski et al. have prepared poly(MMA-g-βBL) copolymers via anionic grafting of βBL from a modified PMMA backbone [85]. PMMA chains were partially saponified by potassium hydroxide and complexed by 18C6 crown ether so as to act as multifunctional mac-

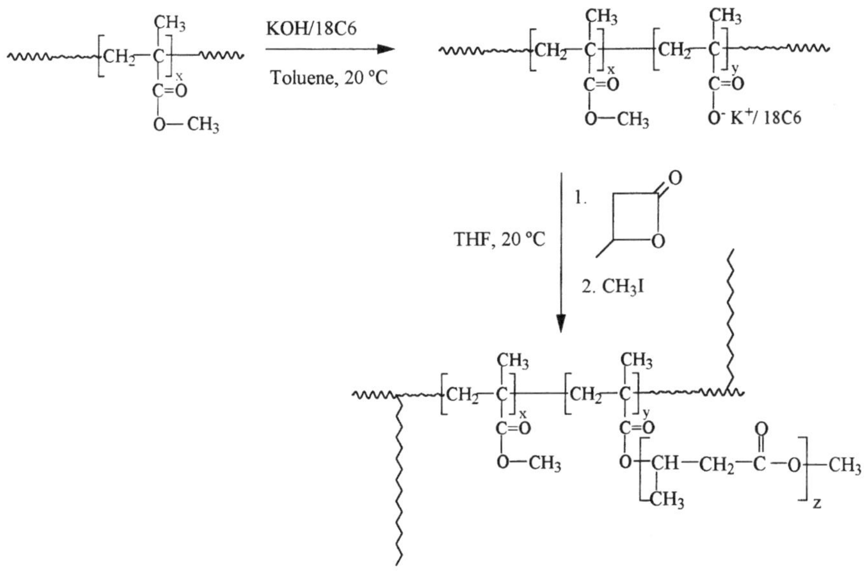

Scheme 27

roinitiators for the βBL anionic polymerization (Scheme 27). Although the graft efficiency was high and the branch density easily predetermined by the degree of saponification, the use of poorly defined PMMA backbones, actually obtained by a conventional non-controlled free radical polymerization, yielded graft copolymers with broad MWD. In a similar approach, Caywood et al. prepared poly(ethylacrylate-g-pivalolactone) copolymers which displayed interesting thermoplastic elastomer properties [86].

Some of us [46] and Egiburu et al. [87] have studied the "macromonomer technique" as a method for producing PCL- and PLA-containing graft copolymers. Purposely, PLA and PCL macromonomers have been prepared by εCL and LA coordinative ROP as initiated by the reaction product of 2-hydroxyethyl methacrylate (HEMA) and AlEt$_3$ (see Scheme 5). These polyester macromonomers have been copolymerized with styrene and methacrylate derivatives by using AIBN as a free radical initiator. Nevertheless, due to the free radical process used in the copolymerization step, there was a lack of control over the formation of the main backbone. Graft copolymers polydispersed in compositions and molecular weights were accordingly obtained. In order to overcome this drawback we have examined again both metal- and nitroxyl-mediated controlled/"living" radical polymerization [88]. According to the synthetic strategies depicted in Scheme 28, a variety of graft copolymers were prepared among different styrenic and (meth)acrylic monomers and PCL and PLA grafts. The graft copolymers were characterized by controlled compositions, predetermined molecular weights for backbone and grafts, and narrow MWD (PDI=1.2–1.4). Fur-

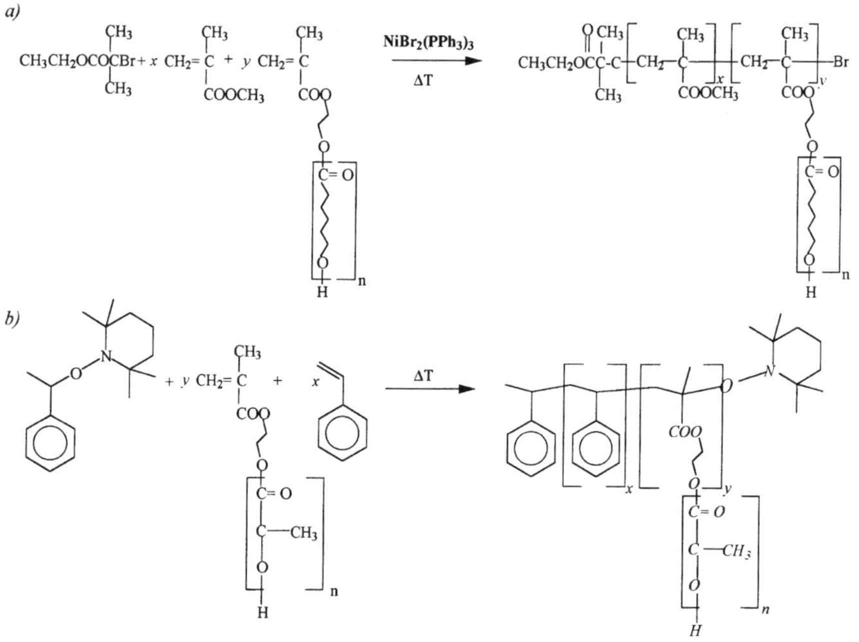

Scheme 28

thermore, the radical process allowed the introduction of functional groups into the graft copolymers by copolymerization of HEMA, methacrylic acid, etc.

3.5
Ring Opening Metathesis Polymerization

The recent developments of ring-opening polymerization (ROMP) have opened new avenues for synthesizing new polymeric materials [89]. Of particular importance is the emergence of ruthenium-based catalyst due to its versatility and compatibility with different polar functionalities and a diminished sensitivity to atmospheric oxygen and water [90]. Our laboratory has pursued the synthesis of graft copolymers by combining coordinative ROP and ROMP processes. To this end, α-norbornenyl PCL macromonomers were prepared by using a purposely functionalized aluminum monoalkoxide as initiator (see Scheme 5) [50]. Once again, the molecular parameters of the PCL macromonomers were perfectly controlled. Copolymerization of norbornene and the α-norbornenyl PCL macromonomers was promoted by a ruthenium based catalyst developed by Noels and Demonceau (Scheme 29) [90]. The active catalyst is generated in-situ by reaction of trimethylsilyldiazomethane (TMSD) and [RuCl$_2$(p-cymene)]$_2$ in the presence of tricyclohexylphosphine (PCy$_3$) in chlorobenzene at 60 °C. Under these experimental conditions, graft copolymers with a quite narrow MWD

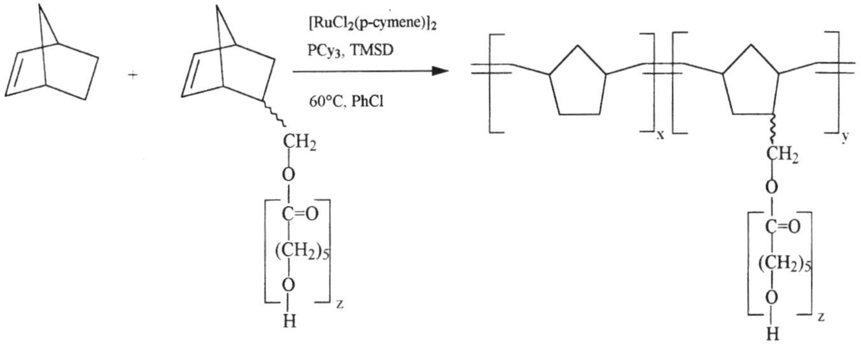

Scheme 29

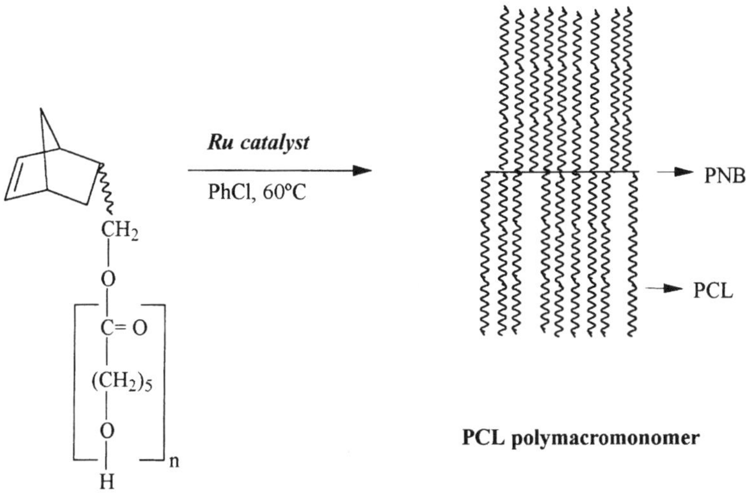

Scheme 30

PCL polymacromonomer

(PDI=1.2–1.4) were prepared at the strict condition to control kinetically the copolymerization step. Actually, long reaction times are much more favorable to transfer side reactions with a concomitant broadening of the MWD and the formation of cyclic chains.

Very recent work has demonstrated the ability of ROMP to homopolymerize α-norbornenyl PS and PEO macromonomers in a controlled way and with high yields [91]. This new strategy has led to the preparation of PCL polymacromonomers by homopolymerization of the above-mentioned α-norbornenyl PCL macromonomers using the same Ru based catalyst (Scheme 30) [50]. It is worth pointing out the quantitative recovering of high molecular weight polymacromonomers with very narrow MWDs (PDI=1.10) as observed by SEC (Fig. 4).

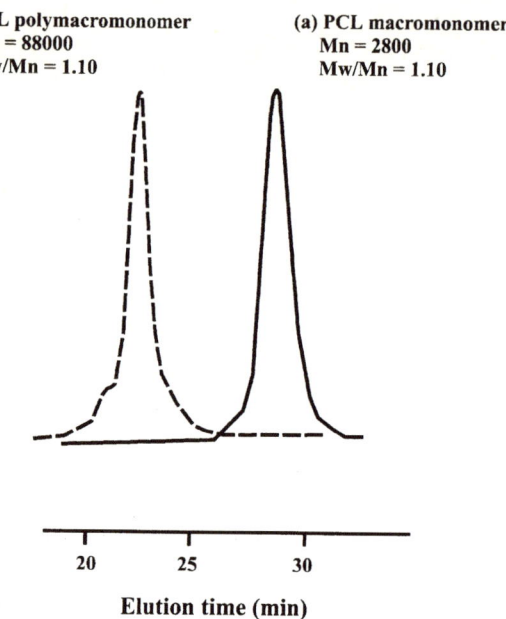

Fig. 4. SEC traces for PCL macromonomer (*curve a*) and PCL polymacromonomer (*curve b*) For synthetic conditions, see Scheme 30

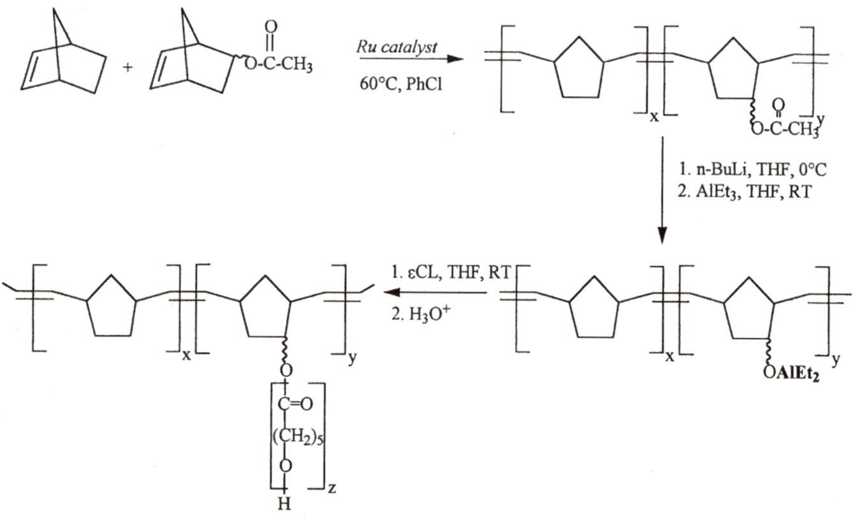

Scheme 31

Moreover, alcohol functionalities have been introduced into the polynorbornene (PNB) backbone by copolymerization of norbornene with a few percent of 5-acetate norbornene and subsequent acetate reduction. After transformation of the pendant hydroxyl functions into diethyl aluminum alkoxides, εCL has been ring opening polymerized (Scheme 31). Owing to the controlled/"living" character of both polymerization processes the isolated poly(NB-g-CL) graft copolymers were characterized by well-defined composition, controlled molecular weight and branching density, and narrow MWD (PDI=1.2–1.4) [92].

3.6
Polycondensation

Although polycondensation does not lead to well-defined polymers with precisely controlled molecular weight and narrow MWD, it offers easy access to polymer families and structures difficult or impossible to be obtained through a polyaddition process. For instance, the synthesis of multiblock copolymers is readily achieved by the step-growth process. Despite their commercial interest, only a few examples of the introduction of aliphatic polyester oligomers as building blocks in polycondensates have been reported in the scientific literature. As an example, Maglio et al. used hydroxy-terminated PL-LA oligomers, coded as HO-(PLA)-OH, together with sebacoyl dichloride and 1,6-hexanediamine in the synthesis of random multiblock PL-LA-polyamide copolymers by means of a two-step polycondensation (Scheme 32) [93].

Another approach was attempted by Seppala and Kylma who reported the synthesis of poly(ester-urethane)s by condensation of hydroxyl terminated telechelic poly(CL-co-LA) oligomers with 1,6-hexamethylene diisocyanate (Scheme 33) [94]. The diisocyanate acts as chain extender producing an increase in molecular weight of the preformed oligomers. The authors claim that some of the copolymers present elastomeric properties. Using a similar method, Storey described the synthesis of polyurethane networks based on D,L-LA, GA, εCL,

x HO-(PLA)-OH + y ClC(O)-R-C(O)Cl $\xrightarrow[(-2xHCl)]{}$ ClC(O)-R-C(O)O-(PLA)-O(O)C-R-C(O)Cl

x ClC(O)-R-C(O)O-(PLA)-O(O)C-R-C(O)Cl + (y - 2x) Cl(O)C-R-C(O)Cl + (y-x) H$_2$N-R'-NH$_2$

$\downarrow$

—[C(O)-R-C(O)O-(PLA)-O(O)C-R-C(O)-NH-R'-NH]$_n$-[C(O)-R-C(O)-C-NH-R'-NH]$_m$—

with R = -(CH$_2$)$_8$-
R'= -(CH$_2$)$_6$-

Scheme 32

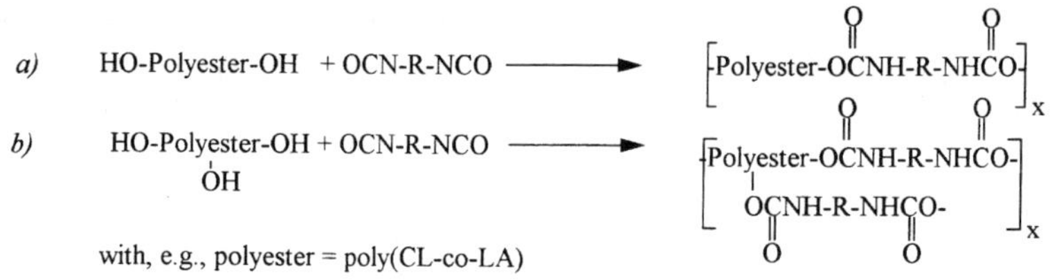

a) HO-Polyester-OH + OCN-R-NCO ⟶ [-Polyester-OCNH-R-NHCO-]$_x$

b) HO-Polyester-OH + OCN-R-NCO ⟶
       ÓH

$$\begin{bmatrix} \text{-Polyester-OCNH-R-NHCO-} \\ \text{OCNH-R-NHCO-} \end{bmatrix}_x$$

with, e.g., polyester = poly(CL-co-LA)

Scheme 33

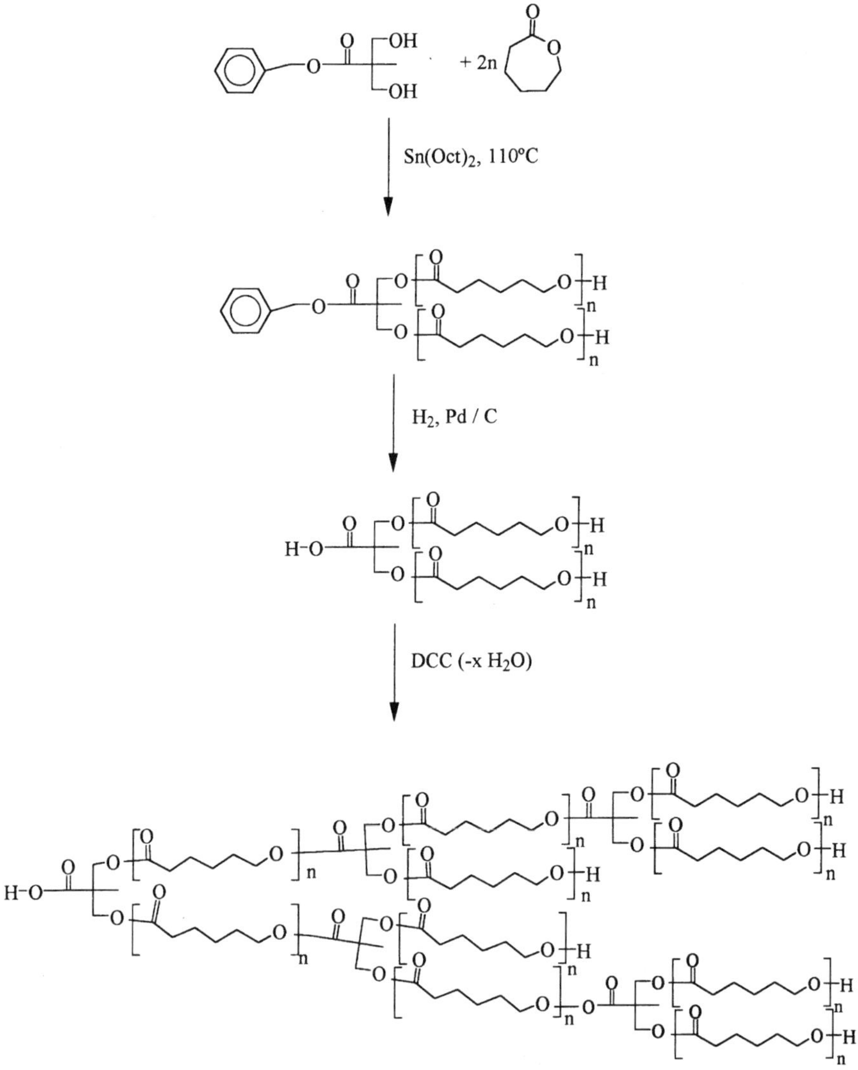

Scheme 34

and TMC (Scheme 33b) [95]. The condensation of preformed trifunctional oligomers with tolylene-2,6-diisocyanate triggered the network formation.

In an original way, hyperbranched aliphatic polyesters (PCL) have been prepared by condensation of AB_2 macromonomers (Scheme 34) [96]. The PCL macromonomers were prepared by living ROP of εCL coupled with a sequence of protection-deprotection reactions. The AB_2 macromonomers were self-polymerized in the presence of dicyclohexylcarbodiimide (DCC) producing the expected hyperbranched polyesters. Other AB_2 macromonomers have also been synthesized by copolymerization of εCL and a few molar percent of TOSUO. After self-polycondensation, the hyperbranched polyesters so obtained have been treated successively with $(C_6H_5)_3CBF_4$ and $NaBH_4$ in order to remove the ethylene ketal protection and to reduce the resulting ketone functions, respectively. Pendant hydroxyl groups are therefore made available along the PCL branches and can initiate further polymerizations of εCL catalyzed by $Sn(Oct)_2$, for instance. Highly branched polycaprolactones, i.e., dendri-graft PCL, with significant additional functionalities have been accordingly synthesized [121].

Last but not least, some of us have recently synthesized polyimide-aliphatic polyester triblock and graft copolymers in collaboration with Hedrick and his coworkers [97, 98]. Well-defined aminophenyl or diaminophenyl end-functional polyester oligomers have been synthesized on purpose and used as end-cappers or macromonomers leading to the aforementioned triblock or graft copolymers, respectively. The polyimide-polyester copolymers so obtained proved to be highly efficient promoters of polyimide nanofoams (for more details see Sect. 4.2).

3.7
Dendritic Construction

Pioneer works by Frèchet and coworkers have reported on the synthesis of polymers with hybrid linear-globular architectures [99]. Generally, the synthesis has been accomplished by end-capping a preformed linear polymer with a complementary monofunctionalized dendrimer. An alternative approach consists of using convergent dendrimers with a single reactive group at their focal point, as macroinitiator. For instance, poly(aromatic ether) dendrimers bearing a potassium alcoholate at the focal point appeared to be able to initiate the polymerization of εCL and to offer a better kinetic control by comparison to anionic ROP as initiated by simple low molecular weight potassium alkoxides [100]. The authors explain that the bulky dendritic moiety is able to screen, at least temporarily, the active anionic centers and to delay the occurrence of inter- and intramolecular transesterification reactions. Nevertheless, prolonged reaction times again proved to be favorable for promoting the formation of cyclic oligomers. More recently, we have examined the use of similar polyether dendrons bearing a diethylaluminum alkoxide at their focal point, as macroinitiators for the coordinative ROP of εCL and LA in toluene at 25 °C and 70 °C, respectively (Scheme 35; shown for εCL ROP) [101]. Hybrid linear-globular AB-type block

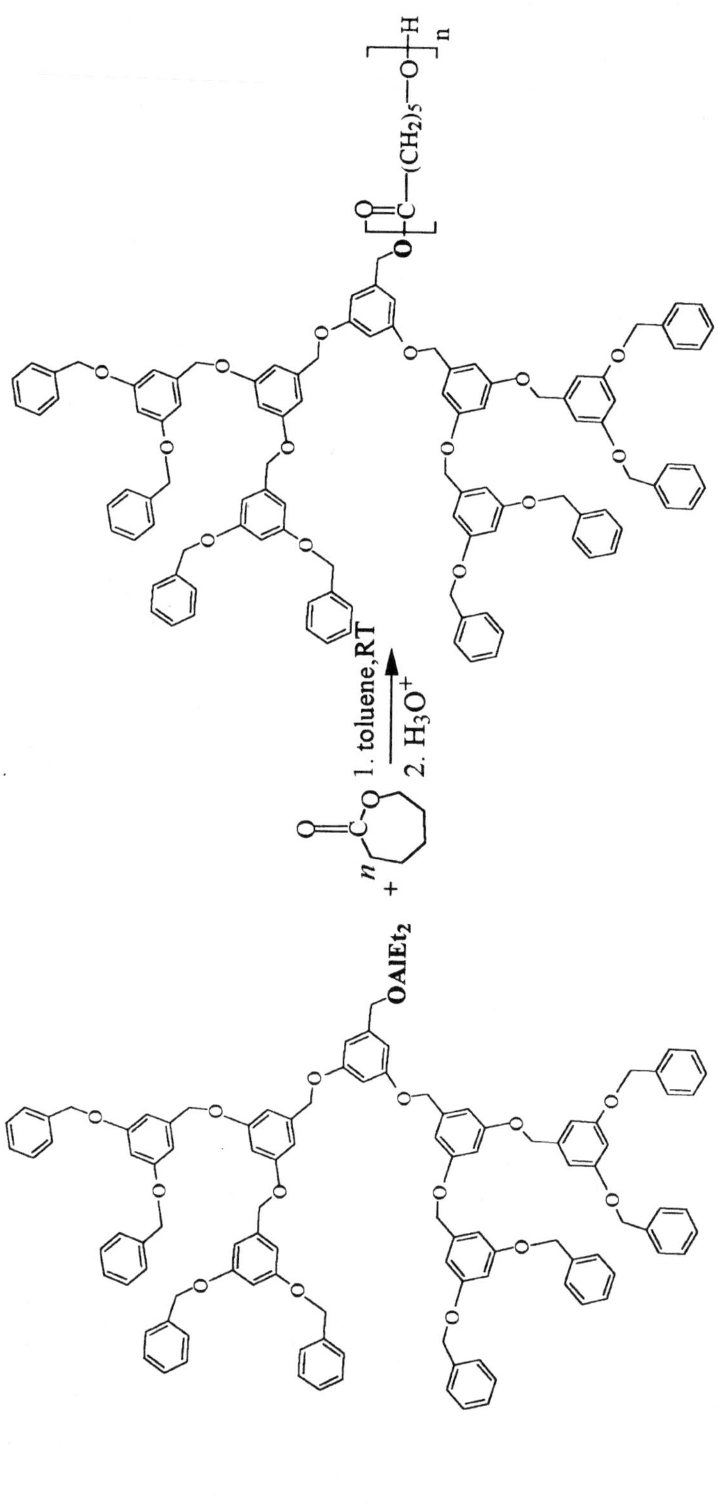

Scheme 35

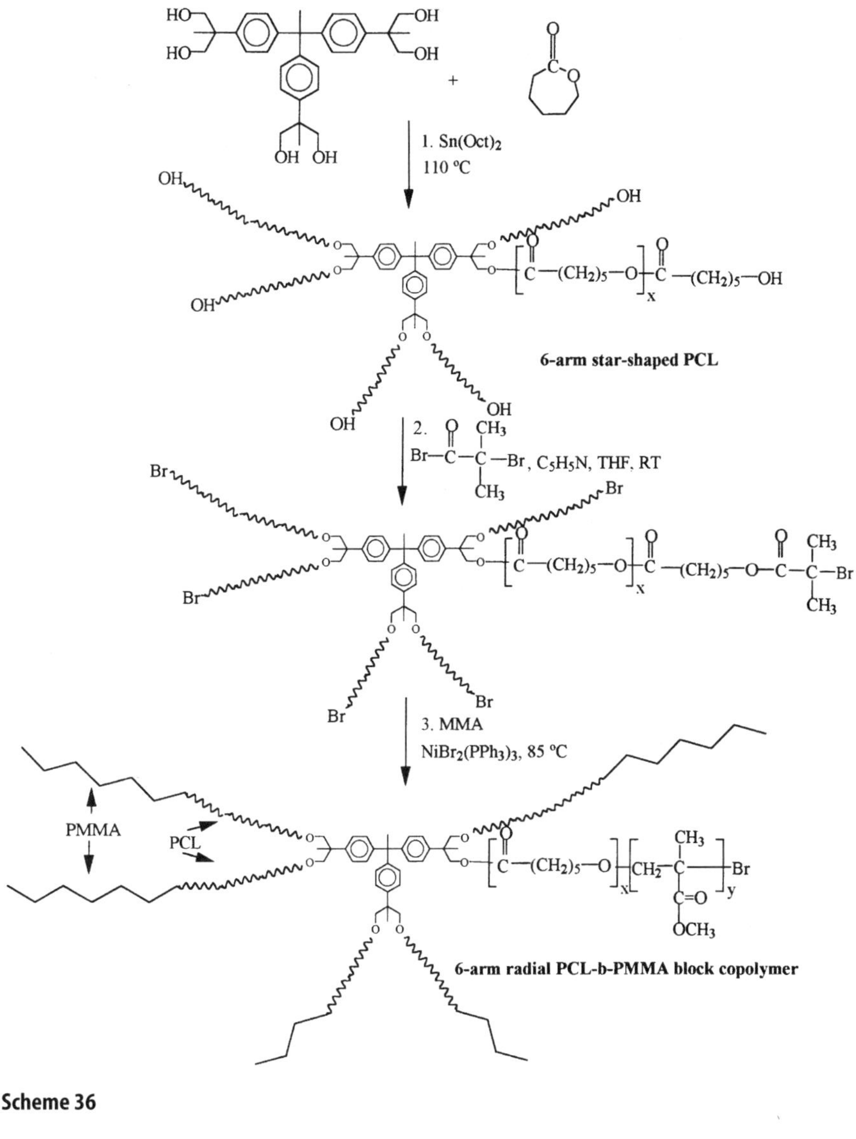

Scheme 36

copolymers with well tailored substructures and narrow MWD have been synthesized with high yields under mild conditions.

In a different approach, Hedrick et al. have studied multifunctional dendritic initiators for the synthesis of multiarm star-shaped copolymers [102]. Several dendritic initiators with hydroxyl functionality ranging from 2 to 48 have been prepared according to the method developed by Hult et al. [120]. The bulk polymerization of εCL initiated by these multifunctional macroinitiators and activated by either Sn(Oct)$_2$ or Al(OiPr)$_3$ as catalyst (see Sect. 3.3) yielded star-shaped copolymers with unimodal and narrow MWD (PDI=1.1–1.3) (Scheme 36, shown for a 6-arm star). Whatever the hydroxyl functionality of the starting dendritic initiator the initiation efficiency evaluated by ^{13}C NMR analysis was very high (>90%). Furthermore, the resulting hydroxy end-functionality of the star-shaped copolymers was explored to prepare novel radial block copolymers. Reaction of every hydroxyl end-group with 2-bromo-2-methyl propanoyl bromide provided a multifunctional macroinitiator very efficient in MMA ATRP (see Sect. 3.4). Following this strategy, an ω-bromo 6-arm PCL star as ATRP macroinitiator was quantitatively converted into a 6-arm radial poly(CL-b-MMA) block copolymer with controlled molecular weights for each block and narrow apparent MWD (PDI=1.16) (Scheme 36).

3.8
Coordination Polymerization

Recent advances in the development of well-defined homogeneous metallocene-type catalysts have facilitated mechanistic studies of the processes involved in initiation, propagation, and chain transfer reactions occurring in olefins coordinative polyaddition. As a result, end-functional polyolefin chains have been made available [103]. For instance, Waymouth et al. have reported about the formation of hydroxy-terminated poly(methylene-1,3-cyclopentane) (PMCP-OH) via selective chain transfer to the aluminum atoms of methylaluminoxane (MAO) in the cyclopolymerization of 1,5-hexadiene catalyzed by di(pentamethylcyclopentadienyl) zirconium dichloride (Scheme 37). Subsequent equimolar reaction of the hydroxyl extremity with AlEt$_3$ afforded an aluminum alkoxide macroinitiator for the coordinative ROP of εCL and consecutively a novel poly(MCP-b-CL) block copolymer [104]. The diblock structure of the copolymer

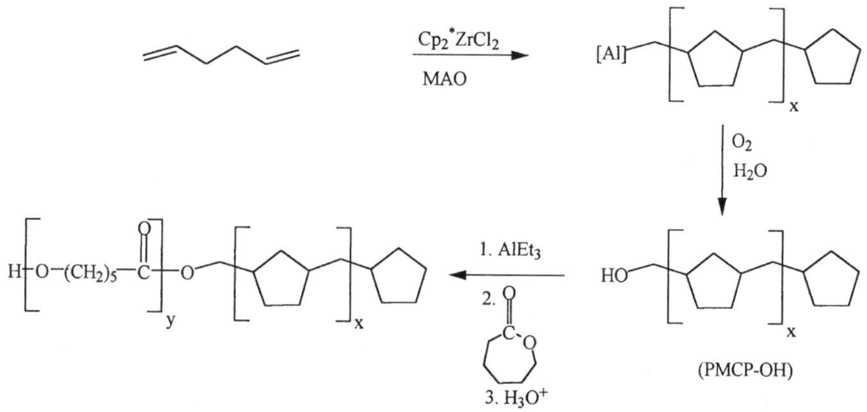

Scheme 37

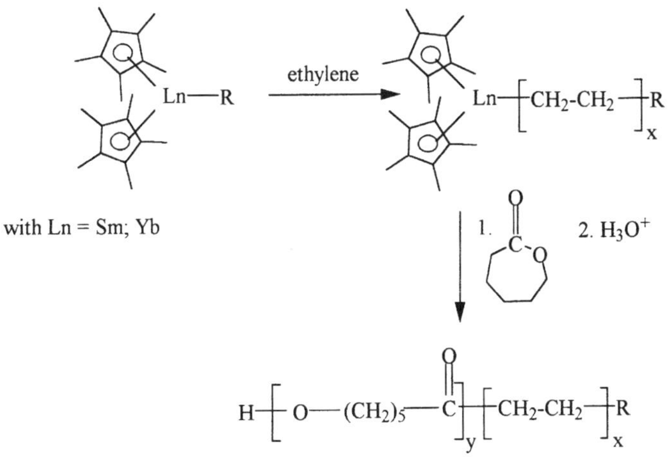

Scheme 38

has been confirmed by selective fractionation and ^{1}H NMR; however the MWD was quite broad as a result of the lack of control in the first 1,5-hexadiene polymerization (PDI=2–5).

Finally, interesting research work has recently been published on the ability of organolanthanide complexes to promote the polymerization of monomers such as (meth)acrylates, lactones, and epoxides in a "living" manner, and olefins but with rather poor control [103]. Accordingly, Yasuda has synthesized block copolymers of MMA and lactones, e.g., εCL and δVL, by sequential addition of the two comonomers [103]. When MMA was added first the copolymerization proceeded smoothly and gave copolymers with narrow MWD (PDI=1.1–1.3). Upon reversed addition of the respective monomers, the copolymerization did not proceed at all, i.e., the polylactone active end-species were totally unable to add an MMA unit. Similarly, original poly(ethylene-*b*-caprolactone) block copolymers have been prepared by sequential addition of ethylene and εCL (Scheme 38). Again, ethylene needs to be added first. Even though the block copolymer structure has been confirmed by GPC-FTIR, the MWD so obtained was relatively broad and one may not exclude the formation of homopolymers (PDI=1.5–2).

4
Aliphatic Polyesters as Building Blocks for New Materials

4.1
Biodegradable and Biocompatible Thermoplastic Elastomers

Since their commercial introduction by Shell in 1965, the poly(styrene-*b*-(butadiene or isoprene)-*b*-styrene) triblock copolymers (SBS or SIS) have attracted a

Table 4. Tensile properties of the poly(CL-*b*-DXO-*b*-CL) ABA triblock copolymers (see Scheme 39)

Block lengths A/B/A	15 K/50 K/15 K	15 K/70 K/15 K
Tensile modulus (MPA)	31	21
Stress at yield (MPA)	3.0	2.1
Elongation at yield (%)	15	18
Strength at break (MPA)	52	53
Elongation at break (%)	1070	1210

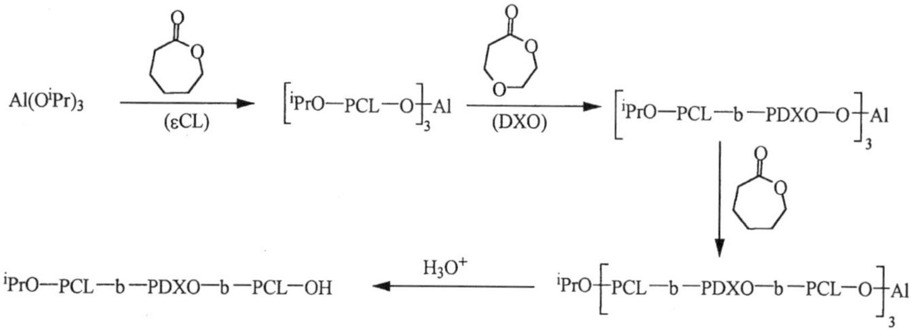

Al(O^{i}Pr)$_3$ $\xrightarrow{(\varepsilon CL)}$ [^{i}PrO—PCL—O]$_3$Al $\xrightarrow{(DXO)}$ [^{i}PrO—PCL—b—PDXO—O]$_3$Al

^{i}PrO—PCL—b—PDXO—b—PCL—OH $\xleftarrow{H_3O^+}$ ^{i}PrO[PCL—b—PDXO—b—PCL—O]$_3$Al

Scheme 39

lot of attention regarding their elastomeric properties. The unique thermomechanical properties of these SBS or SIS thermoplastic elastomers relies upon the microphase separation of the PS hard blocks into glassy microdomains dispersed in a continuous rubbery polydiene matrix. Such a phase morphology provides a physical network of flexible chains cross-linked by thermoreversible glassy microdomains. Clearly, these materials combine the mechanical performances of vulcanized rubbers and the straightforward processing of thermoplastics. SBS and SIS triblock copolymers are currently synthesized by a three-step anionic living copolymerization, typically using butyllithium as initiator. Albertsson and some of us have investigated the synthesis of all aliphatic polyester based ABA triblock copolymers and characterized their performances as original biodegradable and biocompatible thermoplastic elastomers. In fact, poly(CL-*b*-DXO-*b*-CL) symmetrical triblock copolymers with high molecular weights and narrow MWD (PDI=1.20) have been synthesized by sequential coordinative ROP of the monomers, i.e., εCL and DXO, as initiated by Al(O*i*Pr)$_3$ in toluene at RT (Scheme 39) [26, 106]. For low PCL contents, the PCL teleblocks form separate semi-crystalline microdomains dispersed in a continuous PDXO rubbery matrix, the glass transition temperature of which is well below RT (T_g= –36 °C, see Table 1). A temperature increase above 60 °C, which is the melting temperature of the PCL physical cross-linking points of the three-dimensional network, allows for an easy melt material processing.

Of prime interest are the tensile properties summarized in Table 4, and typical of stress-strain curves exhibited by thermoplastic elastomers. The elongation and strength at break were measured above 1000% and 50 MPA, respectively. Both the tensile modulus and the stress at yield increased by increasing the PCL relative content whereas, as expected, the ultimate elongation at break slightly decreased.

To sum up, the "living" character of the aluminum alkoxide mediated ROP of lactones has permitted the synthesis of novel ABA triblock copolymers, the composition and molecular weight of which can purposely be tuned up for displaying excellent elastomeric properties. Interestingly, the inherent biodegradability of each partner, PCL and PDXO, would open up new applications for these novel thermoplastic elastomers.

4.2
Polyimide Nanofoams

Polyimides are currently used as interlayer dielectrics in *microelectronic packaging*, since they have the requisite properties to survive the thermal, chemical, and mechanical stress of manufacturing and operation. The main advantage realized by the use of polyimides over inorganic alternatives is the lower dielectric constant. A reduction in the dielectric constant of the medium reduces pulse propagation delay, allowing for faster machine time, and minimizes the noise between lines. The most common approach for modifying the dielectric properties of polyimides has been via the incorporation of fluorinated substituents, such as hexafluoroisopropylidene linkages or pendant trifluoromethyl groups. While this approach reduces water absorption and the dielectric constant significantly, the mechanical and thermal properties are often compromised. Another means of reducing the dielectric constant while maintaining the desired thermal and mechanical properties is by the generation of a foam. The reduction in the dielectric constant is simply achieved by replacing the polymer with air, which has a dielectric constant of 1. However, it is obvious that the pore size must be much smaller than the film thickness or any microelectronic features. In order to meet the last statement, a new method of generating a polyimide foam with pore sizes in the nanometer regime has been developed by Hedrick and his coworkers at IBM [107]. This approach involves the use of phase separated block or graft copolymers comprised of a high thermally stable polymer and a second component, which can undergo clean thermal decomposition with the evolution of volatile by-products to form a closed-cell, porous structure. Block copolymers are well known for presenting different morphologies, interestingly on the nanometer scale and which depend on their nature and composition. By designing the block copolymers such that the matrix material is a thermally stable polymer with a high T_g and the dispersed phase is a labile polymer that undergoes thermolysis at a temperature below the T_g of the matrix, one can prepare foams with pores in the nanometer dimensional regime. The foam formation takes

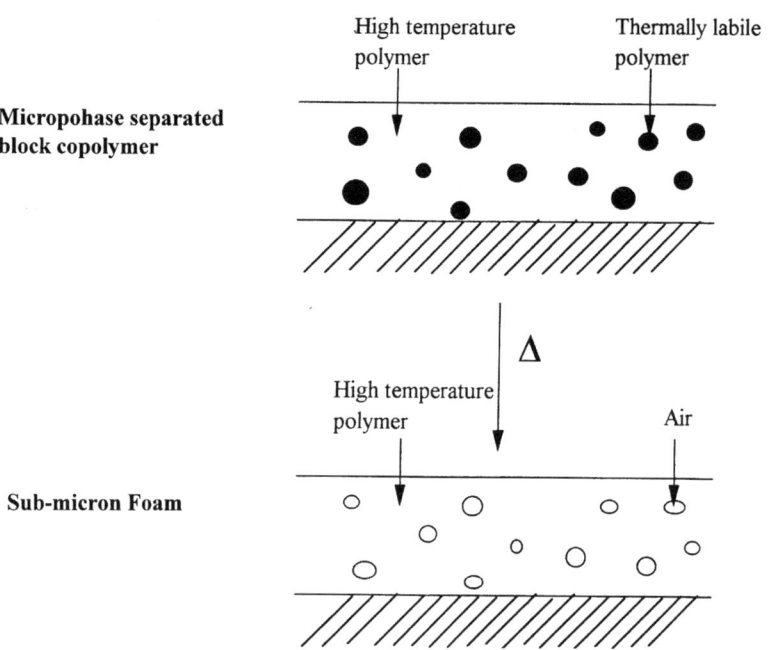

Fig. 5. Illustration of the block copolymer approach to nanofoam formation

place by thermal degradation of the labile block, leaving behind pores the size and shape of which depend on the initial block copolymer morphology (Fig. 5).

Nanofoams have been prepared first by using poly(propylene oxide) as the labile coblock with different polyimides for the continuous matrix [108]. The degradation of the poly(propylene oxide) component can be achieved at 300 °C in an oxygen or air environment via a thermooxidative degradative mechanism which is, however, unacceptable in microelectronic fabrication. In order to overcome this difficulty, aliphatic polyesters have been studied as labile coblocks [97, 98]. Aliphatic polyesters degrade quantitatively in an inert atmosphere into cyclic monomers and other products of low molecular weight. In addition, the decomposition temperature is much lower than the T_g of many polyimides. Concerning the synthetic requirements, the first important step was the availability of preparing well-defined oligomers having controlled molecular weight and functionality. Like the previously studied polyether blocks, polyesters end-capped with a single or double aryl-amine functionality were needed to be amenable towards polyimide copolymerization. The synthesis of such ω-functional polyesters has been accomplished by reacting monofunctional living aluminum alkoxide terminated oligomers with 4-nitrophenyl chloride or 3,5-dinitrophenylchloride in the presence of pyridine (Pyr) and further modification by hydrogenation over Pearlman's catalyst, giving the corresponding and desired (di)amine (Scheme 40a,b). Alternatively, the single aryl-amine functionality has

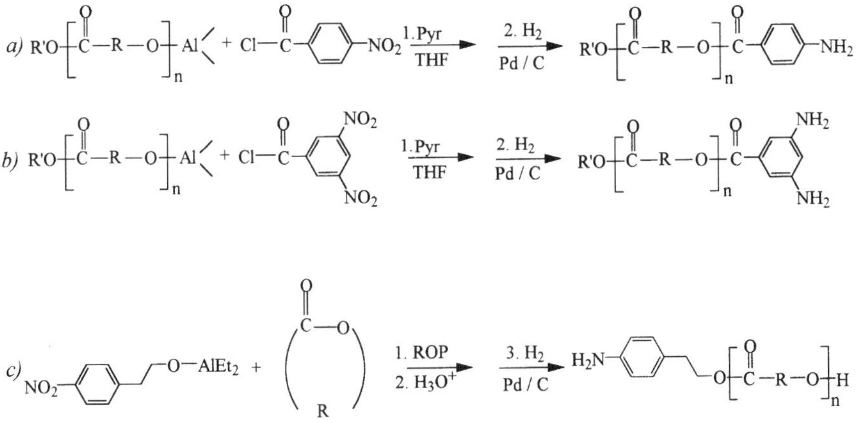

Scheme 40

been incorporated by using diethyl aluminum 2-*p*-nitrophenyl-ethoxide as initiator of the lactone ROP (Scheme 40c). It is worth remembering that the nitro protecting group was required instead of the primary amine due to its ability to co-initiate the coordinative ROP (see Sect. 2.3).

The next step concerned the synthesis of the polyester-polyimide copolymers. Triblock copolymers have been prepared by a step-growth copolymerization of stoichiometric amounts of an aromatic diamine and dianhydride (e.g., PMDA and 3FDA, as depicted in Scheme 41a) added with the single ω-amino polyesters as chain end-cappers. Graft copolymers can be prepared as well. In this case, the diamine end-functionalized oligomeric macromonomers are copolymerized with the polyimide condensation comonomers (Scheme 41b).

Block and graft copolymers based on PMDA-3FDA polyimide and either PVL, PCL or PD,L-LA have been accordingly prepared. The copolymers were characterized by ^{1}H NMR, DSC, TGA, SAXS, and DMTA. Microphase separation was observed in all the studied copolymers and the polylactone blocks were not able to crystallize. Films of the polycondensates were prepared by solution casting from *N*-methyl pyrrolidone followed by a slow heating increase up to 300 °C. This thermal treatment removed the residual solvent and imidized the prepolymer without degrading the labile block [97]. Then different cure treatments were applied in order to degrade efficiently the labile polyester coblocks. The use of PCL as the thermally labile coblock was a successful route to polyimide nanofoams, although the PVL and PD,L-LA based copolymers did not show the expected nanopore formation. The explanation was found in the plasticization of the polyimide matrix by the PVL and PD,L-LA degradation products which produce the collapse of the porous structure. Although some collapse was also observed in the case of PCL-based copolymers, the foam efficiency was high specially for cross-linked polyimide matrix. The nanofoams were characterized by a variety of experiments including TEM, SAXS, density, and refractive index

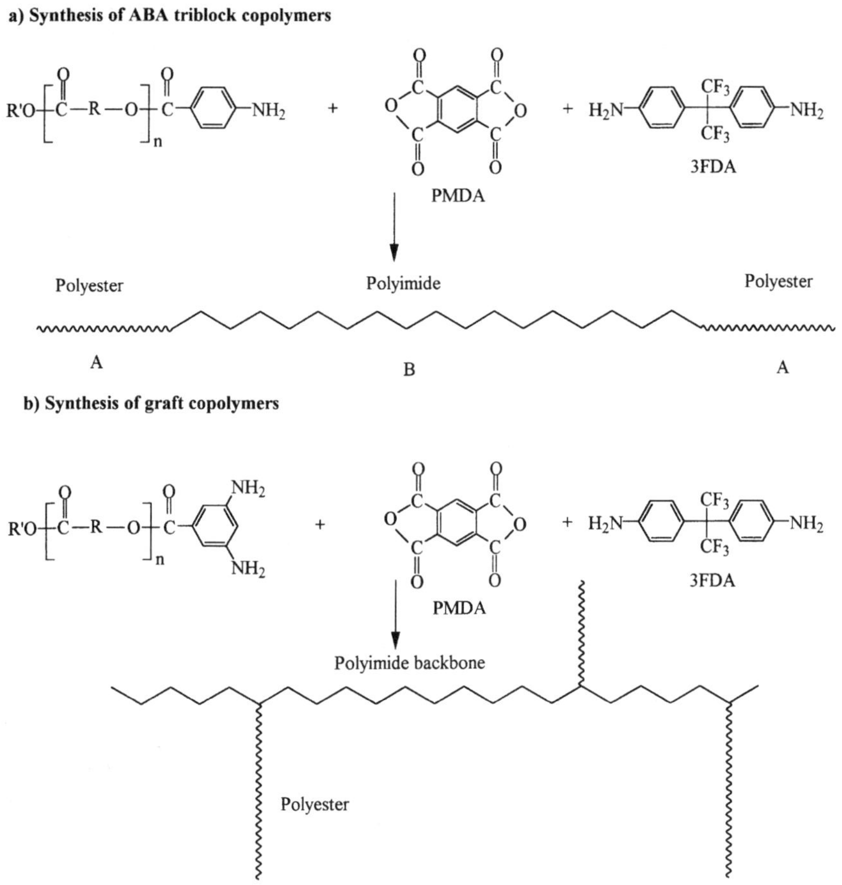

a) Synthesis of ABA triblock copolymers

b) Synthesis of graft copolymers

Scheme 41

measurements. Thin organic film with pores of 10–20 nm and volume fraction of voids 10–20% were obtained by this method and were characterized by dielectric constants as low as 2.4.

4.3
Organic-Inorganic Nanocomposites

During the last few years, special attention has been paid to the design and characterization of inorganic-organic hybrid materials prepared by the sol-gel process. This interest has been driven by the opportunity offered by this method to combine in a quite controlled way the most remarkable properties of inorganic glasses and organic polymers. In order to obtain a successful incorporation and a nanoscopic dispersion of the organic polymers into the hybrid materials, a

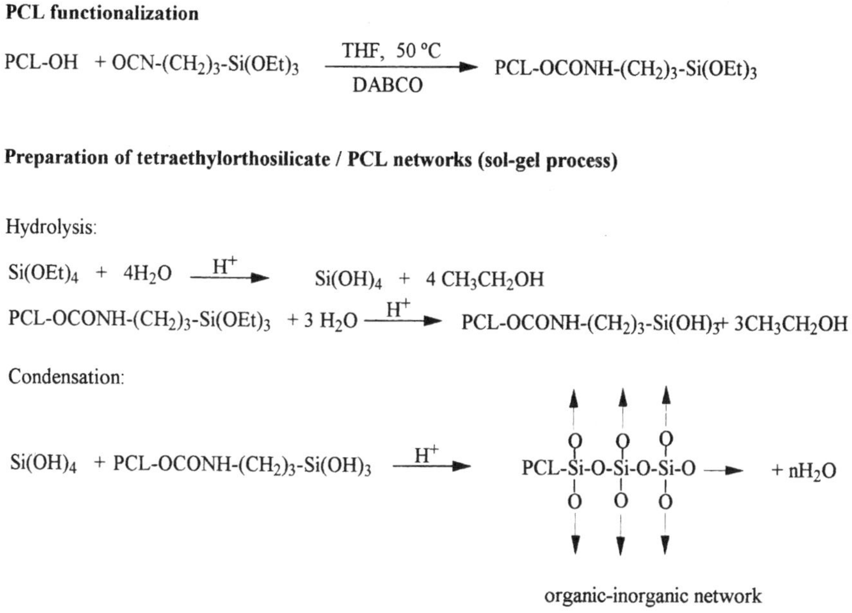

PCL functionalization

PCL-OH + OCN-(CH$_2$)$_3$-Si(OEt)$_3$ $\xrightarrow[\text{DABCO}]{\text{THF, 50 °C}}$ PCL-OCONH-(CH$_2$)$_3$-Si(OEt)$_3$

Preparation of tetraethylorthosilicate / PCL networks (sol-gel process)

Hydrolysis:

Si(OEt)$_4$ + 4H$_2$O $\xrightarrow{\text{H}^+}$ Si(OH)$_4$ + 4 CH$_3$CH$_2$OH

PCL-OCONH-(CH$_2$)$_3$-Si(OEt)$_3$ + 3 H$_2$O $\xrightarrow{\text{H}^+}$ PCL-OCONH-(CH$_2$)$_3$-Si(OH)$_3$+ 3CH$_3$CH$_2$OH

Condensation:

Si(OH)$_4$ + PCL-OCONH-(CH$_2$)$_3$-Si(OH)$_3$ $\xrightarrow{\text{H}^+}$ PCL-Si-O-Si-O-Si-O $\longrightarrow$ + nH$_2$O

organic-inorganic network

Scheme 42

prerequisite is needed – the existence of strong interactions between the inorganic and organic components. Therefore, polymers prone to hydrogen bonding with residual protic moieties, e.g., Si-OH, have been successfully incorporated into silicon oxide networks. Another method consists of attaching covalently the organic component to the inorganic material by using tailored-functionalized polymers, which can participate in the sol-gel process via a condensation with the in-situ generated Si-OH groups, for instance. Aliphatic polyesters are ideally suited for being incorporated into hybrid materials because they can be readily end-capped with functional groups and their ester carbonyl functions can interact strongly and form hydrogen bonds with the inorganic component. On that basis, the preparation of biodegradable and biocompatible inorganic-organic hybrid materials by sol-gel process between tetraethoxysilane (TEOS) and PCL oligomers has been investigated in our laboratory [109–112]. Depending on the coordinative ROP conditions (nature of the aluminum alkoxide initiator, type of termination reaction, copolymerization of εCL with TOSUO, etc.), PCL can be end-capped with hydroxyl groups at one end, at both ends, or with pendant hydroxyl functions along the chain. Reaction of those hydroxyl groups with 3-isocyanatopropyltriethoxysilane has proved to be an easy and direct access to the introduction of inorganic functionalization into the PCL (Scheme 42). The oversimplified reaction pathway of the sol-gel process for the synthesis of PCL containing ceramers is also schematized. The network formation is a complex process where hydrolysis and condensation interplay depending on the reaction conditions. The extent of the PCL incorporation into the silica network depends on the PCL relative content and molecular weight, and the number and reactivity of the PCL functional groups, e.g., HO- or (EtO)$_3$Si- functions.

Totally transparent hybrid materials incorporating up to 50 wt% PCL into TEOS-based silica networks have been prepared [110]. PCL is so intimately incorporated into the polymer that it remains completely amorphous as confirmed by DSC and DMA analysis. TEM and image analysis achieved on thin films of ceramer containing 46 wt% PCL shows a co-continuous two-phase morphology with microdomains ca. 5 nm in size. These materials have a number of potential applications in the biomedical field. For example, the hybrid ceramers containing ca. 50 wt% PCL may be envisaged as a novel type of degradable bioglass. They could also be used as coating films for bone implants and prosthetic devices and, due to an unexpectedly high scratch resistance, they would be a valuable coating for organic polymers as well, particularly for polymers compatible with PCL, such as polycarbonate and PVC. On the other hand, bioglass can serve as support for enzyme immobilization and culture. To confirm their biocompatibility, fibroblast cultures were attempted to evaluate the cytotoxicity of the new hybrid materials. It has been observed that the extent of the cellular attachment depends on the PCL content of ceramers; the density of cells attached to the ceramer surface actually decreases when the PCL content is increased. In other words, the fibroblast adherence decreases with decrease of hydrophilicity of the substrate. As presumed, "in-vitro" biodegradation studies have clearly shown that the inherent biodegradability of PCL is preserved while incorporated in the nanocomposites [111].

Hedrick et al. have investigated the preparation of nanoporous inorganic oxides by sol-gel process carried out on hyperbranched polyesters (see Sect. 3.6) and polysilesquioxanes [96]. Interestingly, the organic species are used to template the inorganic counterpart and, after the completion of the polycondensation reaction, they can be selectively removed through thermolysis, leaving behind a nanoporous inorganic oxide. Similar investigations have been carried out by Chujo et al. using "starburst" dendrimers to generate porosity in silica gel [124]. Hyperbranched polyesters are again ideally suited for this study since they are capable of significant interaction with the inorganic alkoxides and they thermally decompose quantitatively into non-reactive species via an unzipping mechanism. Nanoporous silica films have great potential as low dielectric materials for semiconductor applications [113].

4.4
Biodegradable Amphiphilic Networks

Biodegradable amphiphilic networks have been synthesized by free radical copolymerization of PCL or PD,L-LA dimacromonomers with a hydrophilic comonomer, i.e., 2-hydroxyethylmethacrylate (HEMA) [45, 46, 114]. Well-defined α,ω-methacryloyl PCL and PD,L-LA oligomers were synthesized by the living ROP of the parent monomers initiated by diethyl aluminum 2-hydroxyethyl-

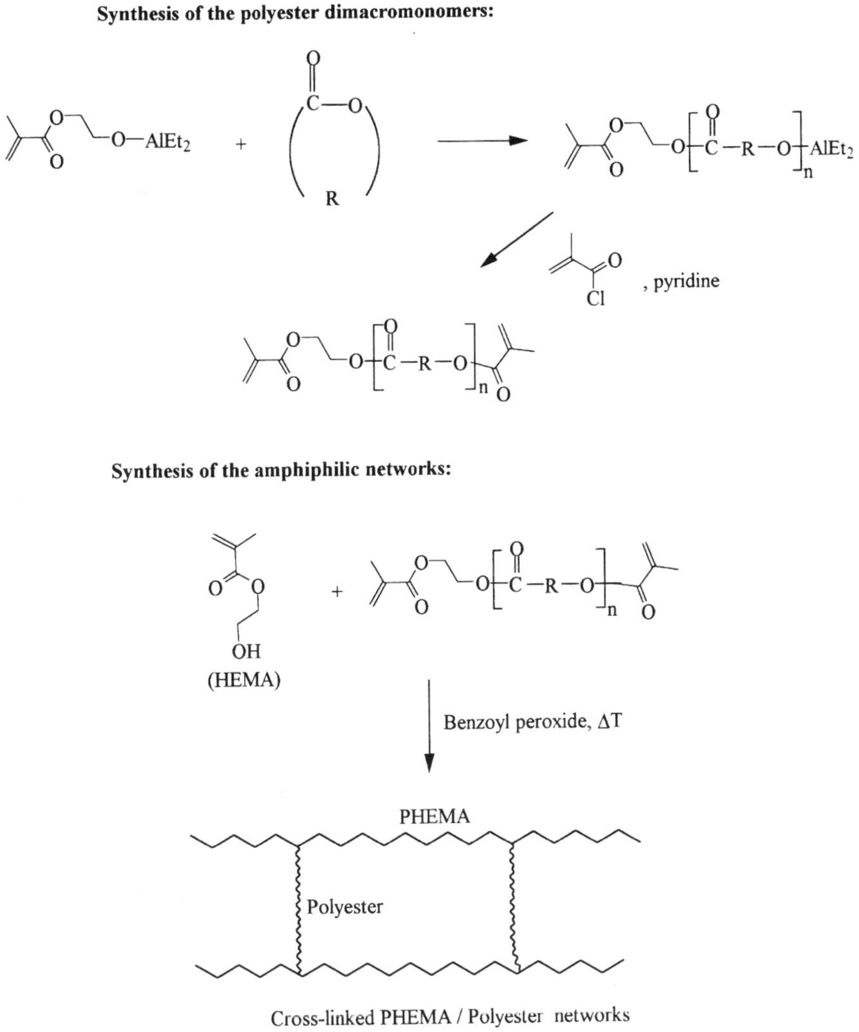

Synthesis of the polyester dimacromonomers:

Synthesis of the amphiphilic networks:

Scheme 43

methacrylate and then terminated by a controlled reaction of the propagating aluminum alkoxide groups with methacryloyl chloride (Scheme 43). The free radical copolymerization of those dimacromonomers with HEMA was performed in bulk at 65 °C using benzoyl peroxide and yielded homogeneous networks with high comonomer conversions (ca. 90%).

The amphiphilic nature of the polyester/PHEMA networks provides gels able to swell in both organic and aqueous media, which prompted us to study them as controlled drug delivery systems [114]. The most valuable characteristic feature of these amphiphilic networks is their potential to sustain the drug delivery of both hydrophilic and hydrophobic drugs. First, the swelling characteristics of the binary networks have been studied in relation to the network composition, the length, and the nature of the dimacromonomers. Further, dexamethasone was chosen as a drug model, since it can be handled as either a hydrophilic (sodium phosphate salt) or a lipophilic (acetate) compound. The drug was incorporated into the network directly during the free radical synthesis and cross-linking step or by swelling of the preformed network in water or in $CHCl_3$, depending on the drug formulation. It has been concluded that the kinetics of the drug release were mainly governed by the network swelling rate.

4.5
Nano- and Microspheres for Biomedical Applications

Biodegradable and biocompatible microspheres of PCL, PLA, PGA, and blends thereof have currently been prepared by different methods, such as emulsion/evaporation, precipitation processes, and dispersion polymerization, and have found applications as carriers for sustaining the release of bioactive molecules. In this section our attention will be exclusively focused on the advantages offered by new molecular architectures available from the macromolecular engineering of aliphatic polyesters in micro- but also nanocarriers. For instance, Slomkowski et al. have prepared microspheres by the straightforward dispersion polymerization of L-LA and εCL in a solvent mixture in which the generated polyesters are poorly soluble. The polymerizations were carried out in a 1,4-dioxane/heptane solvent mixture and were promoted by either $Sn(Oct)_2$ catalyst or aluminum alkoxide initiators [115]. The dispersion of the growing polyester particles was actually stabilized by poly(dodecylacrylate-g-caprolactone) graft copolymers. The graft copolymers were prepared by the macromonomer technique, as reported previously (see Sect. 3.3). The appropriate selection of these polymeric surfactants allows one to obtain, in a one-step dispersion polymerization, PLA and PCL microspheres with controlled mean particle size and narrow size polydispersity. In a different approach, Egiburu et al. have prepared microspheres of preformed PL-LA grafted copolymers by the emulsion/evaporation method [116]. Three types of copolymers were reported: poly(MMA-g-LA), poly(MA-g-LA), and poly(VP-g-LA). The authors monitored by UV spectroscopy the encapsulation loading level and release kinetics of ibuprofen, an anti-inflammatory drug, from small microspheres (mean size=1–5 μm) in buffered water solution. Interestingly, by varying the composition and thus the hydrophilic character of the copolymers, good control over the release rate was reached.

Biodegradable polyester-based nanoparticles have also been studied, especially in the biomedical domain. Like microelectronics, biomedical research follows the rule: "smaller is better". A typical example of nanoparticles based on the aliphatic polyester engineering by living ROP is provided by the poly(CL-b-GA) copolymers which form stable colloidal dispersions in organic solvents such as toluene and THF without the need of any additional surfactant [27]. The poly(CL-b-GA) particles form a new class of stable non-aqueous dispersions in which a PGA core is stabilized by a PCL shell [122]. The copolymer micelles were characterized by TEM and PCS that show a bimodal distribution. The smallest particles usually dominate the size distribution curve. Their average size is in the range 10–40 nm when the diblocks are dispersed in toluene at a 0.1 wt% concentration. The average size increases as the concentration of the dispersion increases, which might be due to the aggregation of the smallest particles. These colloidal particles of diblocks consisting of two biocompatible polyesters with drastically different physical properties, such as melting temperatures and biodegradability, are very attractive for biomedical applications. Our group also studied the use of poly(HEMA-g-(D,L-LA)) graft copolymers (for their synthesis, see Sect. 3.3) in nanoparticles formulation [123]. In a DMSO/H_2O (10/90 v/v) solvent mixture, these graft copolymers formed stable micelles as observed by PCS, the average size of which was in the range 40–70 nm, depending on the copolymer composition, molecular weight, and concentration. It is worth noting the high efficiency of these copolymers for immobilizing somatotrophine without losing protein activity. Finally, as already discussed in Sect. 2.5, the poly(εCL-co-TOSUO) copolymers were able to form stable aqueous dispersions with a mean particle size below 100 nm. Surprisingly enough, even though the functional comonomer was incorporated on a statistical basis at quite a low content (12 mol%), the copolymer was able to stabilize the hydrophobic PCL segment in water. The size of the colloidal dispersions decreases from 213 nm in the case of ethylene acetal pendant groups to 72 nm and 74 nm upon successive deprotection and reduction steps which lead to more polar ketone and hydroxyl groups, respectively. A decrease in the dispersion concentration from 0.10% to 0.01% results in smaller colloidal particles (from 213 to 72 nm). The suspensions are stable for more than 48 h at room temperature as checked by PCS [74]. This novel family of copolymers has potential for biomedical applications, particularly as tailored drug colloidal vectors with a core-shell-like structure. Reactive groups on the surface of the nanoparticles are indeed available to the binding of species selected for molecular recognition and drug targeting.

5
Conclusions

In conclusion, a quite complete macromolecular engineering of aliphatic polyesters, recognized as biocompatible and biodegradable materials, has been reached. The perfectly well controlled "coordination-insertion" ROP of cyclic (di)esters as initiated by aluminum alkoxides, functionalized or not, has proved to be essential to this end. The related (co)polyesters can be designed in such a way that α-, or α,ω- functional telechelic polymers, block, graft, comb-like, and star-like copolymers, hyperbranched and dendri-graft (co)polymers, and one- or two-component polymer networks are now readily available. Along with the above topologies, macrocyclic (block) polylactones have been very recently synthesized by using 2,2-dibutyl-2-stanna-1,3-dioacyclalkanes as cyclic initiators

[125]. This method exclusively yields macrocyclic polyesters without any competition with linear polymers. Furthermore, the "coordination-insertion" ROP process can take part in a more global construction set, ultimately leading to the development of new polymeric materials with versatile and original properties. Note that other types of efficient "coordination" initiators, i.e., rare earth and yttrium alkoxides, are more and more studied in the framework of the controlled ROP of lactones and (di)lactones [126–129]. These polymerizations are usually characterized by very fast kinetics so as one can expect to (co)polymerize monomers known for their poor reactivity with more conventional systems. Those initiators should extend the control that chemists have already got over the structure of aliphatic polyesters and should therefore allow us to reach again new molecular architectures. It is also important to insist on the very promising enzyme-catalyzed ROP of (di)lactones which will more likely pave the way to a new kind of macromolecular control [6, 130–132].

Acknowledgments. DM would like to thank the "Gobierno Vasco" for a fellowship "Becas para formacion de investigadores". PhD is "Chercheur Qualifié du Fonds National de la Recherche Scientifique FNRS". Authors would also like to thank all their (international) collaborators whose names can be found in the list of references, and more particularly, J.L. Hedrick and C.J. Hawker from IBM Research Center, San Jose. Acknowledgment is also due to the "Service Fédéraux des Affaires Scientifiques, Techniques et Culturelles" (PAI 4-11: Chimie et catalyse supramoléculaire).

References

1. Vert M, Feijen J, Albertsson AC, Scott G, Chiellini E (1992) Biodegradable polymers and plastics. Royal Society
2. Lofgren A, Albertsson AC, Dubois P, Jérôme R (1995) Recent advances in ring opening polymerization of lactones and related compounds. JMS Rev Macromol Chem Phys 35:379
3. (a) Steinbuchel A, Gorenflo V (1997) Macromol Symp 123:61; (b) Kessler B, Witholt B (1998) Macromol Symp 130:245
4. Yamashita Y, McGrath JE (eds) (1981) Anionic polymerization: kinetics, mechamism and synthesis, ACS Symposium Series 66:199
5. Jonte JM, Dunsing R, Krickeldorf HR (1986)J Macromol Sci Chem 23:495
6. (a) Bisht KS, Svirkin YY, Henderson LA, Gross RA, Kaplan DL, Swift G (1997) Macromolecules 30:7735; (b) Matsumura S, Mabuchi K, Toshima K (1998) Macromol Symp 130:285
7. Lundberg RD, Cox EF (1969) Ring opening polymerization. Marcel Dekker, New York London, chap 6, p 266
8. Cox EF, Hostettler (1972) Union Carbide Corp US Pat 3,021,309
9. Kricheldorf HR, Sumbel M (1989) Eur Polym J 25:585
10. Dubois P, Degée P, Ropson N, Jérôme R(1997) In: Hatada K, Kitayama T, Vogl O (eds) Macromolecular design of polymeric materials. Marcel Dekker, New York Basel Hong Kong, chap 14, p 247
11. Kricheldorf HR, Kreiser-Saunders I, Scharnagl N (1992) Makromol Chem Macromol Symp 32:285
12. Nijenhuis AJ, Griypma DW, Pennings AJ (1992) Macromolecules 25:6419
13. Hamitou A, Ouhadi T, Jérôme R, Teyssié P (1977) J Polym Sci Polym Chem Macromol Symp 32:285

14. Ouhadi T, Stevens C, Jérôme R, Teyssié P (1976) Macromolecules 9:927
15. Ouhadi T, Stevens C, Teyssié P (1975) Makromol Chem Suppl 1:1991
16. Endo M, Aida T, Inoue S (1987) Macromolecules 20:2982
17. Hofman A, Slomkowski S, Penczek S (1987) Makromol Chem Rapid Commun 3:387
18. Baran J, Duda A, Kowalski A, Szymanski R, Penczek S (1997) Macromol Symp 123:93
19. Yasuda H, Ihara E (1995) Macromol Chem Phys 196:2417
20. Stevens WM, Ankone M, Dijkstra P, Feijen J (1996) Macromolecules 29:6132
21. LeBorgne A, Pluta C, Spassky N (1995) Macromol Rapid Commun 15:955
22. Dubois P (1991) PhD Thesis, University of Liège, Belgium
23. Duda A, Penczek S (1995) Macromol Rapid Commun 16:67
24. Ropson N, Dubois P, Jérôme R, Teyssié P (1995) Macromolecules 28:758
25. Kurcok P, Dubois P, Jérôme R (1996) Polym Int 41:479
26. Lofgren A, Albertsson AC, Dubois P, Jérôme R, Teyssié P (1994) Macromolecules 27:5556
27. Barakat I, Dubois P, Jérôme R, Teyssié P, Mazurek M (1994) Macromol Chem Macromol Symp 88:227
28. Dubois P, Jacobs C, Jérôme R, Teyssié P (1991) Macromolecules 24:2266
29. Degée P, Dubois P. Jérôme R (1997) Macromol Chem Macromol Symp 123:67
30. Degée P, Dubois P, Jérôme R (1997) Macromol Chem Phys 198:1973
31. Degée P, Dubois P, Jérôme R (1997) Macromol Chem Phys 198:1985
32. Ropson N, Dubois P, Jérôme R, Teyssié P (1992) Macromolecules 25:3820
33. Ropson N, Dubois P, Jérôme R, Teyssié P (1997) J Polym Sci Polym Chem 35:183
34. Kuhling S, Keul H, Höcker H (1992) Makromol Chem 193:1207
35. Mecerreyes D, Dubois P, Jérôme R (to be published)
36. Tian D, Dubois P, Jérôme R (1997) Macromolecules 30:2575
37. Duda A, Penczek S, Dubois P, Mecerreyes D, Jérôme R (1996) Macromol Chem Phys 197:1273
38. Barakat I (1993) PhD Thesis, University of Liège, Belgium
39. Vanhoorne P, Dubois P, Jérôme R, Teyssié P (1992) Macromolecules 30:1947
40. Tian D, Dubois P, Jérôme R (1997) Macromolecules 30:1947
41. Jacobs C, Dubois P, Jérôme R, Teyssié P (1991) Macromolecules 24:3027
42. Dubois P, Jérôme R, Teyssié P (1989) Polym Bull 22:475
43. Barakat I, Dubois P, Jérôme R, Teyssié P (1993) J Polym Sci Polym Chem 31:505
44. Dubois P, Ropson N, Jérôme R, Teyssié P (1996) Macromolecules 29:1965
45. Dubois P, Jérôme R, Teyssie P (1991) Macromolecules 24:977
46. Barakat I, Dubois P, Jérôme R, Teyssie P, Goethals E (1994) J Polym Sci Polym Chem 32:2099
47. Hawker CJ, Hedrick JL, Malmstrom E, Trollsas M, Mecerreyes D, Moineau G, Dubois P, Jérôme R (1998) Macromolecules (31:214)
48. Hedrick JL, Trollsas M, Mecerreyes D, Dubois P, Jérôme R (1998) J Polym Sci Polym Chem (36:3187)
49. Carter KR, Richter R, Hedrick JL, HcGrath JE, Mecerreyes D, Jérôme R (1996) ACS Polym Prepr 37:607
50. Mecerreyes D, Lecomte P, Dahan D, Dubois P, Jérôme R, Demonceau A, Noels A (1999) Macromol Chem Phys (in press)
51. Dubois P, Degée P, Jérôme R, Teyssié P (1992) Macromolecules 25:2614
52. Dubois P, Zhang JX, Jérôme R, Teyssié P (1994) Polymer 35:4998
53. Tian D, Dubois P, Jérôme R, Teyssié P (1994) Macromolecules 27:4134
54. Duda A, Penczek S (1995) Macromolecules 28:5981
55. Duda A, Penczek S (1991) Macromol Chem Macromol Symp 47:127
56. Dubois P, Ropson N, Jérôme R, Teyssié P (1995) Macromolecules 28:1965
57. Biela T, Duda A (1996) J Polym Sci Polym Chem 34:1807
58. Duda A (1996) Macromolecules 29:1399
59. Inoue S, Aida T (1993) Macromol Chem Macromol Symp 73:27
60. Vert M, Lenz RW (1979) ACS Polym Prepr 20:608

61. Gelvin ME, Kohn J (1992) J Am Chem Soc 114:3962
62. Fujino T, Ouchi T (1985) Polym Prepr JPN 25:2330
63. Kimura Y, Shirotani K, Yamane H, Kitao T (1988) Macromolecules 21:3338
64. Barrera DA, Zylstra E, Peter TL, Langer R (1993) J Amer Chem Soc 115:11,010
65. Pitt G, Gu ZW, Ingram P, Hendren RW (1987) J Polym Sci Polym Chem 25:955
66. Tian D, Dubois P, Grandfils C, Jérôme R (1997) Macomolecules 30:406
67. Webster OW (1991) Science 251:887
68. Frèchet JMJ (1994) Science 263:1710
69. Endo M, Aida T, Inoue S (1987) Macromolecules 20:2982
70. Degée P (1991) PhD Thesis, University of Liège, Belgium
71. Kurcok P, Dubois P, Sikorska W, Jedlinski Z, Jérôme R (1997) Macromolecules 30:5591
72. Degée P, Dubois P, Jérôme R, Teyssié P (1993) J Polym Sci Polym Chem 31:275
73. Palmgren R, Karlsson S, Albertsson AC (1997) J Polym Sci Polym Chem 35:9
74. Tian D, Dubois P, Jérôme R (1998) Macromol Symp 130:217
75. Heushen J, Jérôme R, Teyssié P (1981) Macromolecules 14:242
76. Balsamo V, Gyldenfeldt F, Stadler R (1996) Makromol Chem 197:1159
77. Mecerreyes D, Dubois P, Jérôme R, Hedrick JL (1998) Makromol Chem (in press)
78. Harris JF, Sharkey WH (1986) Macromolecules 19:2903
79. Rique-Nurbet L, Schappacher M, Deffieux A (1994) Polymer 35:4563
80. Mecerreyes D, Dubois P, Jérôme R, Hedrick JL, Hawker CJ, Beinat S, Schappacher, Deffieux A (1997) Proceed. ACS Polym Mat Sci Eng 77:189
81. Moad G, Rizzardo E (1995) Macromolecules 28:8722
82. Wang JS, Matyjaszewski K (1995) J Am Chem Soc 117:5614
83. Mecerreyes D, Moineau G, Dubois P, Jérôme R, Hedrick JL, Hawker CJ, Malmstrom E, Trollsas M (1998) Angew Chem 37:1274
84. Mecerreyes D, Moineau G, Dubois P, Jérôme R (1998) (to be published)
85. Kowalczuk M, Adamus G, Jedlinski Z (1994) Macromolecules 24:572
86. Thamm R, Buck WH, Caywood SW, Meyer JM, Anderson BC (1977) Angew Makromol Chem 58:345
87. Egiburu JL, Fernandez-Berridi MJ, San Román J (1996) Polymer 37:3615
88. Hawker CJ, Mecerreyes D, Elce E, Dao J, Hedrick JL, Barakat I, Dubois P, Jérôme R, Volksen W (1997) Macromol Chem Phys 198:155
89. Grubbs RH (1982) Comprehensive organometallic chemistry. Pergamon, Oxford, vol 8, p 499
90. Demonceau A, Stumpf AW, Saive E, Noels AF (1997) Macromolecules 30:3127
91. Heroguez V, Gnanou Y, Fontanille M (1997) Macromolecules 30:4791
92. Lecomte P, Mecerreyes D, Dubois P, Demonceau A, Noels AF, Jérôme R (1998) Polym Bull (40:631)
93. Andini S, Ferrara L, Maglio G, Palumbo R (1988) Makromol Chem Macrom Rapid Commun 9:119
94. Kylma J, Seppälä JV (1997) Macromolecules 30:2876
95. Storey RF, Hickey TP (1994) Polymer 35:830
96. Trollsas M, Hedrick JL, Mecerreyes D, Dubois P, Jérôme R (1997) Proceed ACS Polym Mat Sci Eng 77:208
97. Hedrick JL, Russell TP, Sanchez M, Di Pietro R, Swanson S, Mecerreyes D, Jérôme R (1996) Macromolecules 29:3642
98. Hedrick JL, Carter KR, Richter R, Miller RD, Russell TP, Flores V, Mecerreyes D, Dubois P, Jérôme R (1998) Chem Mat 10:39
99. Gitsov I, Wooley KL, Frèchet JMJ (1992) Angew Chem Int Ed Engl 31:1200
100. Gitsov I, Ivanova TP, Frèchet JMJ (1994) Macromol Rapid Commun 15:387
101. Mecerreyes D, Dubois P, Jérôme R, Hedrick JL, Hawker CJ (1999) (to be submitted)
102. (a) Trollsas M, Hedrick JL, Mecerreyes D, Dubois P, Jérôme R, Ihre H, Hult A (1997) Macromolecules 30:8508; (b) Trollsas M, Hedrick JL, Mecerreyes D, Dubois P, Jérôme R, Ihre H, Hult A (1998) Macromolecules 31:2756

103. Yasuda H, Tamai H (1993) Prog Polym Sci 18:1097
104. Mogstad AL, Waymouth RW (1994) Macromolecules 24:2313
105. McLain SJ, Ford TM, Drysdale NE (1992) ACS Polym Prepr 33:463
106. Lofgren A, Renstad R, Albertsson A-C (1995) J Appl Polym Sci 55:1589
107. Hedrick JL, Carter K, Cha H, Hawker C, Di Pietro R, Labadie J, Miller RD, Russell TP, Sanchez M, Volksen W, Yoon D, Mecerreyes D, Jérôme R, McGraft JE (1996) React Func Pol 30:43
108. Charlier Y, Hedrick JL, Labadie J, Lucas M, Swanson S (1995) Polymer 36:987
109. Tian D, Dubois P, Jérôme R (1996) Polymer 37:3983
110. Tian D, Dubois P, Jérôme R (1997) J Polym Sci Polym Chem 35:2295
111. Tian D, Dubois P, Grandfils C, Jérôme R, Viville P, Lazzaroni R, Brèdas JL, Leprince P (1997) Chem Mat 9:1704
112. Tian D (1997) PhD Thesis, University of Liège, Belgium
113. (a) Smith DM, Anderson JM, Cho CC, Gnade DE (1995) Advances in porous materials; (b) Smith DM, Anderson JM, Cho CC, Gnade DE (1995) Mater Res Soc Symp Proc 371:261
114. Barakat I, Dubois P, Grandfils C, Jérôme R (1999) J Polym Sci, Polym Chem (in press)
115. Slomkowski S, Sosnowski S, Gadzinowski M (1997) Macromol Symp 123:45
116. Egiburu JL (1996) PhD Thesis, University of the Basque Country, Spain
117. Hamaide T, Spitz R, Letourneux JP, Claverie J, Guyot A (1994) Macromol Symp 88:191
118. Miola C, Hamaide T, Spitz R (1997) Polymer 38:22
119. Tortosa K, Miola C, Hamaide T (1997) J Appl Polym Sci 65:2357
120. Ihre H, Hult A, Soderlund E (1996) J Am Chem Soc 118:6388
121. Hedrick JL, Trollsas M, Mecerreyes D, Jérôme R, Dubois P (1998) J. Polym Sci, Polym Chem (submitted)
122. Sobry R, Van den Bossche G, Fontaine F, Barakat I, Dubois P, Jérôme R (1996) J Mol Struct 383:63
123. Nyssen C (1996) Master's thesis, University of Liège, Liège, Belgium
124. Chujo Y, Matsuki H (1994) J Chem Soc Chem Commun 635
125. Kricheldorf HR, Lee S-R, Eggerstedt S, Hauser K (1998) Macromol Symp 128:121
126. Shen Y, Shen Z, Zhang Y, Yao K (1996) Macromolecules 29:8289
127. Simic V, Spassky N, Hubert-Pfalzgraf L (1997) Macromolecules 30:7338
128. Stevels W, Ankoné M., Dijkstra P, Feijen J (1996) Macromolecules 29:8296
129. Stevels W, Dijkstra P, Feijen J (1997) Trend Polym Sci 5:300
130. Uyama H, Takeya K, Hoshi N, Kobayashi S (1995) Macromolecules 28:7046
131. Kobayashi S, Shoda S, Uyama H (1995) Adv Polym Sci 121:1
132. Nobes GAR, Kazlauskas RJ, Marchessault RH (1996) Macromolecules 29:4829

Received: August 1998

Nanoscopically Engineered Polyimides

James L. Hedrick[1], Jeff W. Labadie[1], Willi Volksen[1], Jöns. G. Hilborn[2]

[1]IBM Research, IBM Almaden Research Center, 650 Harry Road, San Jose, CA 95120, USA
[2]Department of Material Science, Swiss Federal Institute of Technology
Lausanne, MX-D Ecublens, CH-1015 Lausanne, Switzerland
E-mail: hedrick@almaden.ibm.com

Polyimides are currently the materials of choice for interlayer dielectrics in microelectronic applications, since polyimides, as a class of materials, best satisfy the requisite properties to survive the thermal, chemical, and mechanical stresses associated with microelectronic fabrication. As more function is demanded of these polymer dielectrics, e.g., low residual thermal stress, adhesion, photosensitivity, and low dielectric constant, it becomes increasingly difficult to design materials with the desired enhancements without compromising existing properties. This article will describe an approach to modify polyimide with minimal sacrifice to its desirable properties. The preparation of block and graft copolymers provides a means of tailoring the morphology and properties of polyimide through the judicious choice of the coblock, coblock composition, molecular architecture, and block lengths. It is the advent of the poly(amic alkyl ester) intermediate to the polyimide that allows for the controlled synthesis of such block copolymers. The hydrolytic stability of the poly(amic alkyl ester) precursor allows for the isolation and characterization of the copolymers prior to imidization. Such systems represent self-assembling arrays with considerable potential for the preparation of nanostructures, and this article will describe the modification of rigid and semi-rigid polyimides through copolymerization to address favorably such issues as residual thermal stress, dielectric constant, auto-adhesion, and other key design criteria.

1	Introduction	62
2	Modification of Auto-Adhesion Characteristics of Polyimide	67
2.1	Synthesis of Heterocycle-Containing Poly (Aryl Ethers)	68
2.2	Synthesis of Polyimide Copolymers	71
2.3	Adhesion, Morphology, and Mechanical Characteristics of Polyimide Copolymers	77
3	Control of Polyimide Dielectric Constant	83
3.1	Imide-Perfluoroalkyl Ether Copolymers	83
3.2	Polyimide Nanofoams	85
3.2.1	Criteria for the Thermally Labile Coblock	88

3.2.2	Synthesis and Characterization of Polyimide Copolymers	91
3.2.3	Foam Formation and Characterization	95
4	**Residual Thermal Stress Control of Polyimide Through the Use of Self-Assembled, Phase Separated Block Copolymers**	103
4.1	Synthesis of Imide-Siloxane Copolymers	104
4.2	Mechanical Properties of Imide-Siloxane Copolymers	106
References		109

1
Introduction

Organic polymeric materials have long played significant roles in the microelectronics industry because of their processability, attractive mechanical properties, and relatively low cost. Traditionally, these materials have found application either as structural components or in comparatively low-performance system elements such as cards, cables, or circuit boards (i.e., epoxy-glass fiber composites). However, the use of polymeric materials in the fabrication of multichip modules (MCM) or on chips has been much less pervasive, since inorganics (alumina or silicon oxides) have typically dominated such components which are critical to system cycle time. Until recently, system enhancement could be driven by increased semiconductor device performance. However, the level of integration of devices has advanced to the point where overall system performance is becoming limited by the speed with which signals can be transmitted from device to device within a given semiconductor chip and then between chips performing related functions. The on-chip wiring which constitutes the device-to-device interconnection is typically designated as the "back end of the line" (BEOL) wiring, while chip-to-chip interconnection is the function of the first level "package" or multichip module (MCM). Among the many approaches to decrease signal propagation delays in either BEOL or module wiring, the reduction of the dielectric constant of the medium is the most common approach. It is in these critical wiring components that high performance organic polymers are finding increasing application as dielectrics. The employment of organic materials in these devices allows lower cost manufacturing because the materials can be spin-coated rather than vacuum processed for deposition. However, the major driving force for the implementation of polymeric insulators is their much lower dielectric constant compared to the inorganic alternatives.

The velocity of pulse propagation in these structures is inversely proportional to the square root of the dielectric constant of the medium [1]. Hence, reductions in the dielectric constant of the insulator materials translate directly into improvements in machine cycle time. The pitch or distance between conductor lines must, therefore, be minimized to improve cycle time. The minimum distance from signal lines is dictated by noise issues or "cross-talk" that result from induced current in conductors adjacent to active signal lines. The lateral extent of this field – induced noise is again directly dependent upon the dielectric constant of the insulator material. A reduction of the insulator dielectric constant permits moving the signal lines closer together, allowing designers to reduce the length of conductor lines and thereby improve machine cycle time accordingly [1].

Although polymeric insulators have several attractive properties (i.e., ease of processing and low dielectric constant), in all other requirements they have serious shortcomings when compared to the inorganic alternatives. Insulating materials in these applications must be able to withstand the high temperature associated with the processes used to deposit metal lines and join chips to modules (i.e., C_4 soldering) [1]. At the bare minimum, they must be able to withstand soldering temperatures without any degradation, outgassing, or dimensional change (i.e., a T_g significantly higher than soldering temperature). Another major factor in the use of organic insulators is control of residual thermal stress, which becomes further exacerbated with each additional level [2]. The stress results primarily from the mismatch in thermal expansion between the ceramic substrate and the insulating material. In addition, the polymeric insulator must have good adhesion to the ceramic substrate and to itself (i.e., self-adhesion) so as to permit reliable fabrication of multilevel structures [3]. The ability to process the polymer from a common organic solvent, and the ability to planarize (both global and local planarization) underlying topography are also critical features. However, once deposited, the polymer should show solvent resistance to allow the fabrication of multilayer structures and possible lithography steps without dimensional change. Finally, cost (i.e., production volumes at reasonable cost) and reliability are also important design factors.

A wide variety of polymers have been evaluated for use in microelectronics applications; however, polyimides have emerged as the favored class of materials [4–8]. The rigid and semi-rigid aromatic combination of thermal and mechanical properties which include low thermal expansion coefficient, high modulus, and tough ductile mechanical properties as judged by high elongations to break. In addition, these properties are largely retained to 400 °C; above this temperature a partial softening or flow is observed in some cases. These desirable properties reflect the high degree of molecular packing [9, 10]. For instance, the polymer chains of PMDA/ODA polyimide assume locally extended conformations in a smectic-like layered order. Furthermore, imidization of PMDA/ODA polyimide films in contact with a substrate produces substantial orientation of the molecules parallel to the surface [11]. This orientation results in physical properties that are anisotropic. This has been shown to produce a high thermal expansion coefficient and low modulus out of the plane of the film as well as anisotropic swelling behavior. These effects are even more pronounced for the BPDA/PDA polyimide. A manifestation of the orientation, responsible for the excellent mechanical properties, is a corresponding anisotropy in the dielectric constant [12]. In particular, the in-plane dielectric constant can be as much as

0.7 to 0.8 higher than the out-of-plane dielectric constant. An isotropic dielectric constant is clearly an unacceptable limitation to device design. Furthermore, the key advantage realized by the use of polymeric materials over inorganics is largely negated.

These rigid aromatic polyimides are generally applied by spin coating onto silicon wafers as dilute solutions. Since most of these structures are insoluble, they are applied as soluble poly(amic-acid) precursors, typically in aprotic dipolar solvents such as N-methyl-2-pyrrolidone (NMP) or dimethyl acetamide (DMAC) [13]. Imidization is then effected by a subsequent heating step denoted as curing. During the cure, significant stress can build up due to solvent evaporation, imidization, and the mismatch in thermal expansion coefficient between the film and the substrate. Furthermore, as the number of polymer layers and overall thickness of the thin film structures increases, controlling the stresses developed in the polymer when coated and cured becomes increasingly difficult. These high stresses can lead to warpage of the substrate, loss of adhesion, or cohesive failure of the polymer or substrate. Currently, stress is controlled at low levels by using polyimides with thermal expansion coefficients which match either the ceramic or silicon substrate. For instance, residual thermal stresses in the rigid and highly ordered polyimides i.e., poly(biphenyl dianhydride-phenylene diamine) (BPDA/PDA polyimide) and poly(pyromellitic dianhydride-phenylene diamine) (PMDA/PDA polyimide) are low due to the similar thermal expansion coefficients of the polyimide and substrate. This approach is very limited with respect to the number of possible rod-like structures obtainable. For instance, many of the rod-like structures (i.e., PMDA/PDA polyimides) have low thermal expansion coefficients but brittle mechanical properties. Additional drawbacks include the synthetic route employed. Corrosion of copper metallurgy has often been observed from polyimides derived from the poly(amic acid) precursor [14], while polyimides derived from the poly(amic alkyl ester) precursors generally show substantially higher thermal expansion coefficients and residual thermal stress than those derived from the poly(amic acid) analogs [15]. Furthermore, as more function is demanded of polymer dielectric layers, e.g., photosensitivity, it becomes increasingly difficult to design materials with the desired enhancements without compromising existing properties.

Additional drawbacks to the use of polyimide insulators for the fabrication of multilevel structures include self- or auto-adhesion. It has been demonstrated that the interfacial strength of polyimide layers sequentially cast and cured depends on the interdiffusion between layers, which in turn depends on the cure time and temperature for both the first layer (T_1) and the combined first and second layers (T_2) [3]. In this work, it was shown that unusually high diffusion distances (~ 200 nm) were required to achieve bulk strength [3]. For $T_2 \geq T_1$, the adhesion decreased with increasing T. However, for $T_2 < T_1$ and $T_1 \sim 400\,°C$, the adhesion between the layers was poor irrespective of T_2. Consequently, it is of interest to combine the desirable characteristics of polyimide with other materials in such a way as to produce a low stress, low dielectric constant, self-adhering material with the desirable processability and mechanical properties of polyimide.

Mixing two polymers is a convenient means of obtaining materials with properties characteristic of the respective homopolymers. Although it has been demonstrated that engineering thermoplastics such as poly(aryl ether ketones) and poly(aryl ether sulfones) are miscible with selected polyimides, these polyimides are amorphous with low T_gs [16–20]. For the case of rod-like polyimide mixtures with flexible polymers, phase separation generally occurs. Slight nonfavorable interactions between the segments far outweight the small entropic gains on mixing [21]. There are reports that some mixtures of such rigid rod and flexible polymers form "molecular composites" [22]. However, the extent of phase separation is kinetically rather than thermodynamically controlled. These reports include mixtures of rigid and flexible polyimides prepared by the respective poly(amic acid) precursors often undergo transamidization leading to block and random copolymers. By the nature of the connectivity of the segments and blocks, miscibility is promoted. Alternatively, Yoon and co-workers [25] have demonstrated the formation of molecular composites of polyimide/polyimide mixtures where at least one of the polyimide precursors was in the stable alkyl ester form, which is not prone to transamidization reactions. Even here, however, the desired level of phase separation was kinetically controlled. The subsequent molecular composites showed enhanced auto-adhesion with the retention of many of the desirable characteristics of the rigid rod polyimide.

An alternative approach to the modification of the characteristics of rigid and semi-rigid polyimides involves the preparation of random copolymers or the combination of different diamines and dianhydrides. For instance, imide-aryl ether phenylquinoxaline and imide-aryl ether benzoxazole statistical or random copolymers have been prepared and their morphology and adhesion characteristics investigated [26, 27]. In each case, the incorporation of the *co*-diamine containing either a preformed phenylquinoxaline or benzoxazole moiety resulted in significantly improved auto-adhesion. However, wide-angle X-ray diffraction measurements showed that the "liquid crystalline"-like ordering observed in the PMDA/ODA polyimide persists in the copolymers where the phenylquinoxaline or benzoxazole compositions were less than 50 wt%. At higher concentrations the ordering vanishes due to hindrance of interchain packing and, hence, a disruption of the ordering. Coincident with this loss in ordering is the clear development of a T_g and loss of the high-temperature dimensional stability. Thus, these copolymers show improved adhesion with the retention of the ordered morphology over a very narrow composition range.

Likewise, the most common approach in modifying the dielectric properties of polyimides has been via the incorporation of comonomers containing perfluoroalkyl groups. Examples include the incorporation of fluorinated substituents either as hexafluoroisopropylidene linkages [28], main chain perfluoroalkyl groups [29], or pendent trifluoromethyl groups [30, 31]. This approach produces films with dielectric constants as low as 2.6 and low water absorption; however, the mechanical properties and solvent resistance are generally sacrificed. The addition of pendent trifluoromethyl groups appears to circumvent this drawback, but the incorporation of sufficient trifluoromethyl groups into the polymer is synthetically limited. Alternatively, the dielectric constant of polyimides can be lowered through the introduction of kinks and separator linkages in the polymer backbone to reduce chain-chain interactions [32]. When these structural modifications were copolymerized with the judicious choice of fluorine-containing comonomers, polyimides with dielectric constants in the range of 2.4–2.8 were achieved.

Here we will describe another approach in modifying polyimide without sacrificing the ordered morphology and properties associated with this morphology. This approach involves the preparation of block copolymers derived from polyimide and a wide variety of coblocks. The use of block copolymers offers numerous advantages over polymer/polymer mixtures and random copolymers since the molecular architecture, block lengths, and composition can be designed to produce materials with a wide range of properties and morphologies [33]. Furthermore, since the two dissimilar materials are covalently bonded, miscibility is enhanced and phase separation, when it occurs, is restricted to dimensions of the order of 100–400 Å. By modifying the copolymer composition, it is possible to control the domain shapes as well as curvature of the domain interfaces. In an AB diblock copolymer, the domain shapes range from spheres of A in a matrix of B for low compositions of A to more interconnected cylindrical, lamella, or even more complex morphologies for higher compositions of A. Such systems represent self-assembling arrays with considerable potential for the preparation of nanostructures depending on the blocks, processing conditions, etc. Polyimides, as a class of materials, have received little attention as a component in the synthesis of block and segmented copolymers [34–36], and among these examples, the imide-siloxane copolymers have been the most widely studied [35, 36]. The general synthetic methodology used for the imide-siloxane copolymers utilized a monomers-oligomers approach via the poly(amic acid) precursor to the polyimide [35]. Bis(amino)siloxane oligomers of various molecular weights were coreacted with either a dianhydride or a mixture of a dianhydride and a diamine to yield poly(amic acid) solutions, which were cast and cured to imidize the polymers producing an $[AB]_n$ multiblock molecular architecture.

We have used an alternative synthetic procedure for the preparation of imide-containing copolymers based on a poly(amic alkyl ester) intermediate to the polyimide (Scheme 1) [13, 37]. In this route, pyromellitic dianhydride (PMDA) is opened by ethanol to yield a meta, para mixture of half esters which can be separated by fractional recrystallization and converted to the representative acid chlorides. Polymerization with a diamine yields the target poly(amic ethyl esters). Through the judicious choice of the ester moiety, further synthetic flexibility is possible in both imidization temperature and solubility. In our work, we have primarily used the poly(amic ethyl ester) precursor to polyimide since it is soluble in a variety of solvents and solvent mixtures and imidization occurs at a significantly higher temperature than for the poly(amic acid) analogue. Using a poly(amic alkyl ester) precursor offers more synthetic flexibility due to improved solubility and greater structural variety in both the polyimide backbone

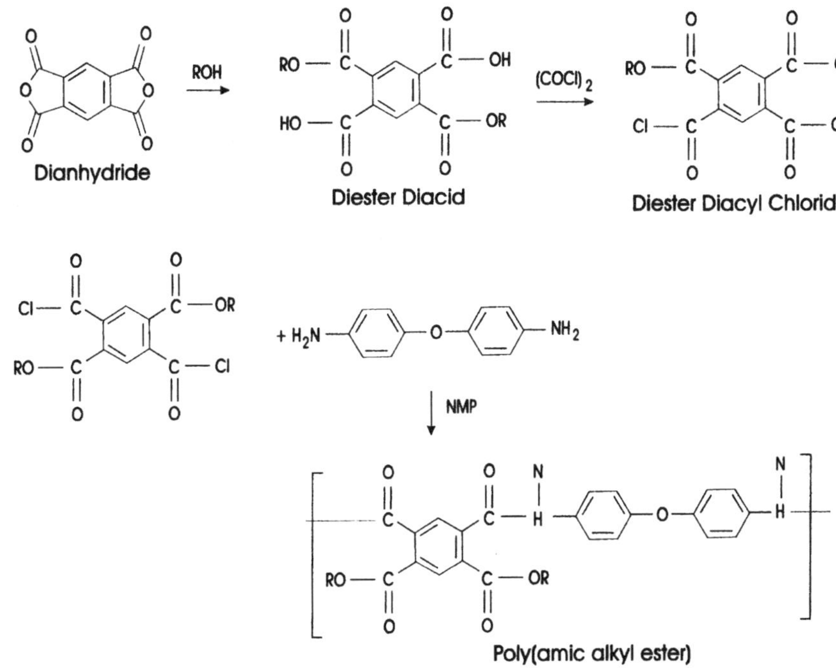

Scheme 1

and the coblock. The hydrolytically stable precursors may be isolated, characterized, and purified. Furthermore, since imidization occurs at substantially higher temperatures, molecular mobility is attained before imidization, thus minimizing the control of the resultant morphology by kinetic factors. Here we will describe the modification of rigid and semi-rigid polyimides through copolymerization to address favorably such issues as residual thermal stress, dielectric constant, auto-adhesion, and other key design criteria.

2
Modification of Auto-Adhesion Characteristics of Polyimide

As previously stated, the rigid polyimides meet many of the requirements for microelectronics applications; however, the presence of an ordered morphology, coupled with the lack of a softening transition results in extremely poor self-adhesion. Alternatively, thermally stable thermoplastics exhibit excellent self-adhesion, but often lack sufficiently high temperature dimensional stability and/or solubility and processability from common organic solvents. For instance, poly(phenylquinoxaline) (PPQ) has a T_g in the 370 °C range, thereby overcoming

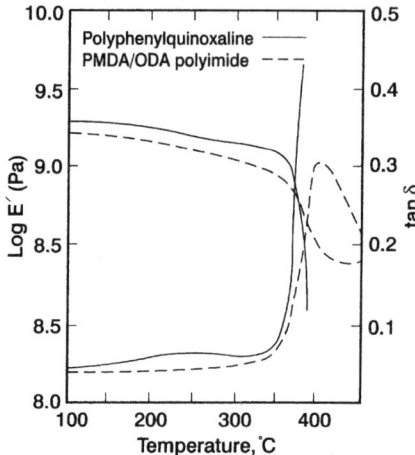

Fig. 1 Dynamic mechanical spectra of PMDA/ODA polyimide and poly(phenylquinoxaline)

the problem associated with self-adhesion at high temperatures [38, 39]. This polymer can best be compared to PMDA/ODA polyimide by their dynamic mechanical spectra (Fig. 1). The PMDA/ODA polyimide shows only a small drop in modulus at ≈ 350 °C, reflecting the absence of a T_g and the retention of the ordered structure, whereas PPQ shows a modulus invariance up to the T_g, at which point the modulus falls dramatically, characteristic of an amorphous melt. Consequently, it was of interest to combine the adhesion characteristics of the thermoplastics with desirable properties of many of the thermoplastics and heterocycle-containing high temperature polymers in solvents amenable towards polyimide copolymerization. To this end, aryl ether linkages were introduced through new synthetic methodologies to impart solubility. Heterocycle-containing monomers were polymerized via nucleophilic aromatic substituting polymerization in which the generation of the aryl ether linkage was the polymer-forming reaction [40–42]. The heterocycle-containing poly(aryl ethers) prepared by this route were excellent candidates for polyimide block copolymers since control of the end-group type and molecular weight was possible, and the polymers were soluble in common organic solvents.

2.1
Synthesis of Heterocycle-Containing Poly (Aryl Ethers)

Three poly(aryl ethers) were prepared and used as coblocks in imide copolymerizations. The first coblock prepared was poly(aryl ether phenylquinoxaline), since this material has the requisite high T_g (~ 280 °C) and thermal stability, and the polymer can be processed from solution or the melt. The synthesis of poly(aryl ether phenylquinoxalines) involves a fluoro-displacement polymerization of appropriately substituted fluorophenylquinoxalines with bisphenols, using conventional condensation techniques [40]. Facile displacement of aryl fluorides, activated by the phenylquinoxaline ring, was demonstrated, and polymerization of aryl fluoro-substituted bis(quinoxalines) with bisphenols led to high polymer yield. This general synthetic route has been extended to the preparation of functional oligomers amenable towards polyimide syntheses with the use of 3-aminophenol to control molecular weight and produce the amino-functionality as described by Jurak and McGrath (Scheme 2) [43]. Bis(amino) aryl ether phenylquinoxaline oligomers of various molecular weights were prepared by the reaction of 1,3-bis(6-fluoro-3-phenyl-2-quinoxalinyl) benzene, 2,2'-bis(4-hydroxyphenyl)hexafluoropropane, 3-aminophenol in N-methylpyrrolidone solvent in the presence of potassium carbonate [44]. This general synthetic route was also extended to an AB-self polymerizable monomer 3-(4-hydroxyphenyl)-3-phenyl-6-fluorophenylquinoxaline, in the presence of 3-aminophenol using an analogous procedure to that described above, to afford new quinoxaline-based macromers (Scheme 3) [45, 46]. The Carothers equation was used to determine the quantity of 3-aminophenol required to control both the molecular weight and the chain end functionality of the oligomers. Table 1 contains the characteristics of the oligomers prepared. The number-average molecular weights, determined by ^{1}H NMR [44] were 6200 and 15,500 g/mol for oligomers **1a** and **1b**, respectively, for the difunctional oligomers and 6200 g/mol for the monofunctional oligomer, **1f**.

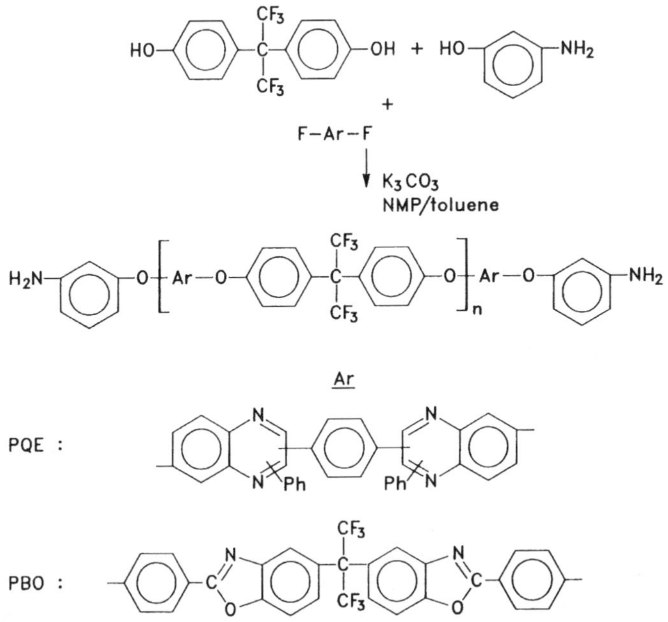

Scheme 2

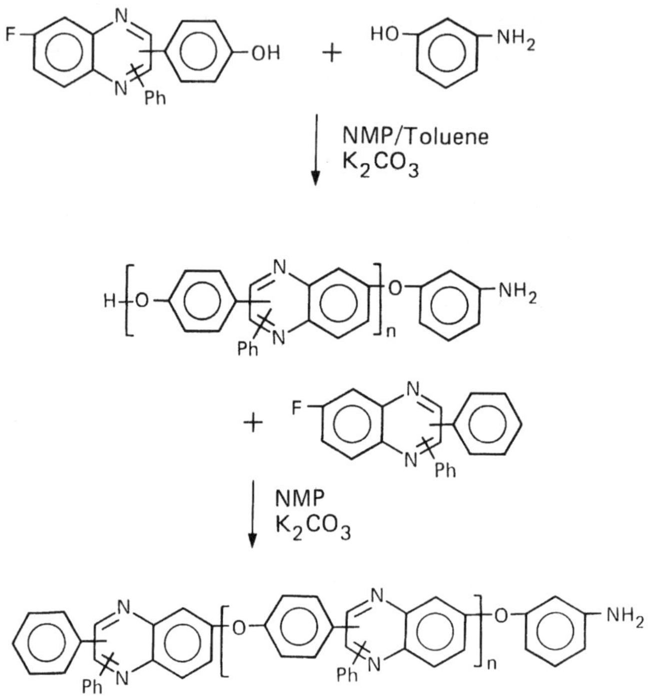

Scheme 3

The heterocyclic-activated fluoro-displacement polymerization was extended to other heterocycles including azoles [47, 48]. Of particular interest was the benzazole-activated fluoro-displacement polymerization as a route to poly(aryl ether benzoxazoles) and functional oligomers amenable towards copolymerization [47, 49]. This class of polymers, like the poly(aryl ether phenylquinoxalines), are tough engineering thermoplastics with the requisite thermal stability and solubility. To this end, functional oligomers were prepared by the reaction of 2,2-bis[2-(4-fluorophenyl)benzoxzol-6-yl] hexafluoropropane, 2,2'-bis(4-hydroxyphenyl) hexafluoropropane and 3-aminophenol using the procedure described above (Scheme 2) [49]. The characteristics of the oligomers prepared **1c** and **1d** are also shown in Table 1.

The third poly(aryl ether) surveyed as a coblock for polyimide copolymerization was poly(aryl ether ether ketone), PEEK, which is a highly crystalline polymer (40–50 % crystallinity) with a T_g of 145 °C and a T_m of 340 °C. However, PEEK is only soluble in diphenylsulfone at temperatures in excess of 300 °C or in strong acids [50]. This insolubility in organic solvents makes the synthesis and

Table 1. Characteristics of amine terminated aryl ether oligomers

Sample Entry	Oligomer Type	<Mn> Theory	(gram/mol) Measured	T_g °C
1a	poly(phenylquinoxaline)	7,000	6,200	225
1b	poly(phenylquinoxaline)	15,000	15,500	235
1c	poly(benzoxazoles)	10,000	10,500	210
1d	poly(benzoxazoles)	25,000	26,000	227
1e	poly(ketimine)	7,000	6,200	145
1f	phenylquinoxaline (monofunctional)	7,000	6,200	240

characterization of function oligomers and subsequent transformations (i.e., copolymerization) difficult. An alternative synthetic route to the preparation of PEEK functional oligomers has been employed which involves a derivatization of one of the monomers to afford oligomer solubility and, after subsequent transformations (i.e., copolymerization), a deprotection step yields the parent polymer. It has been demonstrated that the ketone moiety in 4,4'-difluorobenzenephenone can be derivatized with aniline to produce a ketimine (Scheme 4) [51, 52]. This disrupts the trigonal bond arrangement which provides the planar zig-zag chain packing in the subsequent polymer precluding crystallization. Interestingly, it has been demonstrated that the ketimine moiety is sufficiently electron-withdrawing to activate aryl fluorides towards nucleophilic aromatic displacement. In addition, the ketimine can accept the negative charge developed in the transformation through a Meisenheimer complex which lowers the activation energy for the displacement (Scheme 4) [51, 52]. This allows polymerization in common aprotic dipolar solvents (i.e., NMP). The ketimine group may be quantitatively hydrolyzed in the polymer form to produce the parent structure, PEEK [51, 52].

The synthesis of the bis(amino) aryl ether ketimine oligomer was carried out in an analogous fashion to the poly(aryl ether) oligomers described before (Scheme 5) [43]. The ketimine functional 4,4'-bisfluoride was reacted with hydroquinone and 3-aminophenol in an NMP/toluene solvent mixture in the presence of potassium carbonate. The characteristics of the oligomer synthesized are shown in Table 1 (sample 1e).

2.2
Synthesis of Polyimide Copolymers

The imide-aryl ether copolymers were prepared via the poly(amic alkyl ester) route [53–56]. This intermediate route to polyimide is believed to be more versatile than the poly(amic acid) analogue due to its enhanced solubility. Volksen et al. [53, 54] reported that up to 75 % of a cosolvent could be used along with NMP, which is an important consideration in copolymerization of chemically dissimilar block copolymers. Furthermore, the poly(amic alkyl ester) precursor to the

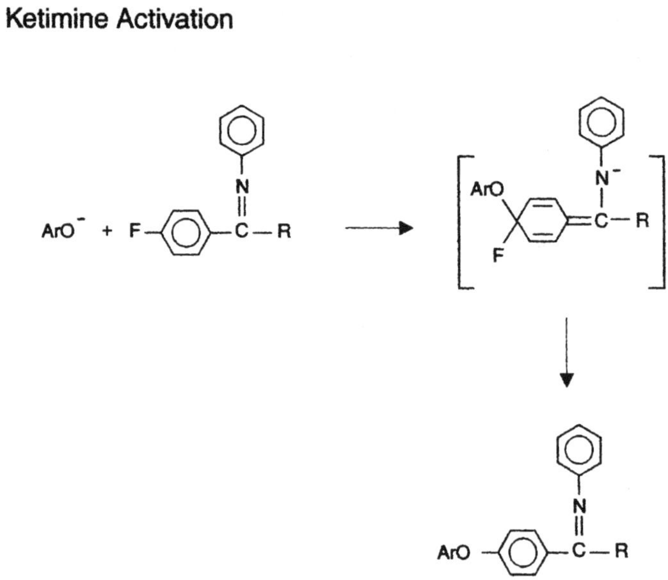

Ketimine Activation

Scheme 4

polyimide can be isolated, characterized, and subjected to selective solvent rinses to remove homopolymer contamination. The copolymer synthesis involved the incremental addition of PMDA diethyl ester diacyl chloride in methylene chloride to a solution of the aryl ether oligomer and ODA in an NMP-based solvent mixture in the presence of an acid acceptor such as N-methylmorpholine or pyridine (Schemes 6–8) [53, 54]. Although both the poly(amic ethyl ester) and the poly(aryl ether) homopolymers are soluble in NMP, mixtures of the two polymers in NMP formed cloudy solutions in some cases, and the subsequent copolymers formed cloudy films, consistent with homopolymer contamination. The addition of a cosolvent to the NMP produced clear solutions and minimized homopolymer contamination observed after copolymerization. For the case of the quinoxaline-based copolymers, N-cyclohexylpyrrolidone was added, and cyclohexane was required for the benzoxazole-based copolymers to produce clear solutions and subsequent films. The solids content for the copolymerization was maintained between 12 and 13 wt%. The acid acceptors used in the polymerization varied, however, N-methylmorpholine was favored since the subsequent salt, N-methylmorpholium hydrochloride, precipitated from the reaction mixture. This is believed to be important since other acid acceptors such as triethylamine form salts which remain in solution and may adversely affect the solubility of the oligomer(s) or subsequent copolymer, leading to homopolymer contamination. However, it should be pointed out that both pyridine and triethylamine have been successfully used even though the salt formed is soluble. High molecular weight polymers were readily achieved as judged by the dramatic increase in viscosities, which is characteristic of most condensation polymer-

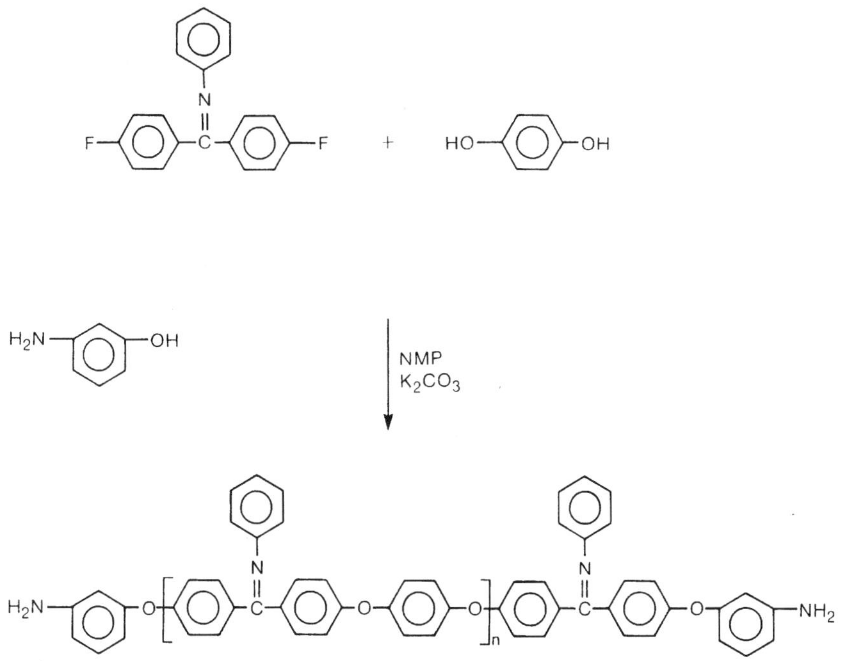

Scheme 5

izations. The resulting amic ester-aryl ether copolymers were precipitated in water and rinsed with methanol to remove salts formed during polymerization.

Table 2 contains the characteristics of the amic ester-aryl ether copolymers including coblock type, composition, and intrinsic viscosity. Three series of copolymers were prepared in which the aryl ether phenylquinoxaline [44], aryl ether benzoxazole [47], or aryl ether ether ketone oligomers [57–59] were co-reacted with various compositions of ODA and PMDA diethyl ester diacyl chloride samples (2a–k). The aryl ether compositions varied from approximately 20 to 50 wt% (denoted 2a–d) so as to vary the structure of the microphase-separated morphology of the copolymer. The composition of aryl ether coblock in the copolymers, as determined by ^{1}H NMR, was similar to that calculated from the charge of the aryl ether coblock (Table 2). The viscosity measurements, also shown in Table 2, were high and comparable to that of a high molecular weight poly(amic ethyl ester) homopolymer. In some cases, a chloroform solvent rinse was required to remove aryl ether homopolymer contamination. It should also be pointed out that both the powder and solution forms of the poly(amic ethyl ester) copolymers are stable and do not undergo transamidization reactions or viscosity loss with time, unlike their poly(amic acid) analogs.

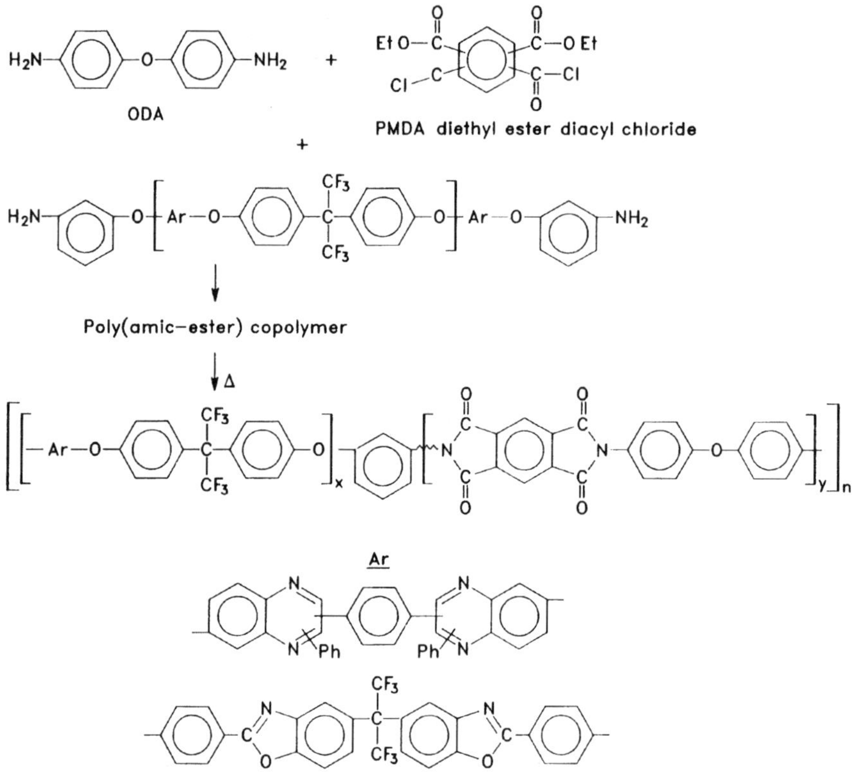

Scheme 6

Table 2. Characteristics of amic ester-aryl ether block copolymers

Sample Entry	Aryl Ether Coblock Type	Aryl Ether Incorporation, wt%		$[\eta]_{25\ °C}^{NMP}$ dL/g
		Charge	Measured	
2a	1a	25	28	0.53
2b	1a	50	50	0.43
2c	1b	25	28	0.43
2d	1b	50	52	0.46
2e	1c	25	21	0.50
2f	1c	50	54	0.36
2g	1d	18	14	0.35
2h	1d	20	16	0.45
2i	1e	20	10	0.52
2j	1f	25	28	0.45
2k	1f	50	49	0.43

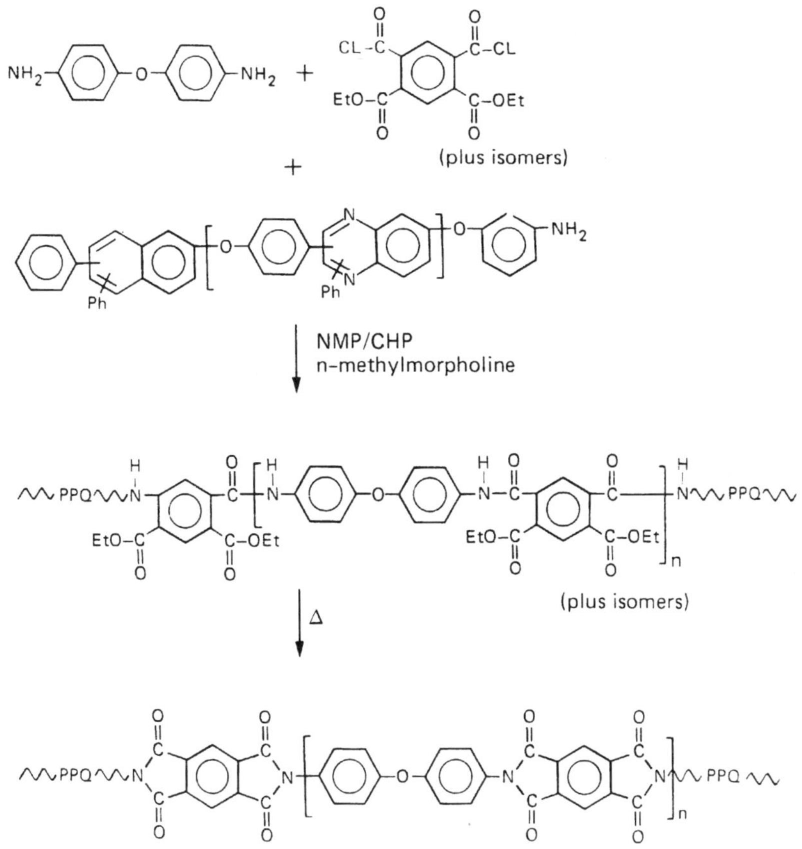

Scheme 7

Prior to imide formation, the imide-aryl ether ketimine copolymers were converted to the imide-aryl ether ketone analogue by hydrolysis of the ketimine moiety with *para*-toluene sulfonic acid hydrate (PTS) according to a literature procedure [51, 52, 57–59]. The copolymers were dissolved in NMP and heated to 50 °C and subjected to excess PTS for 8 h. The reaction mixtures were isolated in excess water and then rinsed with methanol and dried in a vacuum oven to afford the amic ester-aryl ether ether ketone copolymer, **2e** (Scheme 8.)

Solutions of the copolymers (**2a–k**) were cast into thin films and heated to 350 °C to imidize the copolymer (Schemes 6–8), yielding polymers **3a–k**. In each case, clear tough films were obtained indicating minimal homopolymer contamination. The thermal analyses for the copolymers are shown in Table 3 together with that of a polyimide homopolymer for comparison. It is important to note that the aryl ether composition increased after imidization due to the loss of ethanol in the imidization process. No detectable T_g was observed for the polyimide

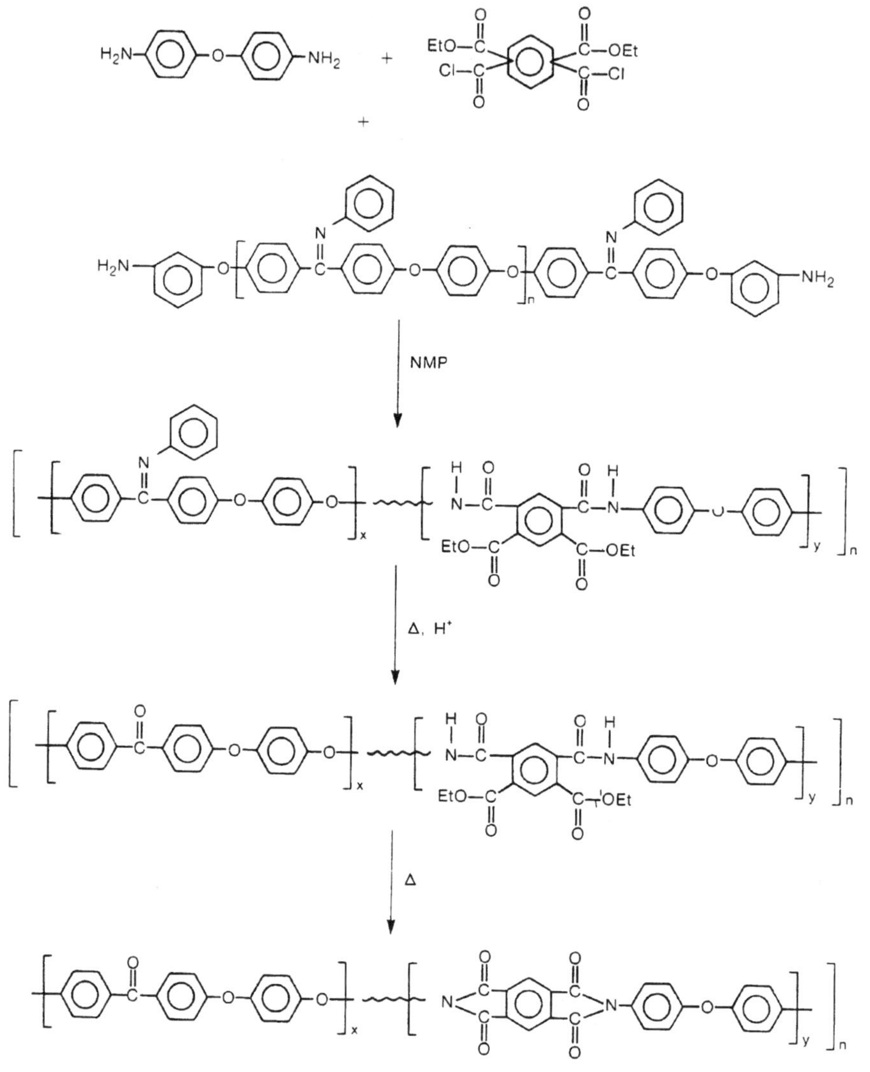

Scheme 8

homopolymer or for the polyimide component in the block copolymer for any of the copolymers surveyed. Thus, the calorimetry measurements provided no insight as to the morphology of the block copolymers. Table 3 also contains the thermal stability, as determined by the polymer decomposition temperature (PDT) and weight loss upon isothermal aging at 400 °C, for the block copolymers and polyimide. The PDTs for the copolymers are comparable to that of the polyimide (~ 480 °C) as was the weight loss upon isothermal aging. This is not entirely unexpected since the aryl ether homopolymers are thermally stable. In-

Table 3. Thermal characteristics of imide-aryl ether block copolymers

Sample Entry	Coblock type	Coblock Composition	Decomposition temperature, °C	Isothermal wt. Loss 400 °C (N_2), wt%	Thermal Expansion Coefficient, ppm
3a	1a	32	490	0.04	19
3b	1a	54	490	0.04	–
3c	1b	29	480	0.04	15
3d	1b	55	480	0.03	–
3e	1c	21	480	0.13	54
3f	1c	21	480	0.13	54
3g	1d	54	460	0.24	90
3h	1d	14	480	0.12	55
3i	1e	19	500	0.04	51
3j	1f	32	490	0.06	18
3k	1f	52	490	0.06	–

terestingly, the thermal expansion coefficients (TEC), also included in Table 3, for the copolymers containing low aryl ether compositions were somewhat lower than that of the parent polyimide. However, at higher aryl ether compositions, the TECs were comparable to that of the parent polyimide.

2.3
Adhesion, Morphology, and Mechanical Characteristics of Polyimide Copolymers

The morphology of block or segmented copolymers is dependent on a number of factors including sequencing, functionality and molecular weights of the blocks, the segmental interaction parameter, and thermal history. One synthetic route utilized a preformed difunctional oligomer-monomer(s) synthetic approach producing an (A–B)$_n$ multiblock architecture where blocks of varying length are randomly placed along the chain [55, 56]. Conversely, the use of monofunctional aryl ether oligomers produces an ABA triblock architecture where the aryl ether block is the A or terminal component. In such cases, the equilibrium morphology based on the volume fraction of the components is difficult to achieve. Furthermore, due to the high T_gs of the components of the copolymer, annealing to refine the morphology is not possible. The average molecular weight of the aryl ether block is identical with that of the preformed oligomer, whereas the average molecular weight of the polyimide block is controlled by the stoichiometric imbalance between the ODA and PMDA diethyl ester diacyl chloride. This is dictated by the aryl ether block length and compositions, where low aryl ether block lengths and high aryl ether compositions generate a larger imbalance between the ODA and PMDA diethyl ester diacyl chloride and, consequently, a lower polyimide block length. This may be very important, since the polyimide block length may control, to a large extent, the phase purity in the subsequent block copolymers.

Shown in Fig. 2 is the dynamic mechanical analysis of a representative copolymer series containing, in this case, varying aryl ether phenylquinoxaline compositions [44]. Two transitions were observed indicative of a microphase-separated morphology. For this copolymer series, the first transition was observed at

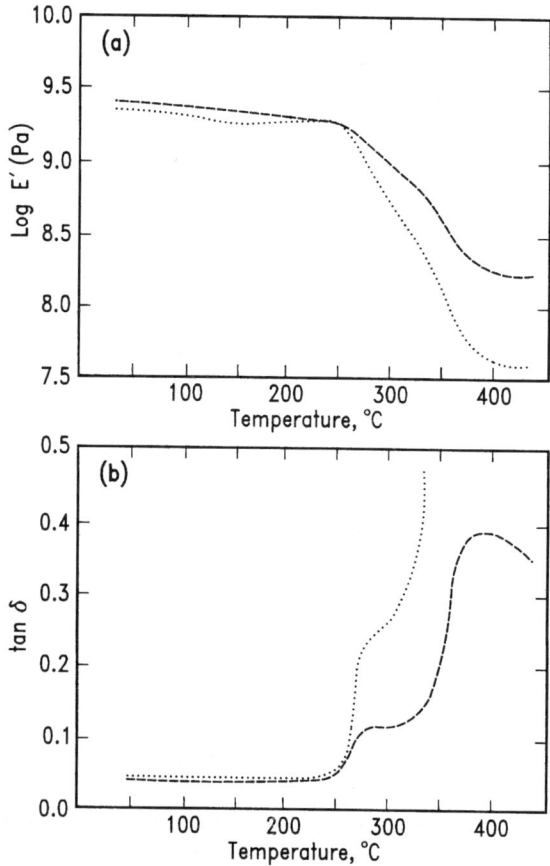

Fig. 2a, b. Dynamic mechanical spectra of imide-aryl ether phenylquinoxaline block copolymers, copolymer 3a (-----) and copolymer 3b (-----)

~ 250 °C resulting from the T_g of the aryl ether phenylquinoxaline phase and the second was observed at approximately 350 °C, which is identical to the transition observed in the dynamic mechanical spectra of the imide homopolymer. For each of the copolymer series surveyed, aryl ether block lengths below 10,000 g/mol showed T_gs in the copolymer somewhat higher than that of the oligomer. This may result from polyimide contamination in the aryl ether phase (i.e., phase mixing), or, alternatively, from the restriction of the aryl ether chain ends since the degree of polymerization in these oligomers is relatively low. Moreover, the region between the transitions, in both the storage modulus and damping, is broad, suggesting a large diffuse interfacial region between the respective domains [44]. The use of higher aryl ether block lengths shifted in the transitions toward those of the respective homopolymers, consistent with either improved phase purity or minimized chain and effects [44, 47]. Markedly different behavior was observed for the imide-aryl ether ether ketone copolymers (Fig. 3) [58]. The T_g of the aryl ether ether ketone component was significantly higher than that of the homopolymer, while the polyimide transition was lower than expected. This behavior is consistent with poor phase purity or phase mixing, which may have resulted from the high temperature processing conditions due to the limited solubility or from an increase in the viscosity of the system due to the rigid nature of 1e. Crystallization of the aryl ether ether ketone component in the block copolymer was not observed.

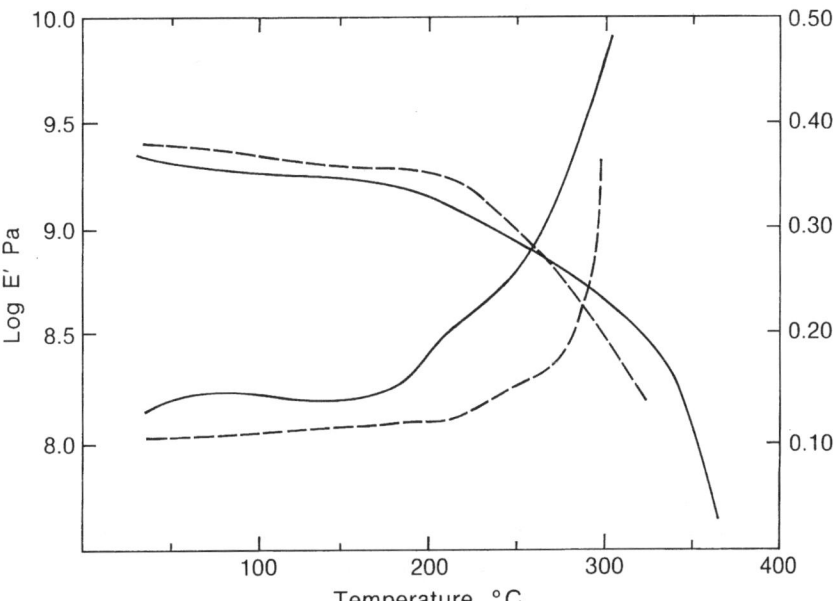

Fig. 3. Dynamic mechanical spectra of imide-aryl ether ether ketone block copolymer (—) and block copolymer/poly(etherimide) blend (-----)

In almost all of the cases investigated, the desirable morphology characteristic of the polyimide was retained in the imide-aryl ether copolymers [9, 10]. The diffraction profiles were determined in a reflection geometry, and consequently the diffraction vector is oriented normal to the surface of the film (Fig. 4). In each case, the diffraction profiles were virtually identical with that of the parent PMDA/ODA polyimide [44]. At ~ 6° a reflection was observed characteristic of the 15,8 Å spacing associated with the projection of the length of the monomer unit onto the chain axis, i.e., the (002) reflection [9, 10]. At ~ 18° and ~ 26° very diffuse reflections were seen characterizing the distances between adjacent PMDA/ODA chains. Thus, at least in terms of the ordering of the PMDA/ODA chains, the introduction of the aryl ether segments has not caused a major perturbation. In addition, the reflection at 6° for both the copolymers and the parent PMDA/ODA is weak in the reflection geometry and of approximately equal intensity. In a transmission geometry, however, this reflection is much more intense. Consequently, the molecular orientation is not altered markedly by the introduction of the aryl ether coblock.

Likewise, the mechanical properties of the copolymers were nearly identical or even somewhat enhanced towards the polyimide homopolymer in terms of the modulus and tensile strength values [44, 47]. For most of the block copolymers, the elongations to break were substantially higher than that of PMDA/ODA polyimide (Table 4). The shape of the polyimide stress-strain curve is similar to that of a "work-hardened" metal with no distinguishable yield point

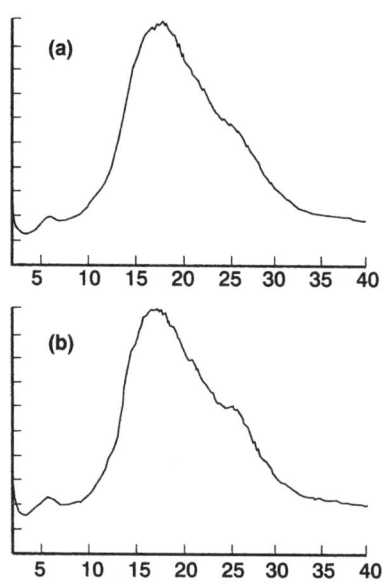

Fig. 4. Wide angle x-ray diffraction profile of imide-aryl ether phenylquinoxaline block copolymers (a) 2c and (b) 2d

Table 4. Mechanical properties of imide-aryl ether block copolymers

Sample Entry	Modulus, MPa	Tensile Strength, MPa	Elongation to Break, %
3a	2600	120	25
3b	2300	115	40
3c	2400	110	30
3d	2500	170	110
3e	2100	110	63
3f	2200	90	22
3g	2300	120	52

characteristic of small-scale or local plastic deformation [60]. The incorporation of aryl ether, particularly at the high compositions, irrespective of the block length used, resulted in a stress-strain curve which appeared more like that of an engineering thermoplastic with a small, yet distinguishable, yield point and necking and drawing. This is characteristic of a larger scale plastic deformation.

The adhesion of polyimide to itself (self- or auto-adhesion) is important in the fabrication of multilayer structures. The interfacial strength of sequentially cast and cured polyimide layers depends on the interdiffusion between layers, which has been shown to depend largely on the cure time and temperature of the first layer [3]. In the most practical application or representative case, the first layer has been cured to 400 °C (T_1 = 400 °C) with a second layer of polyimide deposited (T_2) and the combined layers cured to 400 °C ($T_1 = T_2$ = 400 °C). The peel strength for polyimide under these cure conditions (i.e., ($T_1 = T_2$ = 400 °C) is negligible. The copolymers, on the other hand, showed markedly different behavior (Table 5) [44, 47]. The copolymers with the low aryl ether block lengths and low aryl ether compositions showed significantly enhanced adhesion with peel strengths of ~ 60 g/mm. Peel strengths between 60 and 100 g/mm are considered excellent. As the aryl ether composition increased to 32 and 54 wt%, further improvements in the adhesion were realized. In fact, after the T_2 cure cycle, sequentially cast and cured films were indistinguishable, and the adhesion was characterized as a laminate. The copolymers containing the higher aryl ether block lengths showed the dramatic improvements in adhesion, irrespective of the aryl ether composition. In the last case, the imide aryl ether phenylquinoxaline triblock copolymers, improved adhesion was observed irrespective of block length or composition, and sequentially cast and cured layers were indistinguishable and denoted, once again, as a laminate. This presumably resulted from the enhanced mobility of the aryl ether phenylquinoxaline component for this molecular architecture. For each copolymer, the modulus-temperature profile shows a large drop in modulus at the aryl ether transition and another small drop in modulus at the polyimide transition. Consequently, the moduli of the copolymers at 400 °C are substantially lower than that of the parent polyimide (Fig. 2). Likewise, the damping (tan δ) increases, and this increased mobility in the entire system (melt flow) is believed responsible for the improved adhesion.

Table 5. Adhesion characteristics of imide copolymers: $T_1 = T_2 = 400 \,°C$

Sample Entry	Coblock Type And Composition, wt%	Predicted Morphology	Peel Strength, g/mm
3a	phenylquinoxaline	cylindrical	laminate
3b	phenylquinoxaline	lamellar	laminate
3c	phenylquinoxaline	cylindrical	laminate
3d	phenylquinoxaline	lamellar	laminate
3e	benzoxazole	cylindrical	8
3f	benzoxazole	lamellar	laminate
3g	benzoxazole	cylindrical	1
3h	benzoxazole	lamellar	–
3i	phenylquinoxaline	cylindrical	laminate
3j	phenylquinoxaline	lamellar	laminate

It is important to point out that the improvement in adhesion did not result from an increase in the solubility of the imidized polymer containing the flexible coblock after the T_1 cure cycle. In fact, the block copolymers demonstrated less than 2 % swelling (72 h) in the casting solvent, whereas PMDA/ODA polyimide homopolymer swells approximately 20–30 % (72 h). Clearly these data suggest that the improved auto-adhesion results from melt flow at 400 °C [44].

It is also of interest to adhere or laminate rigid polyimide films to themselves with thermoplastic adhesives, eliminating the use of solvents [51]. Although many of the poly(aryl ether)-based thermoplastics are known to be hot melt adhesives once heated approximately 50–70 °C above their T_gs, adhesion to rigid polyimides is particularly difficult due to both the polymer-polymer immiscibility and the absence of mobility in the rigid polyimide. Harris and co-workers [55] have reported that two such thermoplastics, the amorphous poly(ether imide) and semi-crystalline poly(ether ether ketone), are miscible over their entire compositional range. The imide aryl ether ether ketone block polymers were designed such that the aryl ether ether ketone component would provide the driving force for adhesion with the thermoplastic polyimide. The adhesion, as measured by L-peel test, of sequentially cast and cured layers of poly(ether imide) to PMDA/ODA polyimide is minimal (Table 6). Conversely, the adhesion of poly(ether imide) to the imide aryl ether ether ketone block polymers showed markedly different behavior. The adhesion, once cured to 350 °C, was exceptional. In fact, L-peel adhesion tests could not be performed as the two films were indistinguishable and thus denoted in Table 6 as laminate [59]. The substantial improvements in adhesion presumably resulted from the local miscibility of the poly(ether imide) with the aryl ether ether ketone component of the copolymer at the interface. Mixtures of the imide-aryl ether ether ketone with poly(ether imide) showed a single T_g, where the T_g of the aryl ether ether ketone block was shifted towards the T_g of the poly(ether imide) block, commensurate with the poly(ether imide) content (Fig. 3). These data showed the miscibility of the po-

Table 6. Adhesion studies of poly(ether imide) on polyimide

Sample Entry	Aryl Ether Ketone Composition, wt%	Peel Strength, g/mm
PMDA/ODA polyimide	–	3
copolymer 2i	19	laminate

ly(ether imide) in the aryl ether ether ketone component in the copolymer. Thus, this approach allows a means to laminate rigid high T_g polyimides which have T_gs in excess of the lamination temperature with moderate T_g thermoplastic polyimides.

3
Control of Polyimide Dielectric Constant

An alternative means of reducing the dielectric constant of polyimides is to have a low dielectric constant component dispersed as a second phase within the rigid polyimide matrix. Two key approaches have been pursued to this end. The first involves the preparation of polyimide block copolymers with highly fluorinated coblocks and the second involves the generation of a polyimide foam. The main drawback to the first approach is the solubility of highly fluorinated blocks in organic media which will permit copolymerization with polyimides. This led to the investigation of new semi-fluorinated polymers derived from poly(aryl ethers).

3.1
Imide-Perfluoroalkyl Ether Copolymers

Incorporation of aryl ether groups into semi-fluorinated polymers should give better NMP processability while maintaining the thermal stability, and the techniques known to prepare amine functional poly(aryl ethers) can be applied [43]. This led us to study the synthesis of poly(perfluoroalkylene aryl ethers) from bisphenols and 1,6-(4-fluorophenyl)perfluorohexane using conventional polyether polymerization conditions (Scheme 9) [61]. The resulting polymers displayed good processability in a variety of solvents, excellent thermal stability, T_g = 40 - 100 °C, and dielectric constants in the region of 2.6 (1 MHz). Amine functional oligomers of controlled molecular weight were prepared by proper adjustment of the stoichiometry and addition of 1,3-aminophenol as an end-capping agent (Scheme 9).

The polyimide-perfluoroalkyl ether block copolymers were prepared through the poly(amic alkyl ester) route, where a solution of diester diacyl chloride of PMDA was added to a mixture of ODA and the perfluoroalkyl ether oligomer, analogous to the procedure previously described [55]. Toluene was used as a cosolvent together with NMP to improve the solubility of the fluorinated coblock, and the resulting polymers were isolated by precipitation. The charac-

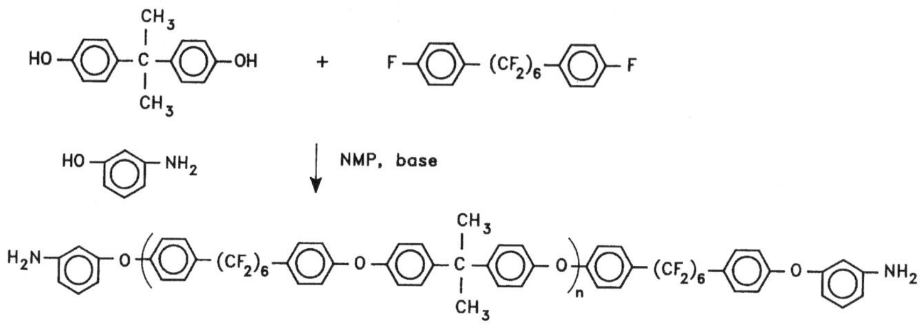

Scheme 9

Table 7. Characteristics of alkyl ester-perfluoroether copolymers

Sample Entry	Perfluoroether Oligomer, Mol. wt. g/mol	Perfluoroether Composition, wt%	
		Theory	Measured
4a	6500	30	33
4b	6500	50	46
4c	6500	75	71
4d	9900	50	47

teristics of selected copolymers are shown in Table 7 (copolymers **4a–4d**). The incorporation of perfluoroalkyl ether coblock is near quantitative, and thermal cure leads to conversion of the poly(amic ester) to polyimide and a concomitant increase in perfluoroalkyl ether content. The thermal stability of the cured copolymers was good at 400 °C (0.20 wt%/h for the 50/50 compositions), albeit lower than the polyimide homopolymer.

The dynamic mechanical analysis of copolymers with perfluoroaryl ether compositions of 35 % (copolymers **4a**) is shown in Fig. 5. The copolymer displayed a transition corresponding to the aryl ether block (100 °C) and a transition at ~ 350 °C, which is identical to that observed for PMDA/ODA. The presence of two transitions at temperatures close to that of the pure homopolymers is indicative of a heterogeneous morphology with good phase purity. The retention of the modulus to high temperature shows the polyimide is the continuous component acting as the matrix. The mechanical properties of selected copolymers are shown in Table 8, and moduli comparable to PMDA/ODA and elongations as high as 100 %. The dielectric constants of the copolymers are in the region of 2.8, which is lower than PMDA/ODA polyimide.

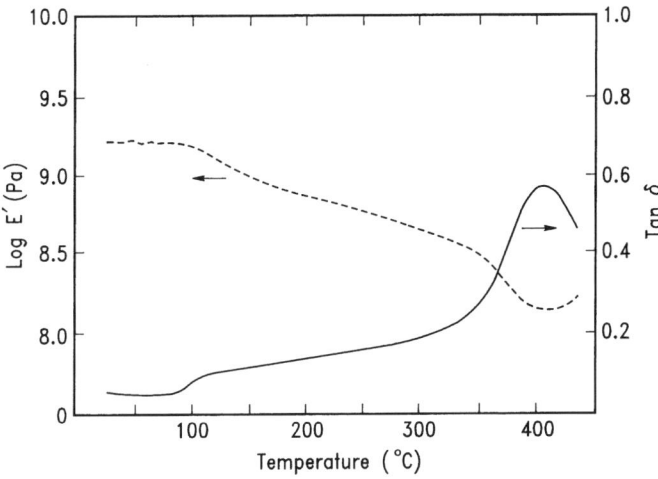

Fig. 5. Dynamic mechanical spectrum of imide-perfluoroalkyl ether block copolymer 4a.

Table 8. Mechanical properties of imide-perfluoroether copolymers

Sample Entry	Modulus, MPa	Stress at Break, MPa	Elongation to Break, %
4b	2030	112	102
4c	2000	83	107

3.2
Polyimide Nanofoams

One alternative to the above procedure is to generate a polyimide foam to reduce the dielectric constant substantially while maintaining the desired thermal and mechanical properties of the aromatic polyimide. The reduction in the dielectric constant is simply achieved by replacing the polymer with air which has a dielectric constant of 1. The advantage of a foam approach is readily apparent by examination of Fig. 6, which shows a Maxwell-Garnett modeling of composite structures based on a matrix polymer, with an initial dielectric constant of 2.8 [62]. Incorporation of a second phase of dielectric constant 2.0, simulating introduction of a second lower dielectric polymer phase, provides some reduction of the overall system dielectric properties. However, when the second phase has a dielectric constant of 1.0, as would be the case on introduction of air-filled pores to form a foam, the reduction of overall dielectric constant is very dramatic. Moreover, it is important to note that the relationship between composition and overall dielectric constant is not linear; in particular, the most significant

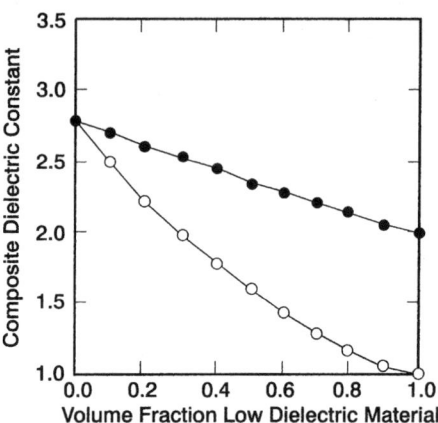

Fig. 6. Maxwell-Garnett theory used for the prediction of dielectric constant containing dispersed regions of low dielectric polymer ($\varepsilon = 2.0$, ○) or air ($\varepsilon = 1.0$, ●)

advantages are realized at modest levels of porosity. However, there are restraints on the materials. It is obvious that the pore size must be much smaller than the film thickness and any microelectronic features. Second, it is necessary that the pores be closed cell, i.e., the connectivity between the pores must be minimal. Third, the volume fraction of the voids must be as high as possible. Each of these can alter the mechanical properties of the film and structural stability of the foam. Failure to meet these restrictions will lead to a collapse of the foam or to a foam with limited use.

Polyimide foams, which have been developed primarily for the aerospace and transportation industries, show high compressive strength, low dielectric constant, low density, good thermal stability, and, unlike other organic foams, polyimide foams tend to be self-extinguishing when burned [63]. The routes to the preparation of polyimide foams include foaming agents [64–66], partial degradation generating a foaming agent [64–73], the inclusion of glass or carbon microspheres [74–78], and microwave processing [79]. The use of foaming agents and reactive systems are the most common routes to polyimide foams and tend to give the most well-defined and controlled pore structures. Alternatively, nadimide functional oligomers have been shown to undergo a rearrangement upon curing to evolve volatiles in which foam structures can be prepared. Most of the high temperature polymer foams reported to date do not satisfy the requirements for applications in microelectronics and, hence, have not found any practical use for thin film applications.

An alternative means of generating a polyimide foam with pore sizes in the nanometer regime has been developed [80–90]. This approach involves the use of block copolymers composed of a high temperature, high T_g polymer and a second component which can undergo clean thermal decomposition with the evolution of gaseous by-products to foam a closed-cell, porous structure (Fig. 7).

- Microphase separated block copolymer systems derived from a high temperature polymer with a thermally unstable polymer that undergoes thermal decomposition at elevated temperature to leave pores in the insulating layer.

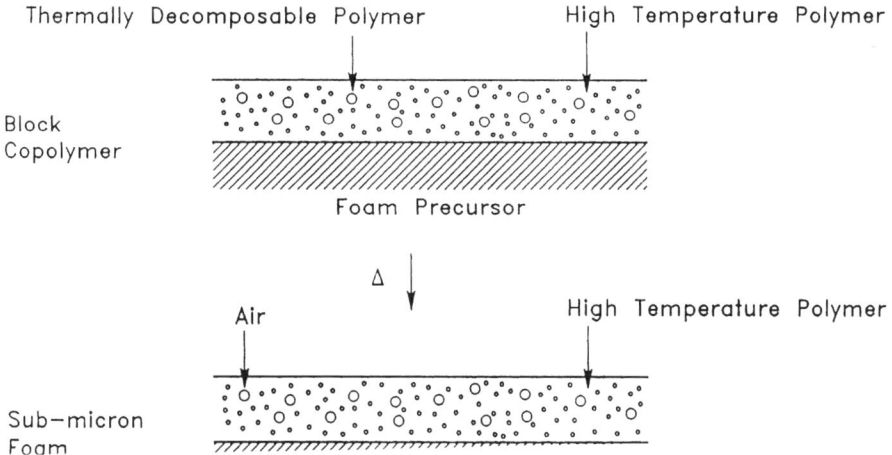

Fig. 7. Approach to the preparation of polyimide nanofoam using microphase separated block copolymers

This concept is similar to that developed by Patel et al. [91] for the preparation of epoxy networks with well-defined microporosity. However, rather than using a phase separation process of two polymers which proceed via a nucleation and growth mechanism and leads to phase separation on the micrometer scale, the work described here takes advantage of the small size scale of microphase separation of block copolymers. These block copolymers can be made to undergo thermodynamically controlled phase separation to provide a matrix with a dispersed phase that is spherical in morphology, monodisperse in size and discontinuous. Furthermore, the molecular structure and molecular weight of the segments allows procise control over both the size and volume fraction of the dispersed phase. By designing the block copolymers such that the matrix material is a thermally stable polymer with low dielectric constant and the dispersed phase is a thermally labile polymer that undergoes thermolysis at a temperature below the T_g of the matrix to give all volatile reaction products, one can prepare foams with pores in the nanometer dimensional regime that have no percolation pathway. That is, they are closed structures containing nanometer size spherical pores that contain air.

The successful implementation of the block copolymer approach to polyimide nanofoams requires the judicious combination of polyimide with the thermally labile coblock. The material requirements for the polyimide block are stringent and require thermally stable, high T_g polyimides which can be readily

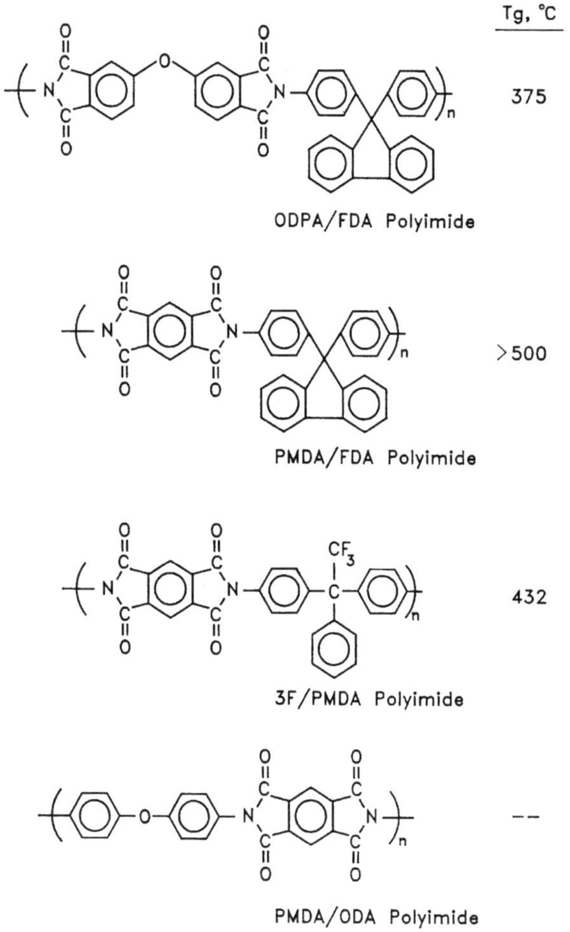

Scheme 10

ODPA/FDA Polyimide — T_g = 375 °C

PMDA/FDA Polyimide — T_g > 500 °C

3F/PMDA Polyimide — T_g = 432 °C

PMDA/ODA Polyimide — --

copolymerized with the appropriate labile block. A number of polyimides were surveyed and the structures of selected polyimides are shown in Scheme 10 [90]. Each of the polyimides had T_gs in excess of 370 °C and decomposition temperatures of ~ 500 °C.

3.2.1
Criteria for the Thermally Labile Coblock

The criteria for the thermally decomposable coblock include the synthesis of well-defined functional oligomers, compared with the synthesis of polyimide. This block must also decompose quantitatively into non-reactive species that can easily diffuse through a glassy polyimide matrix. The temperature at which decomposition occurs is critical; it should be sufficiently high to permit standard film preparation and solvent removal yet below the T_g of the polyimide block to avoid foam collapse. The thermally labile coblocks investigated include poly(propylene oxide), poly(methyl methacrylate), poly(styrene), poly(α-methylstyrene), and poly(caprolactone). Each decomposes quantitatively into small molecules in the appropriate temperature regime. Poly(propylene oxide) is stable in an inert atmosphere up to 300 °C. However, when exposed to oxygen, poly(propylene oxide) decomposes rapidly between 250 and 300 °C. Fig. 8 shows the thermogravimetric analysis (TGA) thermogram of poly(propylene oxide) heated isothermally at 275 °C in air. Within 20 min, a quantitative decomposition is observed. The thermal decomposition temperature of poly(methyl methacrylate) is strongly dependent on the polymerization mechanism used, since the degradation occurs via a depolymerization process. For example, poly(methyl methacrylate) prepared by free radical methods produces a substantial amount of chain ends terminated via a disproportion reaction. Typically, such poly(methyl methacrylate) degrades at relatively low temperatures. Conversely, poly(methyl methacrylate) prepared by anionic or group transfer methods has well-defined end groups and substantially higher decomposition temperatures. In fact, the poly(methyl methacrylate) used in this study has a decomposition temperature of 335 °C. Poly(α-methylstyrene), polystyrene, and styrene/α-methylstyrene copolymers are ideally suited for use as the thermally labile block, since well-defined functional oligomers can be prepared via anionic po-

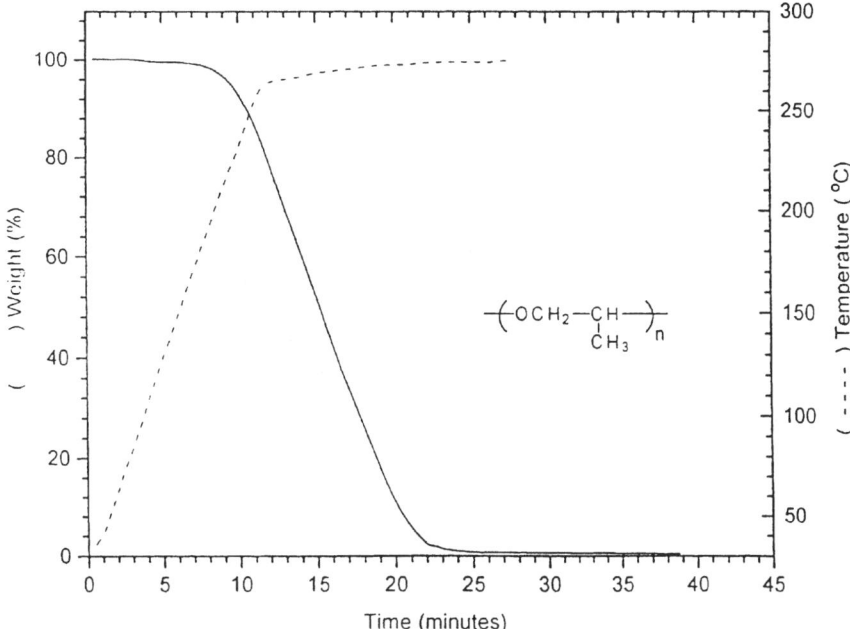

Fig. 8. Isothermal gravimetric analysis of poly(propylene oxide)

lymerization methods, and the polymer depolymerizes to monomer which, in principle, can readily diffuse through the matrix (Fig. 9). Poly(α-methylstyrene) and poly(styrene) undergo rapid thermal decomposition by depolymerization. The average number of monomer units generated per radical formed via initiation or intermolecular transfer defines the zip length [92, 93]. For poly(α-methylstyrene) the zip length is extremely high, ~ 1200, with a nearly quantitative monomer yield [92, 93]. By comparison, poly(styrene) has a zip length of approximately 60 [92]. Thus, the decomposition rate is significantly slower. Alternatively, poly(ε-caprolactone) and related aliphatic polyesters can be prepared by ring opening polymerization methods and decompose in the requisite temperature regime [94–96].

The characteristics of the amine terminated oligomers are shown in Table 9 (**5a–f**). Monohydroxyl terminated oligomers were prepared via anionic, group transfer or "living free radical" polymerization methods and were converted to the amine end groups by a method developed by Hedrick and co-workers [87, 88, 94]. The aminophenyl carbonate end-capped propylene oxide oligomers were prepared by the reaction of the monohydroxyl terminated propylene oxide oligomers with 4-nitrophenyl chloroformate in methylene chloride containing pyridine. The oligomers were then hydrogenated with Pearlman's catalyst (palladium hydroxide) and hydrogen to the amine. The molecular weights ranged from 5000 to 15,000 g/mol.

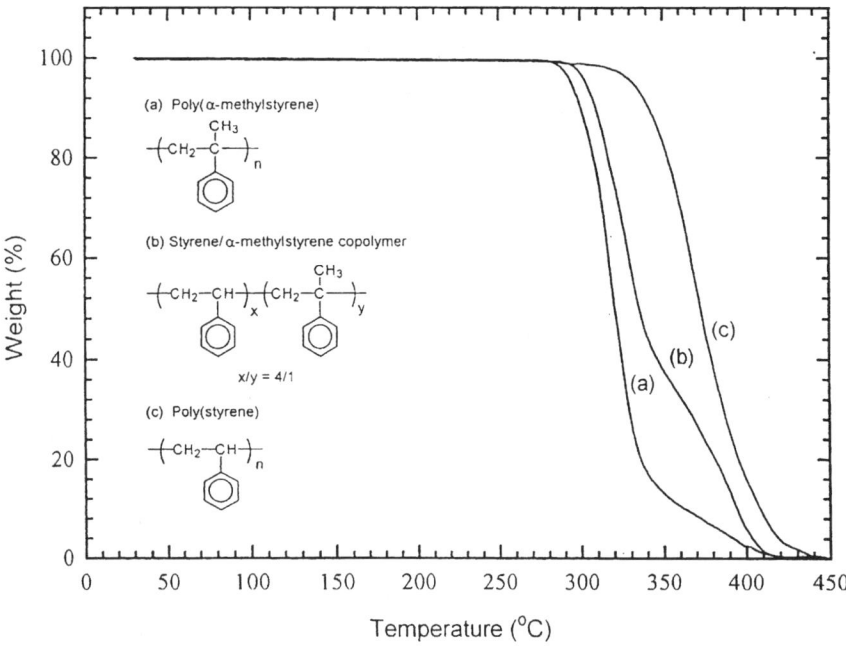

Fig. 9. Thermogravimetric analysis of Poly(styrene), poly α-methylstyrene and copolymers

Table 9. Aromatic amine terminated thermally labile oligomers

Sample Entry	Thermally Labile Block Type	Polymerization Method	Molecular Weight (g/mole)	T_g (°C)
5a	poly(α-methylstyrene)	anionic	12,000	155
5b	poly(styrene)	anionic	14,000	100
5c	styrene/α-methylstyrene	anionic	14,000	100
5d	poly(styrene)	free radical	13,000	100
5e	poly(propylene oxide)	anionic	5,600	-65
5f	poly(methyl methacrylate)	group transfer	15,000	105

3.2.2
Synthesis and Characterization of Polyimide Copolymers

The solubility of the polyimide dictated, to a large extent, the synthetic route employed for the copolymerization. The ODPA/FDA and 3FDA/PMDA polyimides are soluble in the fully imidized foam and can be prepared via the poly(amic-acid) precursor and subsequently imidized either chemically or thermally. The PMDA/ODA and FDA/PMDA polyimides, on the other hand, are not soluble in the imidized form. Consequently, the poly(amic alkyl ester) precursor was used followed by thermal imidization. For comparison purposes, 3FDA/PMDA-based copolymers were prepared via both routes. The synthesis of the poly(amic acid) involved the addition of solid PMDA to a solution of the styrene oligomer and 3FDA to yield the corresponding poly(amic acids). The polymerizations were performed in NMP at room temperature for 24 h with a solids content of ~ 10 % (w/v). Chemical imidization of the poly(amic-acid) solutions was carried out in situ by reaction with excess acetic anhydride and pyridine in 6–8 h at 100 °C. The copolymers were subjected to repeated toluene rinses in order to remove any unreacted styrene homopolymer. The synthesis poly(amic alkyl ester), on the other hand, involved the incremental addition of PMDA diethyl ester diacyl chloride in methylene chloride to a solution of the oligomer and 3FDA in NMP containing pyridine as the acid acceptor (Scheme 11). In these experiments, the meta isomer of PMDA diethyl ester diacyl chloride was used primarily due to its enhanced solubility, and to facilitate comparison with previous studies. The solids composition was maintained at ~ 15 % for each of the polymerizations. The copolymers were isolated in a methanol/water mixture, rinsed with water to remove remaining salts, and rinsed with methanol and toluene followed by drying. Shown in Table 9 are some of the general types of copolymers prepared as foam precursor, showing the scope of combinations of labile with thermally stable blocks (copolymers **6a–n**).

The characteristics of selected copolymers, prepared in the fully imidized form and in the poly(amic alkyl ester) precursor to the polyimide, are shown in Table 10. The weight percentage, or loading, of the labile blocks in the copoly-mers was intentionally maintained low (~ 20 wt%) in order to produce discrete

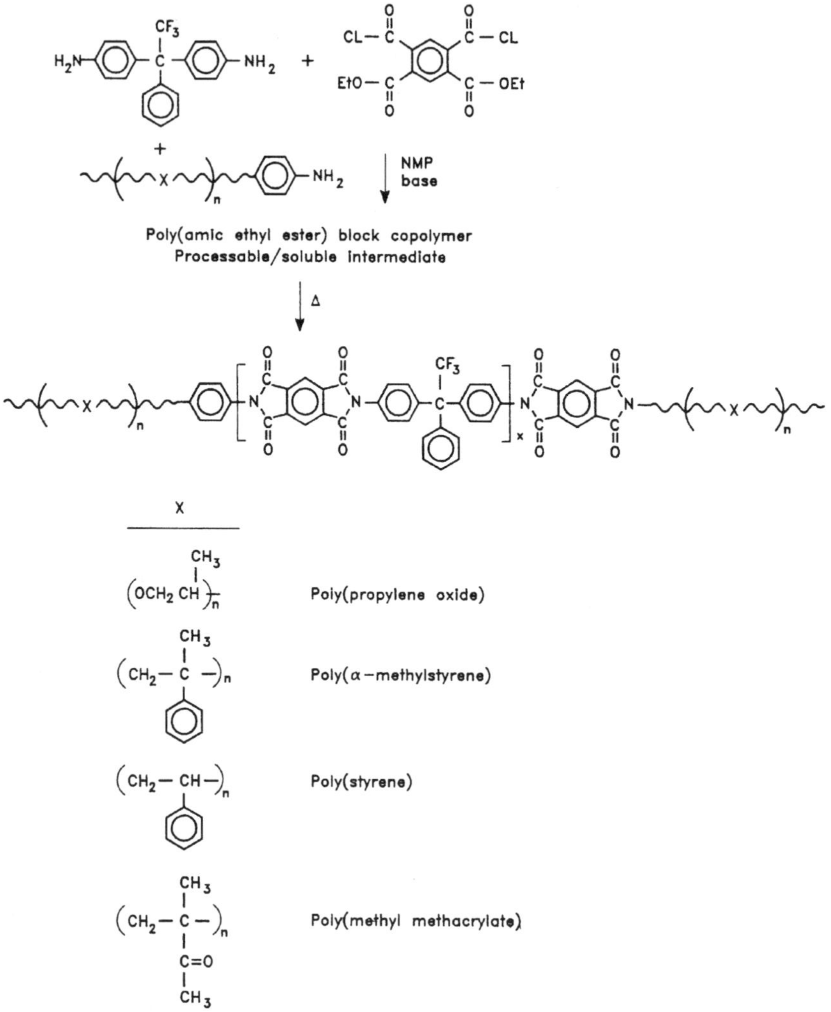

Scheme 11

spherical domains of the block embedded in the polyimide matrix. At higher loadings, phase separation of the block could, in principle, produce cylindrical or more interconnected structures which are undesirable. The loading of the labile block in the copolymer was assessed by thermal gravimetric analysis (TGA) and by ^{1}H NMR. For essentially all cases, the loading of the labile block agreed closely with that theoretically expected from the feed ratios. The use of monofunctional oligomers in the polyimide syntheses described above affords on ABA triblock copolymers architecture, where the thermally labile component comprises the terminal A blocks and the stable polyimide is the B block. It should be noted with such an architecture that, upon thermal decomposition

Table 10. Characteristics of block copolymers

Copolymer Entry	Polyimide Type and Form	Thermally Labile Block Type	Thermally Labile Block Composition, wt%			Volume Fraction of Labile Block, %
			Charge	Incorporated		
				^{1}H NMR	TGA	
6a	3FDA/PMDA (alkyl ester)	poly(propylene oxide)	15	9.9	9	11
6b	3FDA/PMDA (alkyl ester)	poly(propylene oxide)	25	23	22	27
6c	ODPA/FDA (imide)	poly(propylene oxide)	15	13.1	13	–
6d	ODPA/FDA (imide)	poly(propylene oxide)	25	22.5	22.5	–
6e	PMDA/FDA (alkyl ester)	poly(propylene oxide)	15	14	9.2	12
6f	PMDA/FDA (alkyl ester)	poly(propylene oxide)	25	22	18.4	–
6g	ODPA/FDA (imide)	poly(α-methylstyrene)	14	13	14	16
6h	ODPA/FDA (imide)	poly(α-methylstyrene)	25	24	24	29
6i	3F/PMDA (alkyl ester)	poly(α-methylstyrene)	15	14	15	27
6j	3F/PMDA (alkyl ester)	poly(α-methylstyrene)	25	–	24	27
6k	3F/PMDA (alkyl ester)	poly(styrene)	20	18	19	2
6l	3F/PMDA (alkyl ester)	α-methylstyrene/styrene copolymer	20	15	14	18
6m	PMDA/ODA (amic ester)	poly(propylene oxide)	25	22	23	28
6n	PMDA/ODA (amic ester)	poly(methyl methacrylate)	25	20	21	23

of the labile coblock, the molecular weight of the polyimide block remains the same. This is important for mechanical and physical property considerations.

The processing window for film and foam formation was established primarily with the 3FDA/PMDA imide-based copolymers, since these copolymers are soluble and can be isolated and characterized at various stages of processing or imidization. It is critical that the decomposition of the labile block should occur substantially below the T_g of the polymer matrix. Furthermore, the casting solvent must be effectively removed, without labile block degradation, to minimize plasticization of the polyimide matrix, which would further narrow the processing window (i.e., temperature difference between T_g polyimide and decomposition temperature of labile block). Samples were cast from NMP, cured, and the processing window for film and foam formation was established by ^{1}H NMR, TGA, and dynamic mechanical measurements. Since most of the labile blocks are stable at 300 °C in nitrogen, samples cured to this temperature to remove the casting solvent should retain their labile block composition. ^{1}H NMR showed complete solvent removal and the full T_g of polyimide phase was achieved (440 °C). Volksen and co-workers [37] found that the temperature range over which imidization occurred for the poly(amic ethyl ester) derived from PMDA/ODA was 240–355 °C, with a maximum in the rate at 255 °C. I.R., ^{1}H NMR, and calorimetry measurements indicated that the imidization was essentially quantitative under these conditions, with minimal loss to the labile block composition as measured by TGA. However, the processing temperature of the α-methylstyrene-based copolymers was limited to 265 °C to minimize poly(α-methylstyrene) decomposition. In this case, ~ 1–3 % solvent remained and complete imidization was not accomplished, which resulted in a depression in the polyimide T_g.

Dynamic mechanical analysis was one technique used to access the morphology of the copolymers. It is essential that separation occurs with high phase purity in order to obtain a nanofoam while minimizing collapse. Selected dynamic mechanical spectra are shown in Fig. 10. Two transitions were observed in each case, indicative of microphase separated morphologies. For the imide α-methylstyrene copolymer (Fig. 10), the transition occurring near 160 °C is similar to that seen for the α-methylstyrene oligomer used in the synthesis and the damping peaks associated with the α-methylstyrene transition are sharp, indicating that the phases are not only pure but have discrete boundaries. For the α-methylstyrene-based copolymers and other labile coblocks which rapidly degrade, the transition of the imide block shows a strong dependence on the fraction of the α-methylstyrene or other labile block in the copolymer. The copolymer containing ~ 15 wt% α-methylstyrene (copolymer **6g**) shows an imide transition which is nearly identical to that of the polyimide homopolymer. However, the copolymer containing the higher α-methylstyrene fraction shows an imide transition which is substantially depressed (**6h**). Furthermore, the transition appears at nearly the same temperature (~ 320 °C) as the decomposition temperature of poly(α-methylstyrene).

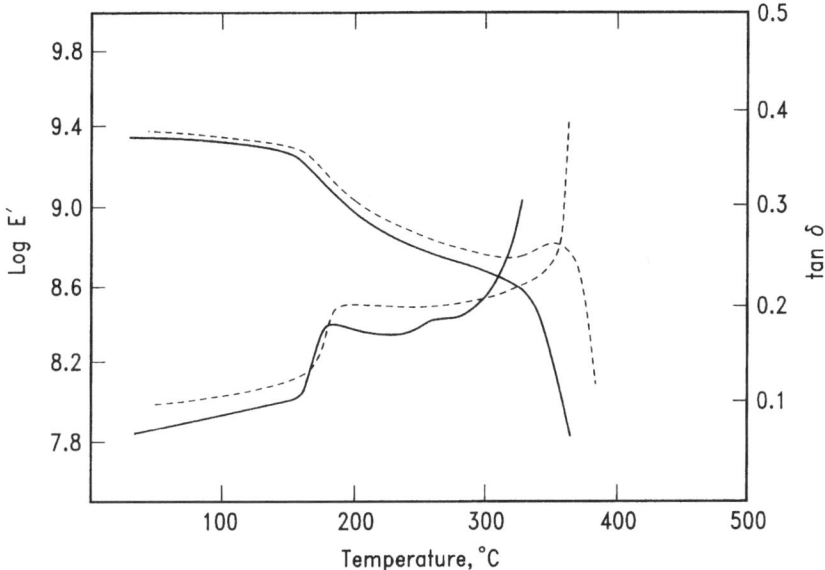

Fig. 10. Dynamic mechanical spectra of copolymers 6g (-----) and 6h (—)

The thermogram of copolymer **6h** in Fig. 11 shows that the α-methylstyrene coblock degrades at ~ 320 °C, and the degradation product, α-methylstyrene, is readily evolved as evidenced by the narrow temperature range for the degradation and monomer evolution. Since poly(α-methylstyrene) thermally degrades by depolymerization, the loss of degradation products is rapid, as shown in Fig. 11. Also shown in Fig. 11 is the dynamic mechanical spectra of copolymer **6h** together with the thermal gravimetric data. A drop in the modulus occurs in the proximity of the α-methylstyrene degradation temperature. This modulus drop was followed by an increase in the modulus just prior to the T_g of the polyimide matrix, consistent with the evolution of degradation products. The copolymer with a high α-methylstyrene content (copolymer **6h**), on the other hand, shows a T_g at a temperature which was commensurate with the degradation of the temperature α-methylstyrene coblock. These data suggest that the α-methylstyrene monomer degradation product strongly interacts with the ODPA/FDA polyimide, which results in plasticization of the matrix.

3.2.3
Foam Formation and Characterization

The balance of these different factors is found in the dependence of nanofoam formation on the labile block content in the copolymer [98]. For copolymers with a low composition of labile block, the degradation is rapid, forming a res-

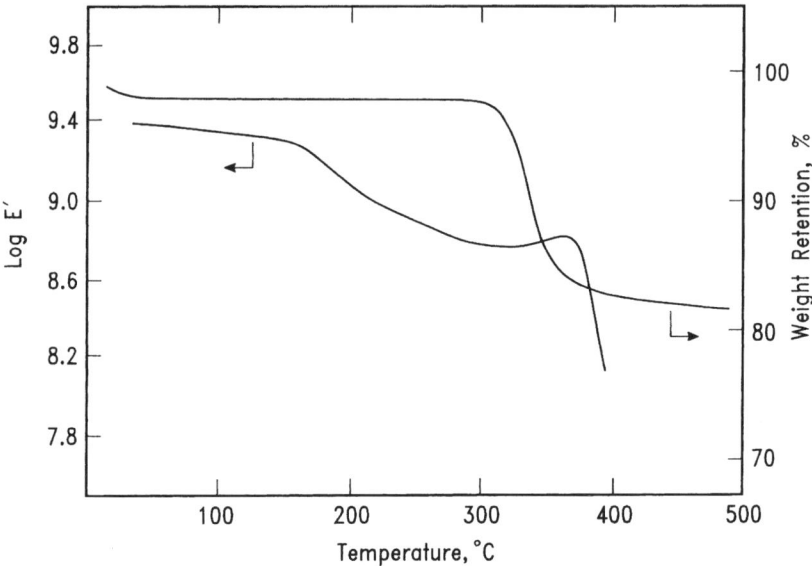

Fig. 11. Dynamic mechanical spectra and thermogravimetric spectra for copolymer 6h

ervoir of degradation products that must be removed in order to form voids. Since the permeability, which takes into account both the solubility and diffusion coefficient, will cause an initial depression (i.e., plasticization) of the modulus followed by an increase in the modulus as the concentration of degradation products diminishes. However, for higher fractions of the labile blocks, a saturation concentration of the degradation products in the imide matrix must be reached. Although the rate at which the degradation products are removed is rapid, within the time scale defined by the relaxation of the plasticized imide matrix, the removal is not rapid enough and leads to a collapse of the nanofoam. Consequently, the retention of the foam structure depends upon a balance between the rate of decomposition of the labile coblock, the solubility of the degradation products in the imide matrix, and the rate at which the degradation products diffuse out the matrix. For each labile coblock surveyed, mild decomposition temperatures were employed to effect the decomposition of the labile block in such a way as to minimize plasticization.

This generation of the nanofoam was accomplished by subjecting the copolymer film to a subsequent thermal treatment to decompose the labile coblock. The temperature range and degradation atmosphere varied depending on the labile block type. For instance, the propylene oxide-based copolymers were heated to 240 °C in air for 6 h, followed by a post-treatment at 300 °C for 2 h to effect the degradation of the propylene oxide component. The degradation process was followed by thermogravimetric analysis (TGA) and ^{1}H NMR, and, under these conditions, quantitative degradation was observed with no evidence of residual by-products nor chemical modification of the polyimide. Conversely, the degradation of the styrenic-based copolymers was accomplished by a step-wise process so as to minimize the degradation rate of these blocks.

The density of the polymer clearly shows the formation of a foamed polymer. The density values for selected foams together with the polyimide homopolymers are shown in Table 11. The density values for the ODPA/FDA and PMDA/FDA polyimides were both 1.28 g/cm^3 and 3FDA/PMDA is 1.34, while most of the propylene oxide-based copolymers showed substantially lower values. The densities of the foamed copolymers derived from these copolymers ranged from 1.09 to 1.27 g/cm^3, which is ~ 85–99 % of that of the polyimide homopolymers. This is consistent with 1–15 % of the film being occupied by voids. From these data (i.e., the comparison of Tables 10 and 11), it appears that the volume fraction of voids or the porosity is substantially less than the volume fraction of propylene oxide in the copolymer (i.e., ~ 70 % or less). Thus the efficiency of foam formation is poor. Conversely, the propylene oxide-based copolymers with PMDA/ODA as the imide component did not show the expected density drop, and the values were essentially identical to that of the homopolymer. In PMDA/ODA-based systems, molecular ordering and orientation were found to be critical in determining the stability of the foam structure. Where the character-

Table 11. Characteristic of polyimide foams

Sample Entry	Initial Labile Block Composition, vol.%	Density, g/cm^3	Volume Fraction of Voids (Porosity) %
3FDA/PMDA polyimide	–	1.35	–
6a	11	1.17	13
6b	27	1.10	18
ODPA/FDA polyimide	–	1.28	–
6c	–	1.20	6
6d	–	–	12
PMDA/FDA polyimide	–	1.28	–
6e	–	1.17	7
6f	–	1.11	12
6g	16	1.23	3.1
6h	29	1.18	7.5
6i	27	1.13	16
6j	27	–	30
6k	22	1.17	14
6l	18	1.18	19
PMDA/ODA polyimide	–	1.41	–
6m	20	1.41	0
6n	25	1.41	0

istic in-plane molecular orientation and molecular ordering of homogeneous PMDA/ODA films were retained, relaxation rates were clearly enhanced in the presence of the pores, leading to collapse of the foam structure well below the matrix T_g. Moreover, there is some suggestion that the presence of pores may enhance stress relaxation rates in step strained imide foams. Thus, the recreation of these effects in a suitably designed block copolymer (where the presence of a second phase should maintain the structural integrity) may be a means of promoting rapid stress relaxation in PMDA/ODA imide films, yielding the sought-after low stress materials.

The porosity values, as measured by the density column, for the foamed copolymers derived from the styrene and α-methylstyrene coblock was difficult. In all cases, the density gradient method yielded results which were consistently lower than those described above, by a substantial amount. During the course of the measurement, the film settled initially at a specific height in the column. With time the sample drifted downwards in the column to higher densities. This, more than likely, resulted from the fluid penetrating into the porous film, which eventually yielded the density of the 3FDA/PMDA polyimide homopolymer. Consequently, an alternative means of measuring porosity was required. It has been shown that infrared spectroscopy provides the means of determining the void content in polymeric materials [95]. Independently measuring the film thickness, the absorbance, calibrated against the bulk polymer, and the refractive index, determined from the interference fringes, yields results that are in quantitative agreement with density gradient methods. The porosities of the foamed copolymers measured by IR are shown in Table 11. The foams derived from the block copolymers comprised of the low block lengths show foam efficiencies comparable to those observed for foams derived from the imidized copolymers. Conversely, the foam derived from the copolymer containing the high molecular weight α-methylstyrene coblock (copolymer **6j**) showed porosities which met and even exceeded the volume fraction of α-methylstyrene coblock. However, many of the foams prepared from α-methylstyrene and styrene labile blocks were somewhat cloudy or even opaque in some cases. This, in turn, led to significant scatter in the IR results. In these samples, transmission electron microscopy (TEM) was used to assess the porosity [96].

Rather convincing evidence that nanofoams can be generated via this process is shown in Fig. 12. Here, an electron micrograph of copolymer **6b** is shown derived from an imide-propylene oxide copolymer. The porous structure of a foam is clearly evident where the white areas are the voids from the degraded labile phase. The average size of the pores is ~ 60 Å which is slightly more than that found by small-angle X-ray scattering. One rather important feature of these data is that the pores are not interconnected. This, as should be remembered, is critical for the end use of these materials. In addition, there is no evidence of very small pores, which is in keeping with the small-angle X-ray scattering data. Furthermore, the periodic nature of the pores is clearly evident. Conversely, the foams derived from the α-methylstyrene- and styrene-based copolymers showed pore sizes ranging from 200 to ~ 1800 Å, which are considerably larger

100 nm

Fig. 12. Transmission electron micrograph of foamed copolymer 6b

than the size of the microdomains of the initial copolymer (Fig. 13). Furthermore, the pores appear to be more interconnected than those obtained from the fully imidized copolymers. Finally, these pores are anisotropic in shape, not spherical. The block copolymers derived from rigid and semi-rigid polyimides with either poly(methyl methacrylate) or poly(propylene oxide) did not show the expected nanofoam formation upon thermolysis of the labile coblock. The rapid collapse of the foam well below the nominal glass transition temperature, T_g, was attributed to the anisotropic mechanical properties, characteristic of

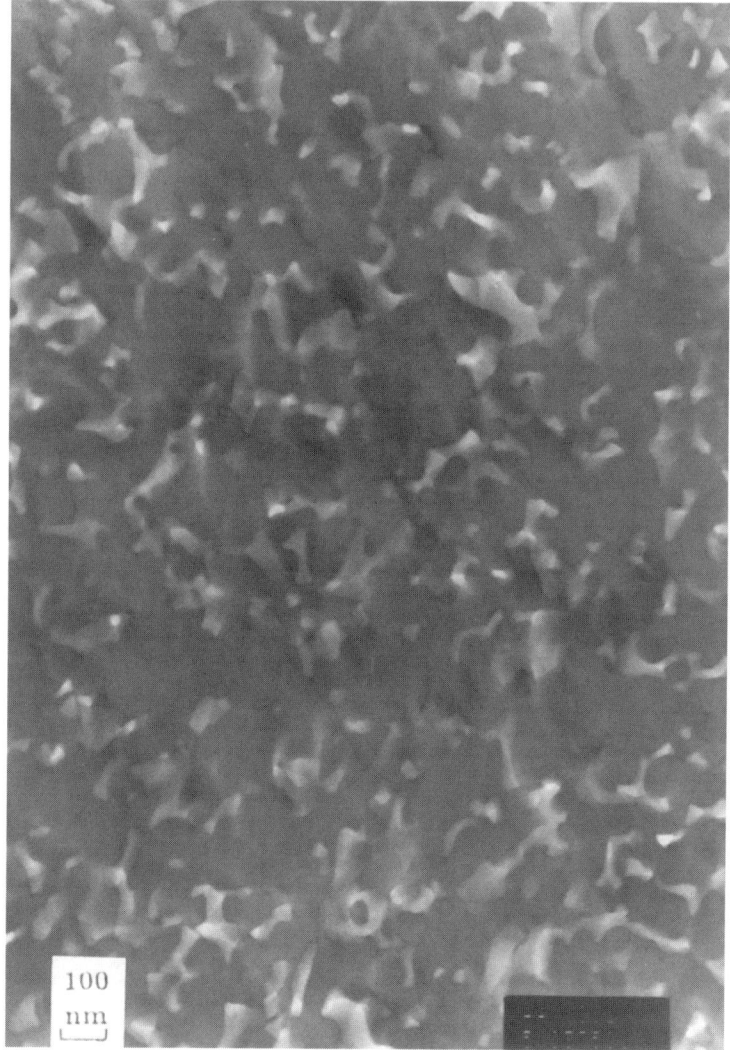

Fig. 13 a. Transmission electron micrograph of: **a** foamed copolymers 6k

thin films of these ordered polyimides. However, in the case of the α-methylstyrene-based copolymers with the rigid polyimides, highly porous materials were prepared as a result of the blowing of the matrix.

Foam formation was possible only in the amorphous high T_g polyimides; however, the volume fraction of voids does not correspond to the volume fraction of propylene oxide in the initial copolymer. A decrease in the volume fraction of voids incorporated into the matrix in comparison to the initial volume fraction of the propylene oxide in the copolymer can be understood by consid-

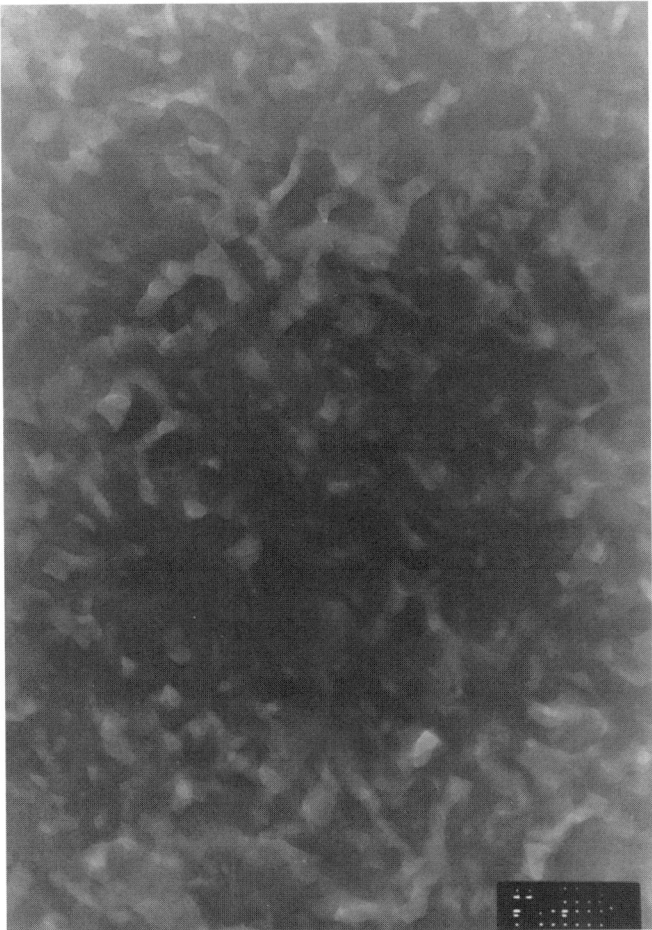

Fig. 13 b. Transmission electron micrograph of: **b** foamed copolymers 6l

ering the distribution of sizes of the propylene oxide phases. Since the copolymers are not monodispersed, a distribution of propylene oxide microdomain sizes is produced. The pressure exerted on a pore will vary as γ/R, where γ is the surface tension and **R** is the radius. Consequently, the higher pressure on the smaller pores will tend to cause them to collapse. In addition to this, there will be a small amount of propylene oxide within the imide phase, which will be removed upon decomposition. However, this cannot account for the larger discrepancies, particularly when the initial propylene oxide content was high. In this case, the large volume fraction of the propylene oxide will necessarily give rise to smaller volumes of the imide phase between the propylene oxide phases. During the degradation of the propylene oxide, the imide can be plasticized by the decomposition products and cause a collapse of the void structure. However,

Fig. 13 c. Transmission electron micrograph of: **c** foamed copolymers 6j

the interaction of the poly(propylene oxide) degradation products with polyimide is not clear. There are approximately 11 major degradation products upon the thermolysis of poly(propylene oxide), where acetaldehyde and acetone comprise nearly 80 % of these products. These polyimides are polar and their solubility parameters are close to those of the major degradation products of poly(propylene oxide). In fact, ODPA/FDA and PMDA/FDA show a significant level of uptake of acetone and acetaldehyde, further suggesting a favorable interaction and the possibility of plasticization. Although the duration of the plasticization may be minimal due to the elevated temperature, this transient mobility coupled with the residual thermal and solvent loss stresses are sufficient to cause a partial collapse of the foam structure. These combined observations suggest that there is an optimal void size and volume fraction that can be incorporated into the imide matrix.

In both styrene and α-methylstyrene coblocks used for the generation of the nanofoam structure, the TEM results demonstrate that the size of the pores generated are much larger than the size of the initial copolymer microdomains. This suggests that the degradation products, α-methylstyrene or styrene, act as blowing agents. As the blocks decompose, some of the monomer diffuses into the polyimide matrix, plasticizing the polyimide. The decomposition of the block, particularly α-methylstyrene, is quite rapid and much more rapid than the diffusion rate. The monomer, at these temperatures, exerts a substantial pressure on the surrounding matrix. Since the matrix is plasticized, it can be effectively blown (Fig. 13). Some of the pores can coalesce with time, giving rise to a partially interconnected pore structure.

To obtain a measure of the dielectric constant and anisotropy of thin films, the refractive index of thin film samples was measured. It has been shown that the measured dielectric constant is approximately the square of the refractive index at 633 nm wavelength [the actual relationship is roughly ε (refractive index)^2 + 02.] and the anisotropy is obtained from the difference between the in-plane and out-of-plane refractive index [97]. The measured anisotropy of foamed polyimides is lower than that observed for non-foamed polyimides. In addition, a drop in refractive index of the samples was observed upon foaming. The polyimide PMDa/3FDA has a measured dielectric constant of ca. 2.9 at 70 °C. A foamed sample of PMDA/3FDA derived from copolymer **6f** showed a drop in dielectric constant of 2.3 [97].

4
Residual Thermal Stress Control of Polyimide Through the Use of Self-Assembled, Phase Separated Block Copolymers

The semi-rigid PMDA/ODA polyimide has a thermal expansion coefficient (TEC) value of approximately 40 ppm, and a measured residual thermal stress in the 30–40 MPa range [15]. However, it should be pointed out that the stress value is significantly less than the value that would be predicted from the TEC and modulus values, and the discrepancy was found to be a result of a significant stress relaxation in the film. However, this relaxation was shown to occur with a large time constant. Thus, while stress relaxation does occur in polyimide, the rate and degree of stress relaxation is insufficient to provide significant relaxation of thermal mismatch stress in a multilayer structure of dissimilar materials such as ceramic or copper. Consequently, it was of interest to modify PMDA/ODA polyimide in such a way as to enhance the stress relaxation process and, subsequently, lower residual thermal stress. Moreover, it would be of significant value to have a generic means of modifying semi-rigid and even amorphous polyimides in such a way as to control stresses which result from thermal expansion coefficient mismatches of adhered films. The incorporation of an elastomeric modifier into polyimide is the most judicious choice, since, in most cases, elastomers relax simultaneously with the applied load. Furthermore, elastomeric modifiers phase separated on the appropriate size scale and should serve as sites of stress concentrations which may enhance the relaxation rate of the polyimide. A number of appropriately designed polyimide block copolymers have been prepared as a means to facilitate the stress relaxation of polyimide and to yield a generic methodology to low stress materials. Poly(dimethylsiloxane) was chosen as the primary coblock, since poly(dimethylsiloxane) is a low T_g, low modulus materials and should, in principle, provide a large modulus mismatch with the polyimide.

4.1
Synthesis of Imide-Siloxane Copolymers

Several series of copolymers were prepared as a route to low stress polyimides [2, 102]. The first series of copolymers prepared were PMDA/ODA imide-dimethylsiloxane copolymers (**7a–c**, Table 12). The copolymers were prepared via the poly(amic alkyl ester) route where PMDA diethyl ester diacyl chloride was added to a solution of ODA and amine functional siloxane in the presence of base (Scheme 12). High molecular weight polymer was obtained for each of the siloxane block lengths surveyed, and the characteristics are shown in Table 12. Since polydimethylsiloxane is nonpolar, microphase separated morphologies were achieved which showed high phase purity while the polyimide block retained the local ordering and orientation, characteristic of the polyimide homopolymer (Fig. 14) [92]. Transmission electron microscopy (TEM) of the imide-dimethylsiloxane copolymer (**7c**) clearly shows the formation of a well-defined two phase structure, where the dark region is the polyimide continuous phase (Fig. 15). In a second series of block copolymers, structural modifications to the poly(dimethylsiloxane) coblock were introduced. PMDA/ODA imide-dimethyldiphenylsiloxane copolymers were prepared with diphenylsiloxane compositions ranging from 25 to 75 wt% (**8a–c**, Table 13). The polar diphenylsiloxane minimized the solubility parameter difference between the imide and siloxane blocks to facilitate phase mixing, as evidenced by the dynamic mechanical behavior (Fig. 14). The characteristics of the copolymers prepared are shown in Table 13. In this study, the siloxane block length and composition remained constant and the diphenylsiloxane composition was varied.

Table 12. Characteristic of imide-dimethylsiloxane block copolymers

Copolymer Entry	Dimethylsiloxane Block Length, g/mol	Dimethylsiloxane Composition, wt%
7a	1000	20
7b	1000	65
7c	5400	20
7d	5400	50

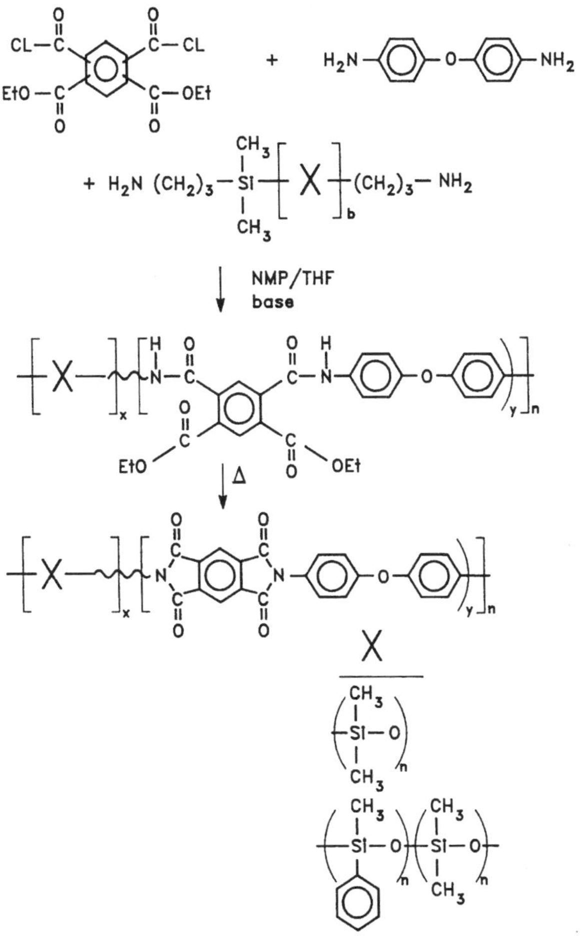

Nanoscopically Engineered Polyimides

Scheme 12

Table 13. Characteristic of imide-dimethyldiphenylsiloxane block copolymers

Copolymer Entry	Dimethyldiphenylsiloxane Block Length, g/mol	Dimethyldiphenylsiloxane in Copolymer, wt%	Diphenylsiloxane Incorporation, wt%
8a	4800	20	25
8b	5400	20	50
8c	5700	20	75

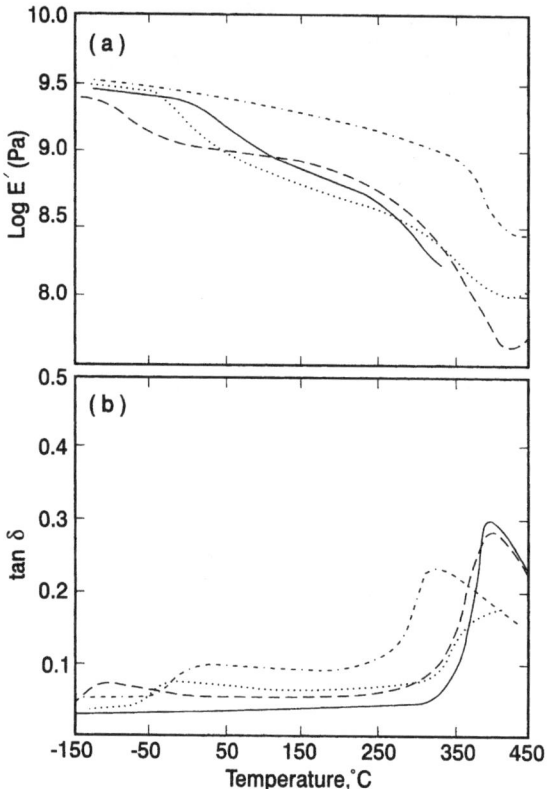

Fig. 14. dynamic mechanical spectra of copolymers 8a (-----), 8b (.....), 8c (—) and polyimide homopolymer (-·-·-·-)

4.2
Mechanical Properties of Imide-Siloxane Copolymers

The residual thermal stress was investigated with a Flexus stress analyzer. The residual stress, σ_F, was calculated from the radii of wafer curvatures before and after polyimide film deposition by the following equation:

$$\sigma_F = \{E_S t_S 2/6 t_F (1 - VS)\}\{1/R_F - 1/R_\infty\}$$

where the subscripts F and S represent the polymer film and substrate, respectively, and E is the Youngs modulus; V is Poisson's ratio, σ is the stress, and t is the thickness. R_F and R_∞ are the radii of a substrate with and without a polymer film, respectively. During thermal curing and cooling, the stress was measured dynamically over the range of 25 to 400 °C. The stress vs temperature profiles for

Fig. 15. Transmission electron micrograph of copolymer 7c

the imide-dimethylsiloxane and imide-dimethyldiphenylsiloxane copolymers with siloxane block lengths of 5400 g/mol are shown in Fig. 16. Each of the samples contained ~ 20 wt% of polysiloxane coblock. The stress vs temperature profile for PMDA/ODA polyimide is also shown to facilitate comparison. Surprisingly, upon cooling from 350 °C, copolymer **7c** shows no buildup in stress. In principle, the residual thermal stress which results from the mismatch in the thermal expansion coefficient between the substrate and the polymer is proportional to the product of the modulus and the thermal expansion coefficient as shown by the following equation:

$$\sigma = \frac{\Delta TE\Delta\alpha}{1-v}$$

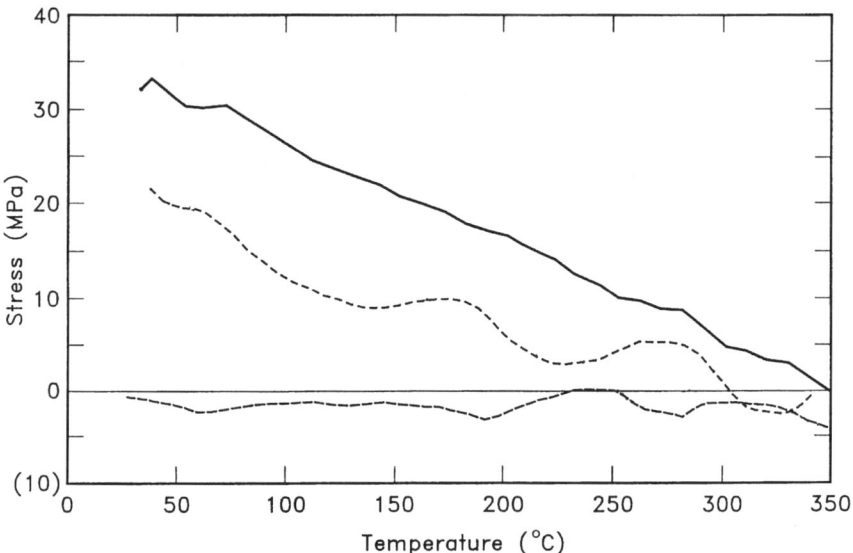

Fig 16. Stress versus temperature plots for copolymers 7c (-----), 7d (--) and polyimide homopolymer (—)

where T is the temperature, E is the modulus, and α is the thermal expansion coefficient. For the case of the imide-dimethylsiloxane copolymer, both the modulus and thermal expansion coefficient are high. Therefore, it is inconsistent to have no stress in the temperature regime studied. The introduction of diphenylsiloxane into the soft block (copolymers **8a–c**) while maintaining the high block length resulted in a stress/temperature profile which is substantially different to that of the imide-dimethylsiloxane copolymer (copolymer **7c**). Although these stress values for the diphenylsiloxane-containing copolymers are between 20 and 25 MPa, which is approximately 25–40 % lower than that of the parent homopolymer, they are still not as low as the copolymer with the dimethylsiloxane block.

The stress invariance with temperature was observed only for copolymer **6c** containing the high dimethylsiloxane block length. Dynamic mechanical analysis and TEM measurements revealed high phase purity for this copolymer. Previous studies have shown that the liquid crystal type morphology, characteristic of the PMDA/ODA polyimide, is retained in such copolymers. A possible qualitative explanation for the enhanced relaxation behavior of the copolymers is the stress concentrating effect of the rubbery inclusions, which results in an enhanced relaxation rate for a given global stress or deformation, since the local relaxation rates in the matrix is strongly stress dependent [2].

References

1. Tummala RR, Rymaszewski EJ (1989) Microelectronics packaging handbook. Van Nostrand Reinhold, New York, Chap 1
2. Hedrick JL, Brown HR, Volksen W, Sanchez M, Plummer CJG, Hilborn JG (1997) Polymer, 38: 605
3. Brown HR, Yan ACM, Russel TP, Volksen W, Kramer EJ (1988) Polymer 29: 1807
4. Numata S, Fujisaki K, Makino D, Kinjo N (1985) Proceedings of the 2nd Technical Conference on Polyimides: Society of Plastic Engineers. Ellenville, New York, p 164
5. Pfeiffer J, Rhode O (1985) Proceedings of the 2nd Technical Conference on Polyimides: Society of Plastic Engineers. Ellenville, New York, p 336
6. Cassidy PE (1980) Thermally stable polymers: Synthesis and properties. Marcel Dekker, New York
7. Mittal KL (ed) (1984) Polyimides. Plenum Press, New York
8. Lupinski JH, Moore RS (eds) (1989) Polymeric materials for electronic packaging and interconnection. ACS Symposium Series 407
9. Takahashi N, Yoon DY, Parrish W (1984) Macromolecules 17: 2583
10. Russel TP (1986) J Polym Sci Polym Phys Ed 22: 1105
11. Gattiglin E, Russel TP (1985) J Polym Sci Part B Polym Phys Ed 27: 2131
12. Boese D, Lee H, Yoon DY, Swalen JD, Rabolt JF (1992) J Polym Sci Part B Polym Phys Ed 30: 1321
13. Volksen W (1994) Adv Polym Sci 117: 111 111
14. Paraszczak J, Edelstein D, Cohen S, Babich E, Hummel J (1993) IEDM Tech. Digest, p 261
15. (a) Czornyj G, Chen KR, Prada-Silva G, Arnold A, Souleotis H, Kim S, Ree M, Volksen W, Dawson D, Di Pietro R (1992) 42nd Electronic Components and Technology Conference, p 682; (b) Ree M, Swanson S, Volksen W (1992) Polymer 33: 1228
16. Chen DH, Chen YP, Arnold CA, Hedrick JC, Graybeal JD, McGrath JE (1989) Int SAMPE Symp 34: 1247
17. Musto P, Karasz FE, MacKnight WJ (1989) Polymer 30: 1012
18. Guerra G, Choe S, Williams DJ, Karasz FE, MacKnight WJ (1988) Macromolecules 21: 231
19. Harris JE, Robeson LM (1986) US Patent 4,609,714 to Union Carbide Corp
20. Harris JE, Robeson LM (1988) J Appl Polym Sci 35: 1877
21. Flory PJ (1978) Macromolecules 11: 1138
22. Hwang W-F, Wiff DR, Benner CL (1983) J Macromol Sci Phys B 22: 231
23. Feger C (1989) In Lupinski JH, Moore RJ (eds) Polymeric materials for electronic packaging and interconnection, p 114
24. Ree M, Yoon DY, Volksen W (1989) PMSE 60: 179
25. Yoon DY, Ree M, Volksen W, Hofer D, Depero L, Parrish W (1988) Presented at the Third International Conference on Polyimides, Mid-Hudson Section of SPE, Nov 2–4, Ellenville, New York
26. Hedrick JL, Labadie JW, Russel TP (1989) In Feger C, Khajastech MM, McGrath JE (eds) Polyimides: Materials, chemistry and characterization. Elsevier Science Publishers, Amsterdam, p 61
27. Hedrick JL, Hilborn J, Labadie J, Russel TP (1990) Polymer 34: 2384
28. Haidar M, Chenevey E, Vora RH, Cooper W, Glick A, Jaffe M (1991) Mater Res Soc Symp Proc 227: 35
29. Critchlen JS, Gratan PA, White MA, Pippett J (1972) J Polym Sci A-1 10: 1789
30. Harris FW, Hsu SLC, Lee CJ, Lee BS, Arnold F, Cheng SZD (1991) Mater Res Soc Symp Proc 227: 3
31. Sasaki S, Matuora T, Nishi S, Ando S (1991) Mater Res Soc Symp 227: 49

32. St Clair AK, St Clair TL, Winfree WP (1988) Proc Amer Chem Soc Div Polym Mater Sci Eng 59: 28
33. Chan R, Haasen P, Kramer EJ (eds) (1993) Materials science and technology, vol 12. UCH Weinheim, p 251
34. Jensen BJ, Hergenrother PM, Bass RG (1989) Proc Polym Mater Sci Eng 60: 294
35. Johnson BC, Yilgor I, McGrath JE (1984) Polym Prepr 25(2): 54
36. Arnold CA, Sumners JD, Chen YP, Boh DH, McGrath JE (1989) J Polym 30: 986
37. Volksen W, Yoon DY, Hedrick JL, Hotopfer D (1991) Mater Res Soc Symp Proc 227: 23
38. Hergenrother PM (1974) J Appl Polym Sci 18: 1779
39. Hergenrother PM (1971) J Macromol Sci Revs Macromol Chem C6: 1
40. Hedrick JL, Labadie JW (1990) Macromolecules 23: 1561
41. Labadie JW, Hedrick JL (1992) Die Macromol Chem Macromol Symp Ser 54/55: 313
42. Hergenrother P, Connell J, Labadie J, Hedrick JL (1994) Recent advances in heterocycle containing poly(aryl ethers). In Hergenrother PM (ed) High Performance Polymers. Springer-Verlag, Berlin Heidelberg New York
43. Jurak M, McGrath JE (1989) Polymer 30: 1552
44. Hedrick JL, Labadie JW, Russel TP (1991) Macromolecules 24: 4559
45. Labadie JW, Hedrick JL, Boyer S (1992) J Poly Sci Polym Chem Ed 30: 519
46. Hedrick JL, Labadie JW (1990) High Performance Polymers 2(1): 3
47. Hilborn J, Labadie JW, Hedrick JL (1990) 23: 2854
48. Hedrick JL (1991) Macromolecules 24: 6361
49. Hedrick JL, Hilborn J, Palmer T, Labadie JW, Volksen W (1990) J Poly Sci Polym Chem Ed 28: 2255
50. Atwood TE, Barr DA, Faasey GG, Leslie VJ, Newton AB, Rose JB (1977) Polymer 18: 354
51. Mohanty DK, Lowry RC, Lyle GD, MacGrath JE (1987) Int SAMPE Symp 32: 408
52. Mohanty DK, Senger JS, Smith CD, MacGrath JE (1988) Int SAMPE Symp 33: 970
53. Volksen W, Yoon DY, Hedrick JL (1992) IEEE Trans on Components, Hybrids and Manufacturing Technol 15: 10
54. Volksen W, Yoon DY, Hedrick JL, Hofer DC (1992) Mater Res Soc 227: 23
55. Labadie JW, Hedrick JL (1990) Proc of the 40th Elec Comp and Tech Conf, p 706
56. Hedrick JL, Labadie JW, Volksen W (1990) Int SAMPE Elec Conf Series 4: 214
57. Hedrick JL, Volksen W, Mohanty DK (1991) MRS Symp Ser 227: 81; "Polyimides Other High-Temp Polym. Proc Eur Tech Symp (1991) Abadie JJM, Sillion B (eds) Elsevier, Amsterdam, Netherlands
58. Hedrick JL, Volksen W, Mohanty DK (1992) J Polym Sci Part A Polym Chem Ed 30: 2085
59. Hedrick JL, Volksen W, Mohanty DK (1993) Polym Bull 30: 33
60. Russel TP, Brown HR, Grubb DT (1987) J Polym Sci Part A Polym Phys Ed 25: 1129
61. Labadie JW, Hedrick JL (1990) Macromolecules 23: 5371
62. Maxwell-Garnett JC (1905) Phil Trans R Soc 205: 237
63. Gaghani J, Supkis DD (1980) Acta Astronaut 7: 653
64. Smearing RW, Floryan DC (1985) US Patent 4,543, 365 to General Electric
65. Krutchen CM, Wu P (1985) US Patent 4,535,100 to Mobil Oil
66. Hoki T, Matsuki Y (1986) European Patent 186308 to Aasl Chem
67. Meyers RA 81969) J Polym Sci Part A-1 7: 2757
68. Carleton PS, Farrissey WJ, Rose JS (1972) J Appl Polym Sci 16: 2983
69. Alvino WM, Edelman LE (1975) J Appl Polym Sci 19: 2961
70. Farrissey WJ, Rose JS, Carleton PS (1970) J Appl Polym Sci 24: 1093
71. Hammermesh CL, Hogenson PA, Tun CY, Sawako PM, Riccitello R (1979) 11th Natl SAMPE Tech Conf, p 574
72. Gagliai J (1980) US Patent 4,241,193 to International Harvester
73. Lee R, Okey DW, Ferro GA (1985) US Patent 4,535,099 to IMI-Tech Corp
74. Alberino LM (1976) Cell Plast Conf 4: 1
75. Narkis M, Paterman M, Boneh H, Kenig S (1982) Polym Eng Sci 22(7): 417

76. McWhirter RJ (1981) Energy Res Abs 6: 2627
77. McIlroy HM (1977) Energy Res Abs 2: 3469
78. Gagliani J, Lee R, Sorathin UAK, Wilcoxson AL (1980) Sci Tech Aerosp Rep 18: 37
79. Gagliani J, Supkis DE (1979) Adv Astronaut Sci 38: 193
80. Hedrick JL, Charlier Y, Di Pietro R, Jayaraman S, MacGrath JE (1996) J Polym Sci Part A Polym Chem Ed 34: 2867
81. Hedrick JL, Di Pietro R, Plummer CJG, Hilborn J, Jerome R (1996) Polymer 37: 5236
82. Charlier Y, Hedrick JL, Russel TP, Di Pietro R, Jerome R (1995) Polymer 36: 4529
83. Hedrick JL, Charlier Y, Russel TP, Swanson S, Sanchez M (1995) Polymer 36: 1315
84. Hedrick JL, Di Pietro R, Charlier Y, Jerome R (1995) High performance polymers 7: 133
85. Hedrick JL, Di Pietro R, Hawker C, Jerome R (1995) Polymer 36: 4855
86. Hedrick JL, Labadie JW, Russel TP, Hofer D, Wakharkar V (1993) High temperature polymer foams. Polymer 34: 4717
87. Labadie JW, Hedrick JL, Wakharkar V, Hofer DC, Russel TP (1992) Proc IEEE Trans-CHMT 15: 925; Proc Electron Compon Technol Conf 42: 688; Hedrick JL, Di Pietro R, Plummer CJG, Hilborn J, Jerome R (1995) Polymer 36: 2485
88. Charlier Y, Hedrick JL, Russel TP, Volksen W (1994) Polymer 34: 4717
89. Hedrick JL, Russel TP, Lucas M, Labadie J, Swanson S (1995) Polymer 36: 987
90. Rogers ME, Moy TM, Kim YS, MacGrath JE (1992) Mat Plas Soc Symp 13: 264
91. Patel N (1990) Multicomponent network and linear polymer system: Thermal and morphological characterization. PhD dissertation, U.P.I. + S.U. (under MacGrath JE)
92. Bywater S, Black PE (1965) J Phys Chem 69: 2967
93. Sinaha R, Wall LA, Bram T (1958) J Chem Phys 29: 894
94. Trollsås M, Hedrick JL, Mecerreyes, Dubois P, Jérôme R, Ihre H, Hult A (199/9 Macromolecules 30: 8608
95. Hedrick JL, Carter KR, Flores U, Miller RD, Russel TP, Mecerreyes D, Jérôme R (1998) Chem. of Mater. 10: 39
96. Trollsås M, Hedrick JL, Mecerreyes D, Dubois P, Jérôme R, Ihre H, Hult A (1998) Macromolecules 31: 2756
97. Hawker CJ, Hedrick JL (1995) Macromolecules 28: 2993
98. Hedrick JL, Miller RD, Hawker CJ, Carter KR, Volksen W, Yoon DY, Trollsås M (1998) Adv. Mater. 10: v
99. Sanchez M, Hedrick JL, Russel TP (1995) J Polym Sci Phys Ed 33: 253
100. Plummer CJG, Hedrick JL, Hilborn JG (1995) Polymer 36: 2485
101. Cha HJ, Hedrick JL, Di Pietro RA, Blume T, Beyers R, Yoon DY (1996) Appl Phys Lett 68: 1
102. Volksen W, Hedrick JL, Russel TP, Swanson S (1997) J. Applied Polym Sci 66: 195

Received: August 1998

Dendritic and Hyperbranched Macromolecules – Precisely Controlled Macromolecular Architectures

Craig J. Hawker

IBM Almaden Research Center, 650 Harry Road, San Jose, CA, 95120-6099, USA
E-mail: hawker@almaden.ibm.com

The development of procedures for the synthesis of dendritic macromolecules by either the convergent or divergent growth approaches is outlined with emphasis placed on the controlled manipulation of three-dimensional structure. The utility of these techniques to prepare a wide variety of different dendritic structures is then discussed in terms of the three distinct regions associated with these novel macromolecular systems – the central core, the interior building blocks, and the chain ends. Control of these regions, coupled with changes in the synthetic approach, gives tailor-made dendritic macromolecules with predetermined physical properties and/or function. The application of current spectroscopic methods to the structural elucidation of dendritic macromolecules is then detailed along with an examination of the influence of dendritic structure on the physical properties of these novel materials. Finally, a comparison is made with the related class of highly branched polymers, hyperbranched macromolecules, and the manipulation of structure/function for these materials is examined.

Keywords. Dendrimers, Hyperbranched macromolecules, Block copolymers, Surface functionalization, Solvatochromism, Intrinsic viscosity

1	Introduction	114
2	Synthesis of Dendritic Macromolecules	114
2.1	Divergent Growth Approach	115
2.2	Convergent Growth Approach	119
2.3	Comparison of the Divergent and Convergent Approaches	124
3	Characterization of Dendrimers	126
4	Structural Control	129
4.1	Interior Building Blocks	131
4.2	Central Core or Focal Point Group	137
4.3	Chain End Groups	144
5	Hyperbranched Macromolecules	153
	References	158

1
Introduction

The construction of complex, three-dimensional macromolecular architectures by the assembly of simple building blocks has long been the sole domain of natural systems. Proteins, enzymes, DNA, etc. are the prime examples of precisely controlled biological macromolecules with their three-dimensional shape being a result of internal branching and non-covalent binding. For enzymes, this three-dimensional structure results in unique active sites, or microenvironments, within the macromolecule that gives rise to extraordinary catalytic ability and is the basis of biological systems. While current synthetic techniques do not allow the preparation of synthetic macromolecules with the same degree of control as found in nature, chemists have achieved a remarkable degree of success in controlling the structure of small molecules. Building on these concepts and methodologies, the construction of a new class of three-dimensional macromolecules, called dendrimers, from simple building blocks has recently attracted considerable attention [1]. While the sophistication is not comparable to natural systems the degree of control over macromolecular structure is significantly greater than for other polymeric systems. This ability to control placement of functional groups and to construct unique microenvironments has stimulated research into such diverse applications such as drug delivery, diagnostic tools, rheology control, nanofabrication, and molecular electronics. The purpose of this review is to detail the methods available for structural and architectural control during the synthesis of dendritic macromolecules. Using these methods it is now possible to engineer specifically dendrimers to give tailor-made nanoscopic macromolecules with predetermined properties and applications.

2
Synthesis of Dendritic Macromolecules

The term "dendrimer" was originally coined by Tomalia et al. [2] to describe a family of regularly branched poly(amidoamines) in which all bonds converge to a single point, each repeat unit contains a branch junction, and a large number of identical functional groups are present at the chain ends. The graphical representation of a dendrimer, **1**, is shown in Scheme 1 and while it is a two- rather than three-dimensional representation it conveys the essential notion that a dendrimer is a highly symmetrical, layered macromolecule which consists of three well defined regions. A central core, or focal point, is connected to a number of layers, or generations of internal building blocks, which are in turn connected to numerous chain ends. The branching geometry of the internal building blocks, coupled with steric considerations, does not allow a dendrimer to assume a flat structure as shown in Scheme 1. Instead the dendrimer fills space and adopts a globular shape that may approximate a sphere in a fully extended configuration, though this conformation may be energetically unfavora-

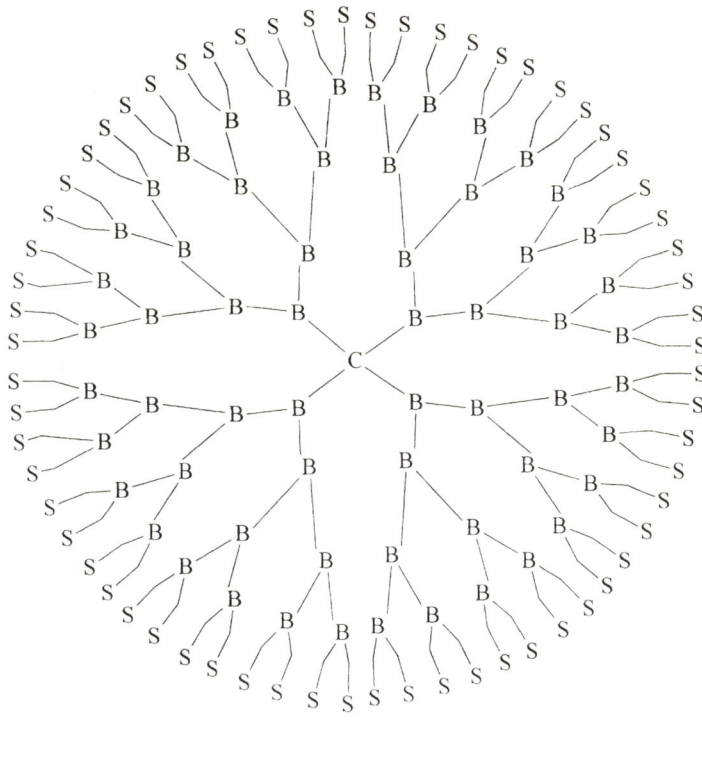

1

Scheme 1

ble. It is this highly symmetrical structure which forms the basis for the development of two distinct synthetic strategies for the preparation of dendrimers.

2.1
Divergent Growth Approach

The pioneering synthetic strategy, developed independently by Tomalia et al. [2] for Starburst dendrimers and Newkome et al.[3] for Cascade molecules, is now recognized as the divergent growth approach. The basis for these methodologies can be traced back to work by Vogtle et al. in 1978 who attempted the preparation of an oligiomeric nitrile-terminated branched polyamine by a series of Michael additions and subsequent reduction steps [4]. This stepwise growth strategy results in an increase in the number of terminal groups and branching units in the structure, although due to the small size and uncontrolled growth of these nitrile-terminated oligomeric amines they cannot be considered to be true dendrimers. Subsequently, Denkewalter et al. applied a similar stepwise repetitive strategy to the synthesis of highly branched poly(lysine)s [5]. In this case growth was continued to the tenth generation, resulting in molecular weights approaching 500,000 Da. However the physical properties of these materials were unremarkable and unlike those subsequently found for dendrimers. A possible explanation for this result is that the unsymmetrical nature of the monomer unit leads to unequal branch segments which disrupts the molecular structure and results in an extended three-dimensional structure and not the pseudospherical structure commonly associated with dendrimers. For optimal dendritic structure and properties it may therefore be necessary to have a symmetrical monomer unit and essentially all subsequent syntheses have utilized this feature.

Following these initial reports it was not until the mid-1980s that a determined effort to prepare and characterize true dendrimers was undertaken. The well-known starburst dendrimers of Tomalia were first reported in the literature in 1985 [2], and almost simultaneously Newkome disclosed the preparation of a cascade molecule which was termed an arborol [3]. In these early publications· the structures were referred to as arborols, cascade or starburst polymers, though the commonly accepted terms at the present time for these materials are dendrimers, or dendritic macromolecules. Although these pioneering efforts employed different chemistries and synthetic strategies the underlying methodology and basic requirements which must be satisfied for the successful preparation of large dendritic structures are found in both cases. Two primary considerations are that the chemistries used for both generation growth and subsequent activation give very high yields with no significant side reactions. The monomer units selected for construction of the dendritic macromolecules are also chosen to afford symmetrical branch segments which alleviates any difficulties due to unequal reactivity or asymmetric shape in the growing dendrimer. These basic principles have subsequently been employed to prepare a wide variety of dendritic macromolecules by the divergent growth approach.

Perhaps the most well known divergent synthesis of dendritic macromolecules is the preparation of poly(amidoamine) (PAMAM or Starburst) dendrimers by Tomalia et al. (Scheme 2) [2]. The basic reaction sequence is similar in concept to that reported by Vogtle et al. [4]; however the troublesome reduction step was circumvented by using an exhaustive amidation process with 1,2-diaminoethane to regenerate the reactive amino groups at the chain ends. As detailed in Scheme 2, the synthesis starts with a polyfunctional core, in this case ammonia, and reaction with methyl acrylate results in the formation of the triester, 2. The second step in this two-step repetitive process is regeneration of the reactive amino groups at the chain ends and this is accomplished by exhaustive amidation with a large excess of 1,2-diaminoethane to give the first generation dendrimer, 3, in which the number of reactive NH groups has doubled from three for ammonia to six for 3. As dictated by the synthetic blueprint, all of the reactive NH groups are located at the chain ends of the dendrimer. Repetition of this two-step process then leads to larger and larger dendritic macromolecules, the structure of which follows a strict geometrical progression. For example, the

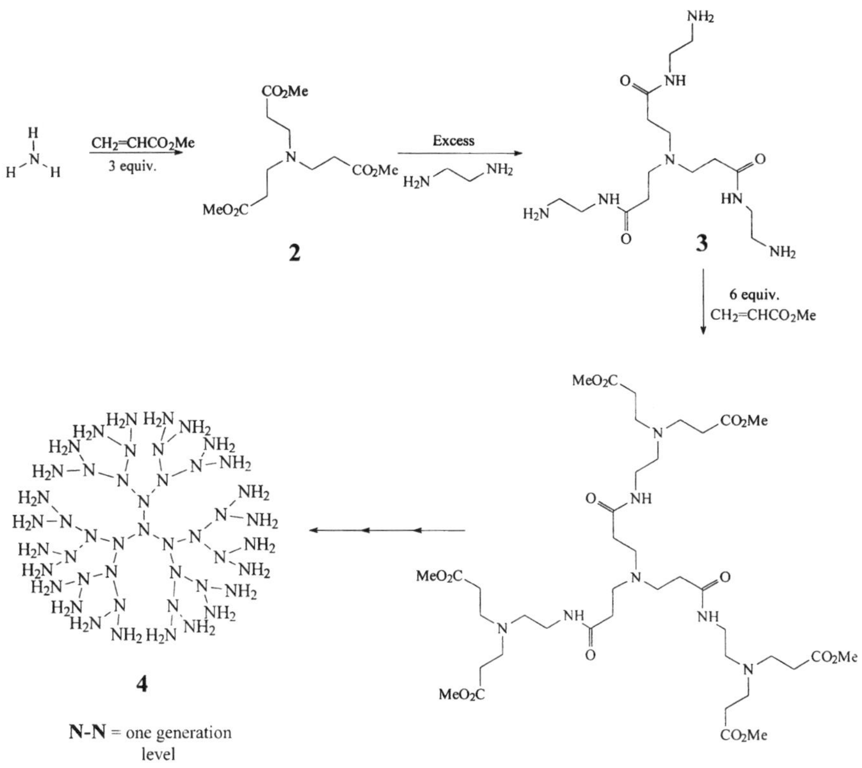

Scheme 2

N-N = one generation level

PAMAM synthesis shown in Scheme 2 utilizes a core moiety and a repeat unit having multiplicities of three and two respectively. Therefore the number of terminal functional groups doubles at each growth step and is related to the generation number by the relationship 3×2^n, where n is the generation number. Examination of the fourth generation dendrimer, 4 (shown schematically), reveals that it has 3×2^4, or 48 reactive NH end groups while the tenth generation dendrimer has 3072 (3×2^{10}) reactive NH end groups. The extremely large number of chain ends groups presents a unique set of opportunities for dendritic macromolecules. As detailed below, functionalization of these chain ends groups can be used to control accurately a variety of physical and chemical properties for dendritic macromolecules. While there were a number of difficulties to overcome in the synthesis of PAMAM-based dendritic macromolecules by the divergent growth approach, careful control of the experimental conditions and purification procedures has allowed Tomalia and co-workers to produce PAMAM dendrimers on a kilogram scale and these materials are now commercially available from a number of sources [6].

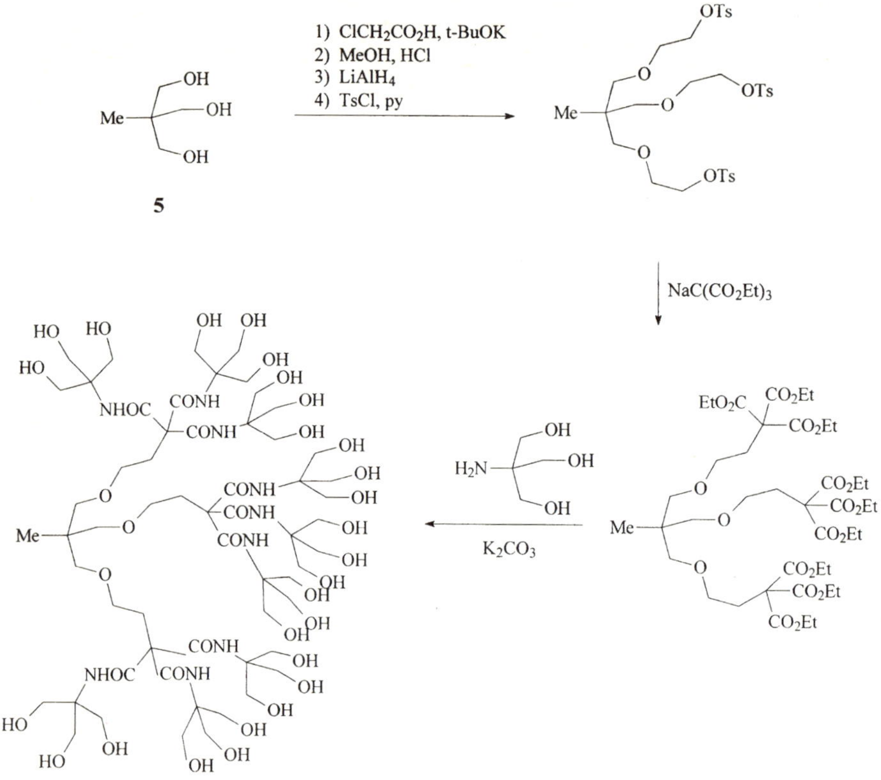

Scheme 3

Recently another family of dendrimers has become commercially available. These polyamines were developed by Meijer and de Brabander-van den Berg of DSM Research and are based on Vogtle's initial synthesis [7]. In this case the troublesome reduction step was performed using a Raney cobalt hydrogenation catalyst and other process improvements have permitted this synthesis to be continued up to the fifth generation with multikilogram quantities available.

The versatility of dendrimer syntheses can be better appreciated if the classical PAMAM synthesis is compared to Newkome and coworker's original "arborol" synthesis of water soluble poly(ether amides) [3]. A divergent strategy was again employed in construction of these macromolecules and in this case the central core was chosen to be tris(hydroxymethyl)ethane, 5. A set of complimentary AB_3 building blocks were employed in the synthesis which greatly increases not only the speed of the synthesis but also the branching density. As shown in Scheme 3, after only two generations, the number of hydroxy chain ends for **6** is 27 (3×3^2) compared to 12 (3×2^2) for the corresponding PAMAM dendrimer. This rapid increase in the number of chain ends, coupled with the compact nature of the repeat units, makes **6** a much more globular structure when compared to PAMAM dendrimers of comparable molecular weight or generation number. The greater number of chain ends in **6** also affords extreme water solubility, which is a desirable feature for a number of applications.

2.2
Convergent Growth Approach

While the divergent growth approach has proved to be highly successful for the preparation of a wide variety of dendritic structures it does have a number of limitations for the preparation of complex globular macromolecules where accurate placement of functionalities at the chain ends, internal building blocks, or central core is desired. The ever increasing number of reactions required to fully functionalize the chain ends during growth of dendrimers by the divergent growth approach also leads to difficulties in maintaining "perfect" growth and/or in removing the large excesses of reagent needed to force these reactions to completion.

To overcome these and other difficulties, an alternative synthetic strategy for the construction of dendritic macromolecules, termed the convergent growth approach, was first demonstrated by Fréchet et al. in 1989 [8]. A different, if conceptually similar, convergent synthesis of smaller dendrimers was outlined by Miller and Neenan in 1990 [9]. From a retrosynthetic viewpoint, the convergent growth approach is essentially the opposite of the divergent growth strategy, in this case growth begins at the periphery of the dendrimer and proceeds inwards with the final reaction being attachment to a polyfunctional core. This approach relies on the fractal and highly symmetrical nature of dendrimers and disconnection leads eventually to the chain ends as the starting point of the synthesis. Therefore growth is begun at the chain ends or "surface" functional groups, S, **7**, and coupling with an AB_x monomer, **8**, where x is 2 or greater, leads to the next generation dendritic fragment, or dendron. Activation of the single functionality at the focal point, P, then gives the reactive dendron, **8**, which can be coupled with the monomer to give the second generation dendron, **9**. The single functionality, P, at the focal point of **9** can now be reactivated to give the reactive dendron, **11**. It should be noted that in going from **8** to **9** to **11** the number of chain ends has doubled at each generation growth step, the molecular weight has essentially doubled, the number of layers of internal building blocks has increased from 0 to 2, while there is still only a single functionality at the focal point (Scheme 4). It is this retention of the focal point group that is a characteristic of the convergent growth approach and is the basis for many of the advantages typically associated with the convergent growth approach. Repetition of this two-step process results in larger and larger dendrons and, if desired, the final reaction can be coupling to a polyfunctional core to give a dendritic macromolecule, **12**, similar to that obtained from a divergent strategy (Scheme 5).

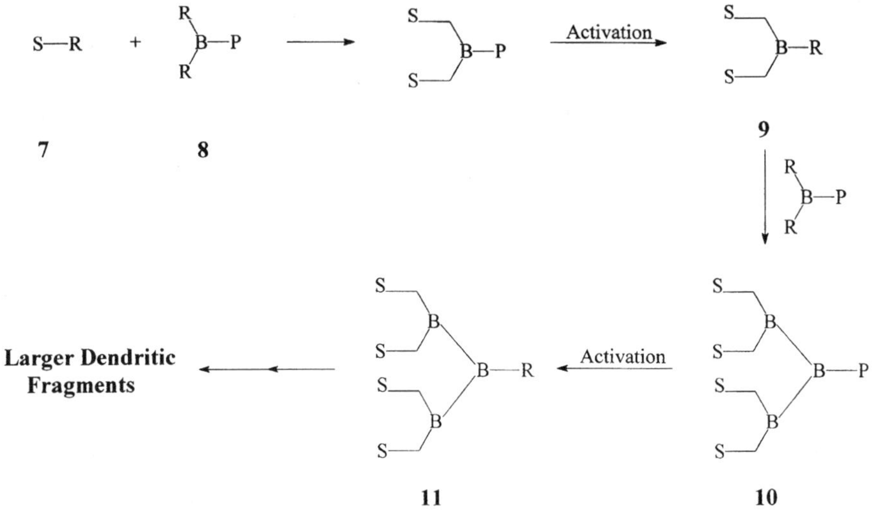

Scheme 4

To demonstrate this new methodology, Hawker and Fréchet have reported the synthesis of a series of dendritic polyether macromolecules based on 3,5-dihydroxybenzyl alcohol, **13**, as the monomer unit (Scheme 6) [10]. As detailed above, the starting material for this convergent synthesis is the surface functional group, which in this case, is the benzylic ether, **14**, derived from benzyl bromide. Reaction of **14** with carbon tetrabromide and triphenylphosphine regenerates the reactive bromomethyl group at the focal point of the first generation dendron, **15**. Alkylation of **15** with the monomer unit under typical Williamson conditions then leads to the next generation dendron, **16**, in which the number of "surface" groups has doubled but there is still only a single functional group at the focal point. The choice of the growth steps in this synthesis was crucial to maintain adequate growth and high dendrimer purity; both the bromination procedure and Williamson coupling chemistry are high yielding reactions, which do not suffer from significant side reactions. Repetition of this two-step bromination/alkylation procedure gives larger monodisperse dendrons such as **17** and **18**, which are characterized by a single functional group at the focal point. In a similar fashion to dendrimers constructed by the divergent growth approach, the structure and functionality of dendritic macromolecules prepared by the convergent growth approach are governed by strict geometrical progressions. For example, Hawker and Fréchet have carried this synthesis to the sixth generation dendron which has 64 (2^6) "surface" benzyl groups, 127 aromatic rings arranged in 7 discrete layers in the ratio 64:32:16:8:4:2:1, a molecular formula of $C_{889}H_{763}BrO_{126}$ and a molecular weight of 13,581 [11].

The same coupling chemistry can be used for attachment of these reactive dendrons to a polyfunctional core. For example, three molecules of a fourth gen-

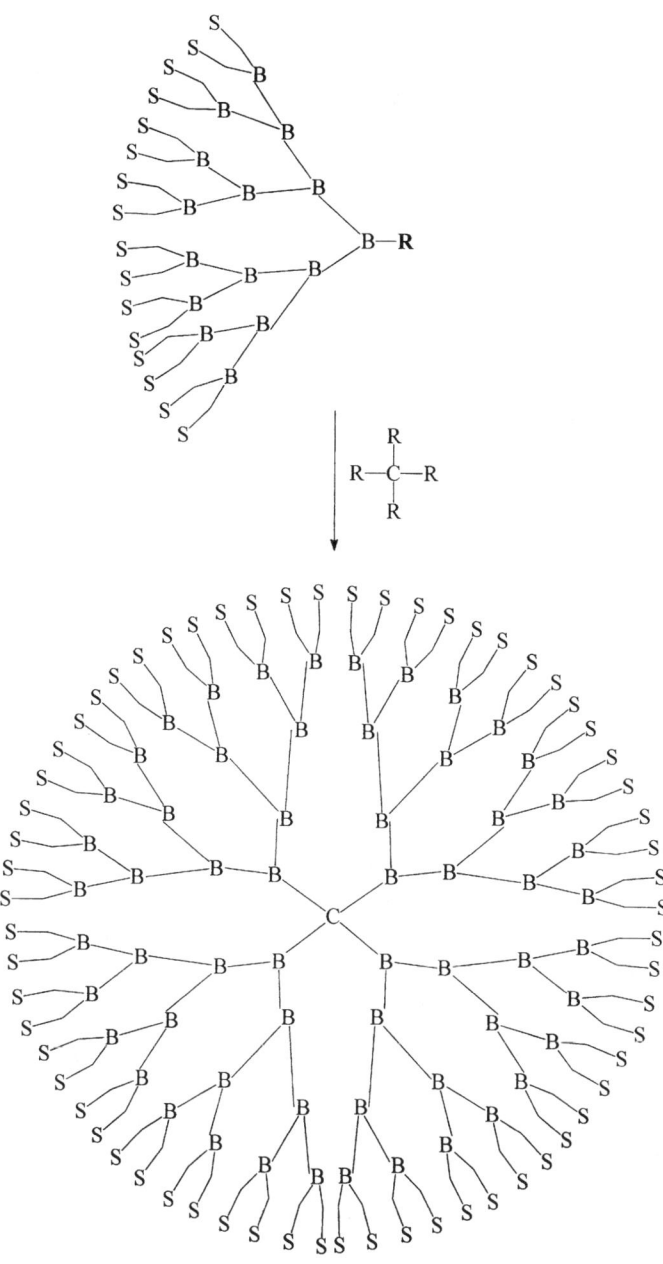

12

Scheme 5

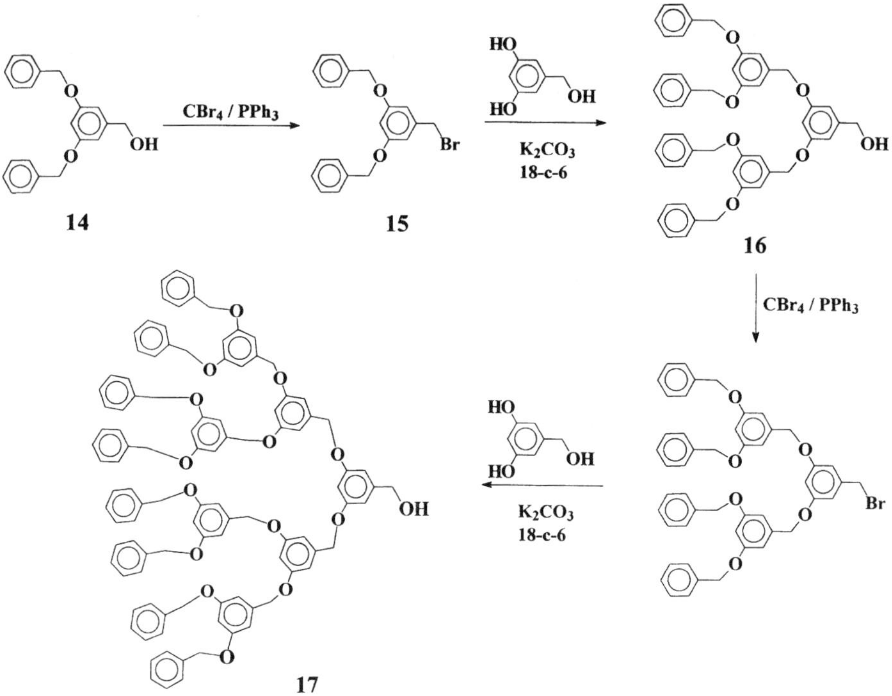

Scheme 6

eration reactive dendron, **18**, can be attached to the trifunctional core, 1,1,1-tris(4'-hydroxyphenyl)ethane **19**, to give the dendrimer, **20**, which no longer has a true focal point group (Scheme 7). Coupling of **19** with the corresponding dendrons with generation numbers ranging from 0 to 6 was also accomplished using this methodology to give a complete series of monodisperse dendrimers with molecular weights ranging from 576 to 40,689. Interestingly, steric inhibition to reaction was not observed in any of these syntheses and only a slight decrease in yield was observed for generation five dendrons and greater. This is surprising given the steric bulk surrounding the single functional group at the focal point and demonstrates that the convergent growth approach is a viable synthetic technique for the synthesis of moderately sized dendritic macromolecules. The degree of control associated with the convergent growth approach can also be gauged by the diverse range of molecules that have been used as cores for attachment of polyether dendrimers such as **18**. Buckminsterfullerene (C_{60}) [12], porphyrins [13], or even large phenol-terminated dendrimers [14], also known as "hypercores', have proved to be interesting core moieties and the importance of such molecules will be discussed in a latter section.

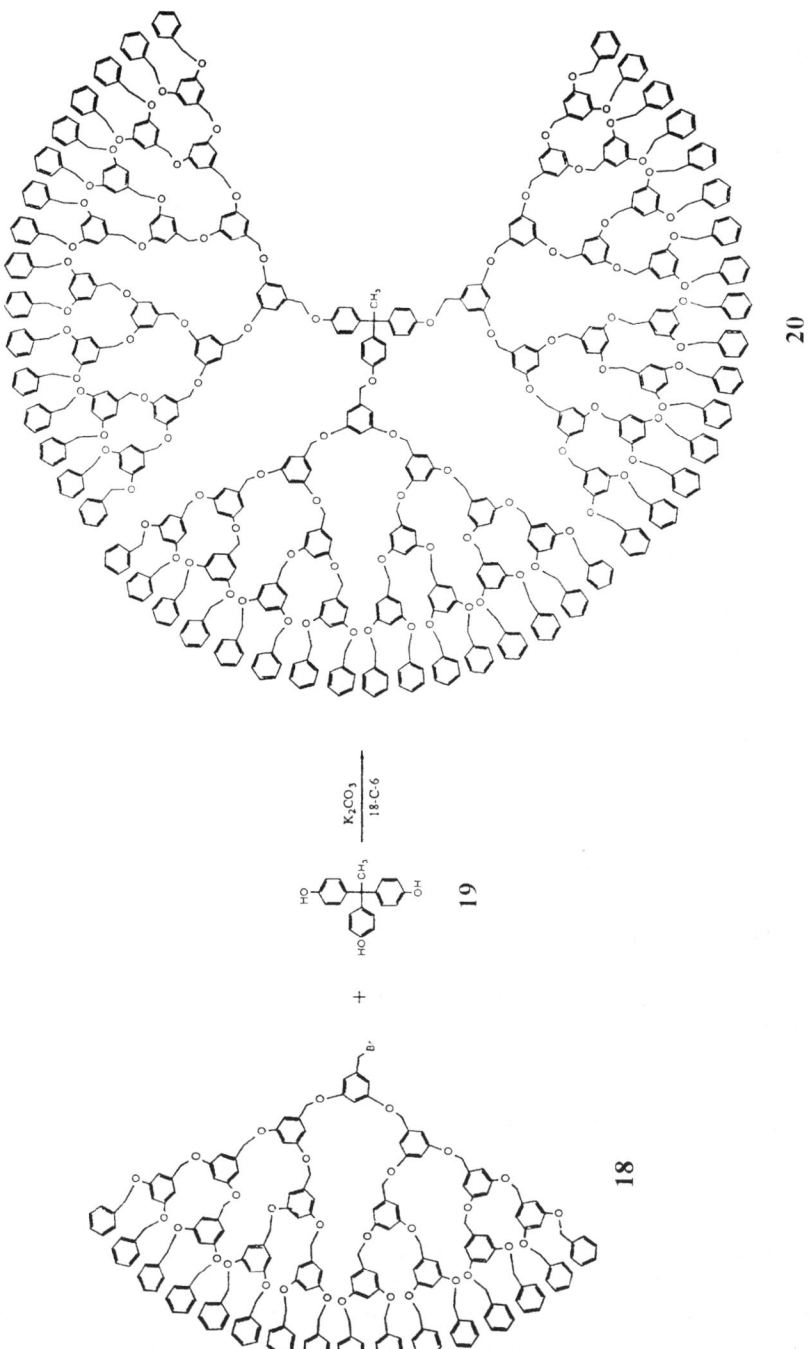

Scheme 7

2.3
Comparison of the Divergent and Convergent Approaches

While both the divergent and convergent growth approaches can lead to the same dendritic structures and involve a repetitive, stepwise growth strategy, there are some fundamental advantages and disadvantages associated with each approach. This results in an almost complimentary relationship between the two synthetic strategies with neither having a distinct advantage over the other. Only when planning the synthesis of a specific dendrimer does the chemistry and functionality involved dictate whether one approach may be more viable than the other.

For the synthesis of moderately sized dendrimers with a very regular structure or accurate placement of functional groups, the convergent growth approach allows a significantly greater degree of control. This increase is due to the unique synthetic blueprint of the convergent strategy, which involves only a limited number of steps for each generation growth, typically two, and the activation of a single functional group at the focal point group. Therefore the possibility of unwanted side reactions is severely reduced and the problems due to incomplete reaction leading to failure sequences minimized. For example, incomplete alkylation of the monomer unit, **13**, with the third generation dendron, **17**, would lead to a mixture of the desired dialkylated fourth generation macromolecule, **21**, and the monoalkylated side product, **22** (Scheme 8). Due to the dramatic difference in size between **21** and **22** and the presence of the polar phenolic group in **22**, purification by simple techniques such as flash chromatography is readily achievable. Also the limited number of alkylation steps combined with this facile separation of partially alkylated products allows control over functional groups placement at various points in the dendrimer. This is in direct contrast to the divergent growth approach, which requires an ever-increasing number of both activation and coupling reactions for generation growth. In this case the possibility of maintaining "perfect" growth becomes increasingly more difficult as the generation number increases and failure sequences, or incomplete reaction, becomes a major factor. The small differences in size and polarity between "perfect" dendrimers prepared by the divergent growth approach and those containing failure sequences make separation essentially impossible. The random nature of the functionalization reactions during generation growth for divergently grown dendrimers also makes accurate functionalization of the chain ends extremely difficult.

In contrast the method of choice for the production of very large dendritic macromolecules (MW>100,000) is the divergent growth approach. This is due to a number of related factors; steric congestion will retard a convergent strategy at a much lower molecular weight than a similar divergent approach. The ability to use large excess of reagents permits the divergent approach to be continued to higher generation numbers and the mass of material produced during the synthesis increases with generation number. The opposite is true for the convergent growth approach which cannot accommodate large excesses of reagents in the

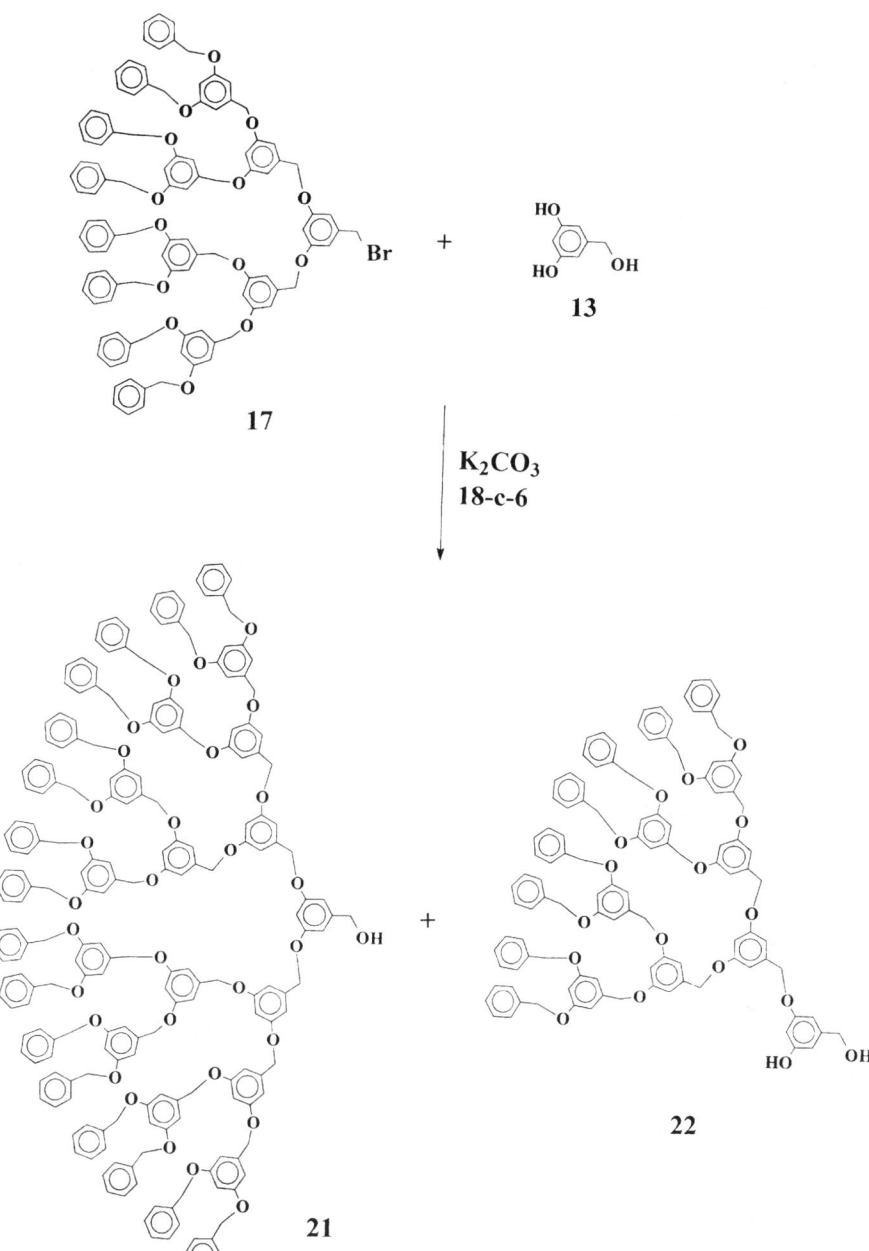

Scheme 8

growth step while the mass of dendrimer obtained after generation growth is actually the same or smaller than the mass of starting materials. Therefore the amount of material actually decreases as the synthesis proceeds making the synthesis of high (ca. tenth) generation dendritic macromolecules difficult.

3
Characterization of Dendrimers

The near monodispersity typically associated with dendrimers prepared by both the divergent or convergent growth approaches presents a number of interesting characterization traits when compared to traditional narrow polydispersity linear polymers. These features are further enhanced by the high degree of symmetry normally associated with dendritic macromolecules and permit an unprecedented amount of structural information to be obtained using standard characterization techniques.

A number of authors have elegantly elaborated the basic requirements and techniques for the structural identification of dendritic macromolecules [15]. Size exclusion chromatography (SEC) has been demonstrated to be an extremely powerful technique for analyzing the purity of dendrimers, particularly those prepared by the convergent growth approach. For both approaches a doubling of molecular weight is observed with each generation growth step and this leads to a gradual shift of the narrow GPC peak to lower elution volume with increasing generation number (Fig. 1). An added benefit in the case of the convergent growth approach SEC allows accurate monitoring of the generation growth step due to the limited number of condensation reactions typically required for generation growth. For the reaction of 3,5-dihydroxybenzyl alcohol with [G-4]-Br, **23** (MW=3351), to give the fifth generation alcohol, [G-5]-OH **24** (MW=6 680), the possible impurities are unreacted starting material, **23**, and the monoalkylated derivative (MW=3410). The dramatic difference in molecular weights between the desired product and side products permits the purity of intermediate dendritic wedges to be accurately determined. It should be pointed out that SEC traces for dendrimers are typically narrower than those for narrow polydispersity standards and while the most carefully purified dendrimer can give polydispersities of 1.01–1.02 by SEC, more sensitive techniques have actually shown them to be monodisperse. It must therefore be recognized that dendrimers can, and often do, exceed the resolution capability of size exclusion chromatography.

A recently developed technique that has found extensive use in the characterization of dendrimers, specifically in determining the purity and monodispersity of these novel materials, has been matrix-assisted laser desorption/ionization (MALDI) mass spectrometry [16, 17]. For dendrimers grown by the divergent strategy, incomplete functionalization of the periphery can lead to subsequent failure sequences and loss of strict dendritic growth. Observation and quantification of these defects is extremely difficult by other techniques, however MALDI mass spectrometry has been successfully employed by a variety of authors to

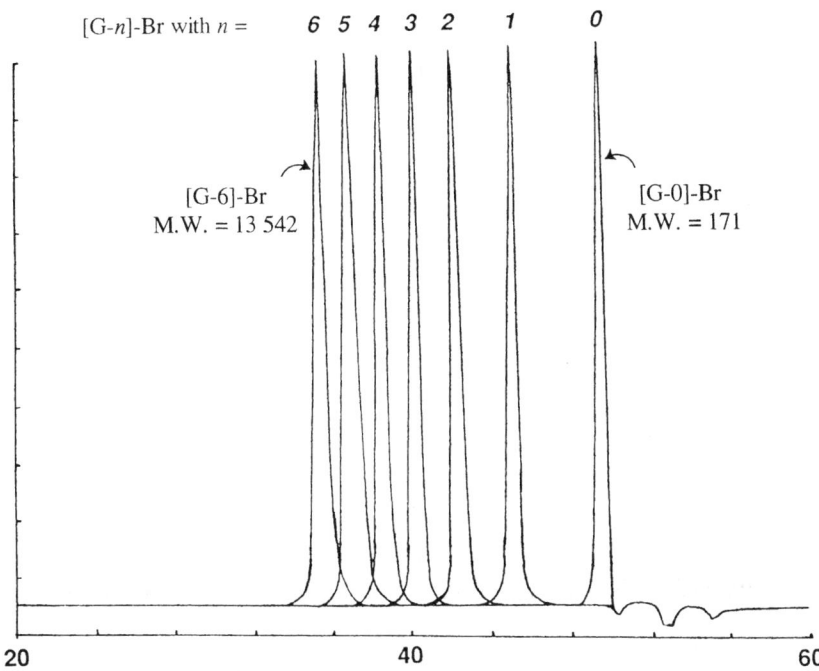

Fig. 1. Overlay of GPC traces for dendritic fragments, [G-0]-Br to [G-6]-Br, based on 3,5-dihydroxybenzyl alcohol as repeat unit

answer this question. An illustration of the power of MALDI mass spectrometry in elucidating the purity of dendritic macromolecules is shown in Fig. 2 where a convergently grown polyether dendrimer, [G-6]-OH, is compared with a narrow polydispersity poly(ethylene glycol) sample. It can be seen that the dendrimer is essentially a single peak at a molecular weight which corresponds to that expected from the synthetic strategy. In contrast the narrow polydispersity linear polymer is composed of an extremely large number of peaks, each representing a different number of repeat units. The combination of SEC and MALDI mass spectrometry permits the purity and to a certain extent the structure of dendritic macromolecules to be accurately determined.

The highly ordered and symmetrical nature of dendrimers permits small and subtle changes in the focal point group, terminal group, and interior building blocks to be readily observed by ^{1}H and ^{13}C NMR spectroscopy. The power of this characterization tool can be better appreciated by the fact that, for select dendritic macromolecules, NMR spectroscopy allows the identification of every unique proton or carbon atom in the structure. For example, the third generation polyether dendrimer, [G-3]-Br **17**, is composed of an outer layer of benzyl ether groups connected to three layers of interior (3,5-dioxybenzyl) building blocks and a final bromomethyl focal point group. Examination of the ^{1}H NMR

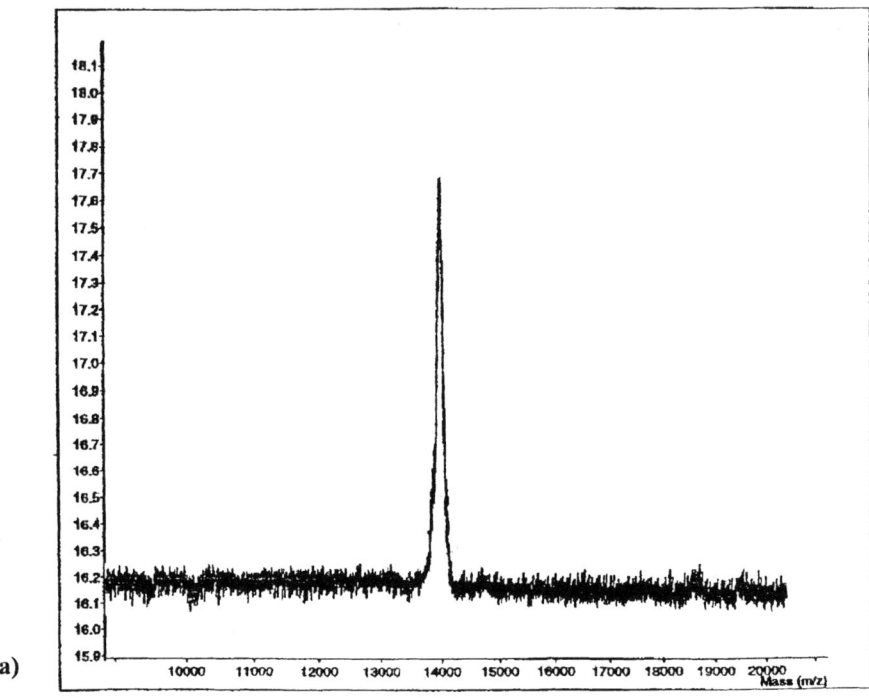

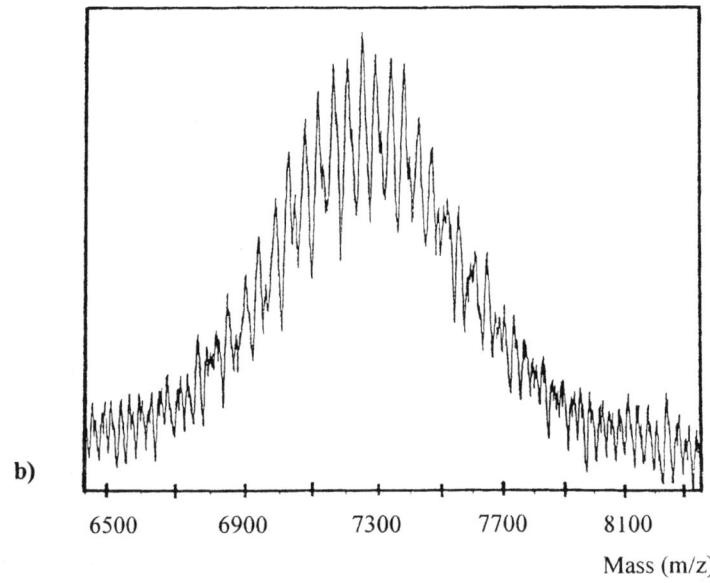

Fig. 2a,b. Comparison of MALDI mass spectra for [G-6]-OH and a narrow molecular weight distribution poly(ethylene glycol), PD=1.05

spectrum of **17** shows a set of resonances between 7.50 and 7.25 ppm corresponding to the aromatic protons of the terminal benzyl groups; a collection of doublets and triplets between 6.45 and 6.70 ppm are due to the aromatic protons of the three layer of interior building blocks. In fact, expansion of these resonances reveals three doublets in the ratio 1:2:4, which correspond to each generation level. Resonances for the benzylic methylene protons are observed at ca. 5.00 ppm and again fine detail is present which correlates with the three different groups of alkyloxymethylene protons. Perhaps the most important resonance in the spectrum is the focal point bromomethyl group at 4.30 ppm (Fig. 3). The sharp and well defined nature of this peak permits its observation even at generation six (MW=13,542) which corresponds to observing 2 protons out of a total of 763. This ability to observe easily the focal point group permits the subtle changes in this group to be readily detected during a typical convergent growth approach. For example, reaction of **17** with 3,5-dihydroxybenzyl alcohol gives the fourth generation derivative, **18**, which results in a change from a bromomethyl group at the focal point to a hydroxymethyl group. This change can be readily followed by ^{1}H NMR spectroscopy since the resonance for the bromomethyl group at 4.30 ppm is observed to disappear completely and is replaced by a doublet at 4.65 ppm corresponding to the hydroxymethyl group. Therefore ^{1}H NMR spectroscopy can also be used to help establish the purity of dendritic macromolecules, as well as to provide an extremely large amount of structural information. Changes in the synthetic strategy, for example introduction of a p-cyano group at the chain ends of the above polyether dendrimers, can also be readily observed by ^{1}H NMR; in this case the set of peaks at 7.25–7.50 ppm is replaced by an AB quartet at 7.80 and 7.20 ppm [18]. In a similar way, exchange of one or more polyether layers by a polyester layer based on 3,5-dihydroxybenzoic acid is also easily observed in the ^{1}H NMR spectrum [19]. Complimentary information can also be obtained from examination of the ^{13}C NMR spectra.

The combination of the above techniques with TLC, HPLC, UV-vis spectroscopy, and other emerging spectroscopic tools demonstrates that the ability to construct a wide variety of different dendritic structures is matched by the ability to determine accurately and confirm those structures. This permits the purposeful design and preparation of tailor-made dendrimers with a degree of structural confidence that is unparalleled in synthetic polymer chemistry.

4
Structural Control

Unlike traditional linear polymers, dendritic macromolecules can be conveniently divided into three distinct regions, *a central core or focal point group, the interior building blocks, and the chain ends*. Each of these regions can be accurately controlled and manipulated to different degrees using either the convergent or divergent approach. It is this feature which makes the engineering of dendritic macromolecules with specific physical properties a reality and dramatically illustrates the potential and versatility of these novel three-dimensional materials.

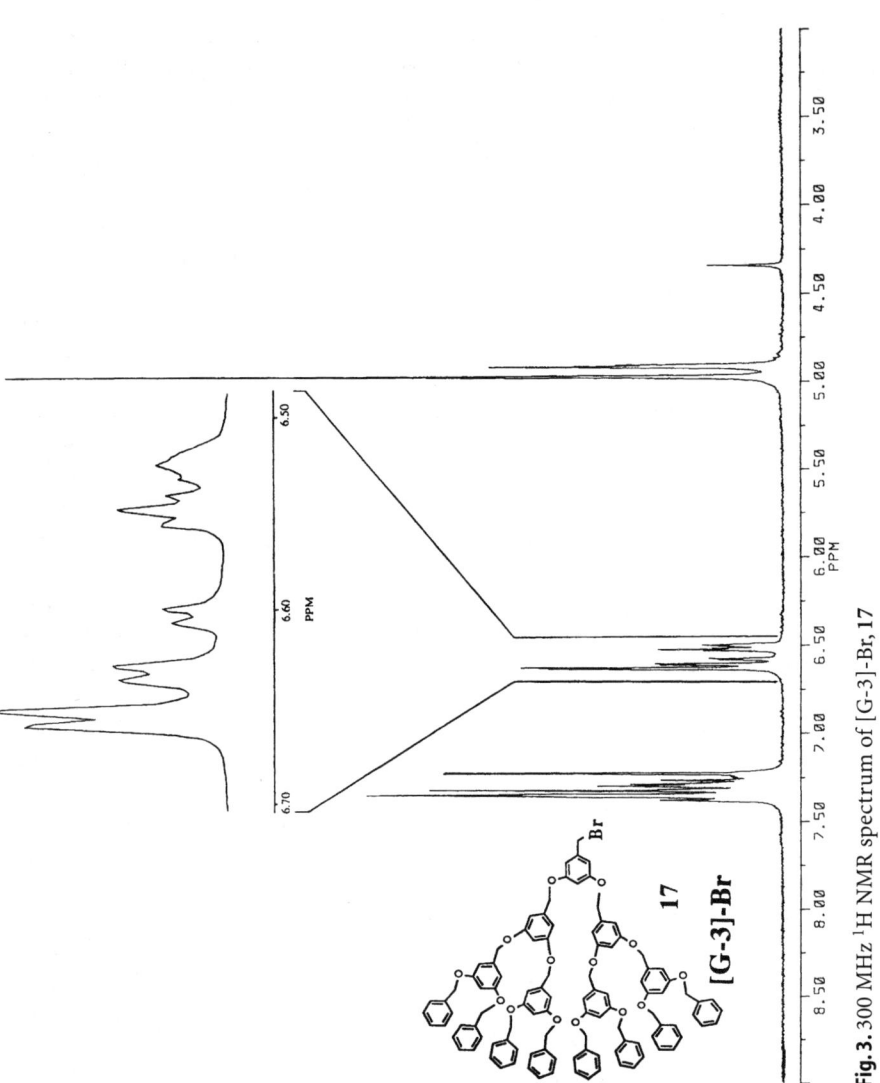

Fig. 3. 300 MHz ^{1}H NMR spectrum of [G-3]-Br, **17**

4.1
Interior Building Blocks

The interior building blocks are essentially the monomer, or repeat units, of the dendrimer and, as such, comprise a significant proportion of the dendritic structure. Therefore the choice of interior building blocks has a dramatic influence on the physical properties and overall structure of the dendrimer. A wide range of building blocks have been employed in the construction of dendritic macromolecules and their choice is governed only by their compatibility with existing functionality and the ability to form linkages in high yields. The early examples of dendrimers published by Tomalia et al. [2] and Newkome et al. [3] focused on linking alkyl units together via amide bonds; subsequently a flurry of activity has seen this selection increased dramatically to include aryl and aliphatic poly(esters) [20–22], poly(phenylenes) [23], poly(arylethers) [8, 10], poly(siloxanes) [24, 25], poly(ethers) [26], poly(arylacetylenics) [27], poly(amines) [28, 29], poly(arylamines) [30], poly(phosphonium salts) [31], poly(crown ethers) [32, 33], poly(carbosilanes) [34], polypyridines [35], etc. Several workers have also introduced chirality into dendritic structures, either at the chain ends [36], core [37], or interior building blocks. Examples of the latter included McGrath and colleagues" use of chiral, non-racemic synthetic monomers [38] and Sharpless and coworkers" use of asymmetric dihydroxylation to give chiral dendrimers based on 1,2-diols [39]. While this is not an exclusive list it does indicate that the divergent and convergent approaches are applicable to the synthesis of dendrimers based on essentially any sub-unit desired [40].

The effect of the dendritic structure on physical properties of dendrimers can be best appreciated when linear polymers, based on a specific repeat unit, are compared with the corresponding dendrimer based on the same repeat unit. Potentially the most dramatic examples of this are the poly(phenylenes) which have been reported by Miller et al. [9, 23]. In this case, the dendritic poly(phenylenes) have exceptional solubility despite the absence of aliphatic or heteroatom linkages. Their preparation utilizes 3,5-dibromo-1-(trimethylsilyl)benzene, **25**, as the monomer unit with generation growth involving palladium-catalyzed Suzuki coupling followed by conversion of the trimethylsilyl group at the focal point to the corresponding boronic acid. In the final step of the synthesis the reactive dendrons are coupled with 1,3,5-tribromobenzene to give a symmetrical, fully aromatic dendritic poly(phenylene) such as **26** (Scheme 9). Significantly, the largest member of this series has a molecular weight of 3510, 46 aromatic rings, is stable to 500 °C, and is freely soluble in toluene. This is in direct contrast to linear poly(phenylene) which is essentially insoluble at molecular weights of less than 700. Since these materials are comprised of the same repeat units this significantly enhanced (ca. 10^5) solubility can only be attributed to the branched dendritic architecture.

The versatility in the construction of dendritic macromolecules, and the unique advantages offered by a controlled step-wise approach, are perhaps best illustrated by the "all hydrocarbon" family of arylacetylenic dendrimers devel-

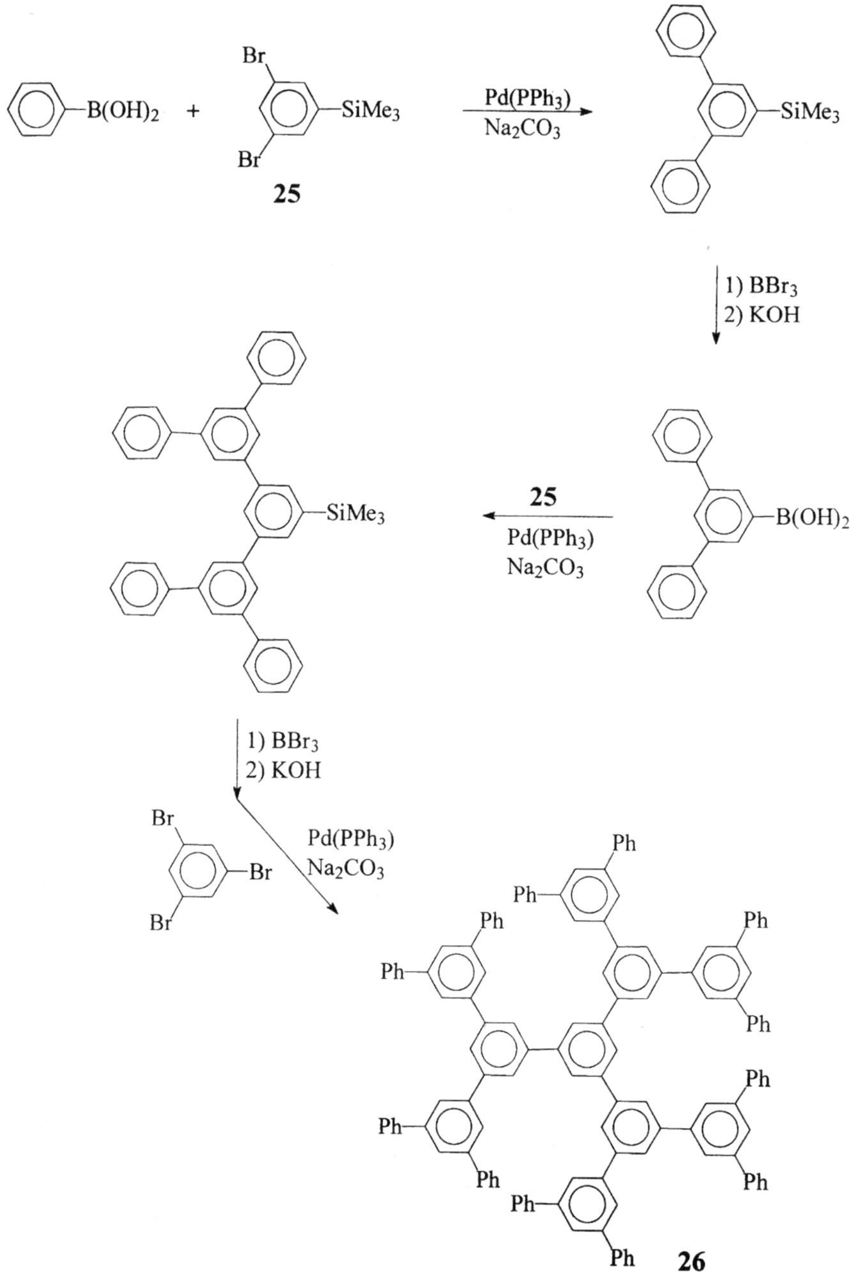

Scheme 9

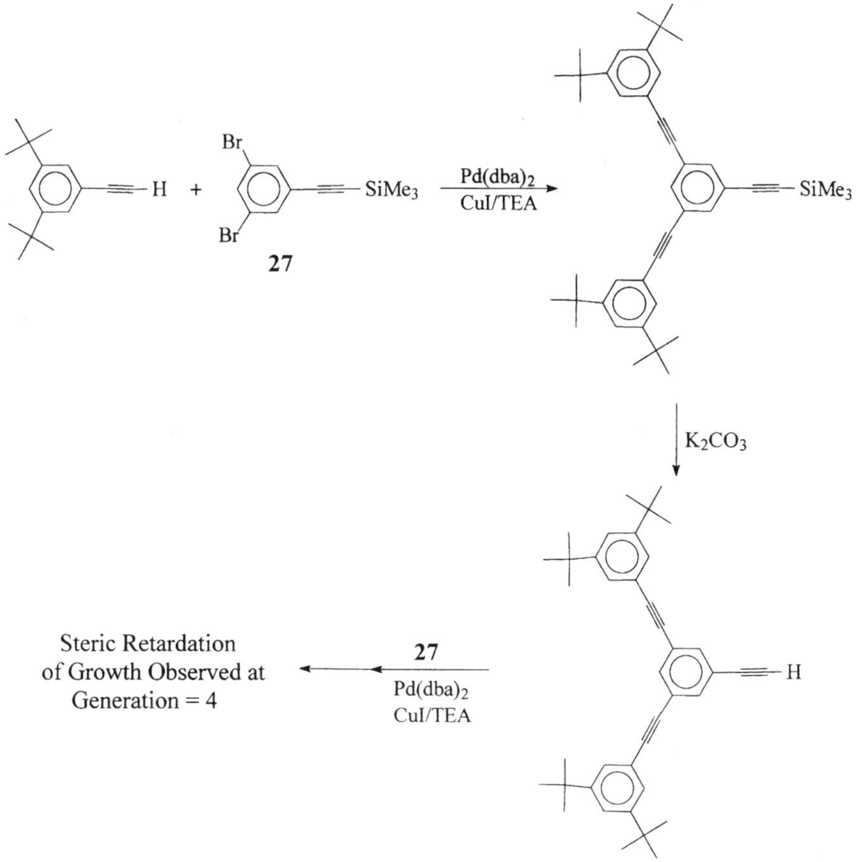

Scheme 10

oped by Moore and Xu [41]. The synthetic blueprint utilizes a convergent strategy based on 3,5-dibromo-1-(2-trimethylsilyl)ethenylbenzene, **27**, as a basic building block with the two-step repetitive procedure involving a palladium-catalyzed cross-coupling reaction between an aryl bromide or iodide and a terminal alkene followed by deprotection of a trimethylsilyl protected terminal alkyne (Scheme 10). In this case the rigidity of the building blocks leads to a substantial steric retardation to growth and dendrimers larger than the third generation could not be prepared using the above building blocks. To overcome these problems the monomer unit was progressively enlarged with rigid spacer units (e.g., **28** and **29**) and these units were incorporated at higher generations when steric retardation became apparent (Fig. 4). Solubilizing t-butyl groups were also included in the terminal units to aid solubility of the larger dendrimers. This dramatic example illustrates how the nature of both the "surface" groups and inter-

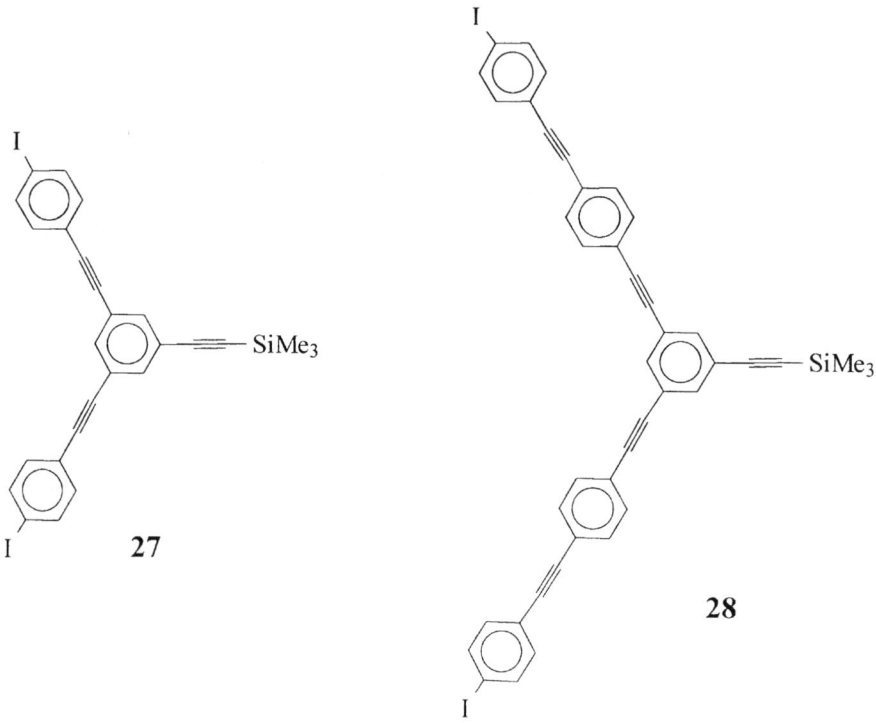

Fig. 4. Extended monomers, 27 and 28, used in construction of dendritic poly(arylacetylenics)

nal building blocks can be manipulated in order to facilitate the synthesis of complex dendritic macromolecules.

Moore has further elaborated on this concept for the synthesis of novel luminescent dendrimers in which the outermost layer of phenylacetylene linkages is replaced by the corresponding triarylamino group [42]. The solubility of the dendrimer, 30, can still be controlled by the introduction of t-butyl groups at the chain ends, although the presence of triarylamino repeat units significantly alters the electronic character of the dendrimer (Fig. 5). The structure of 30 is a perfect example of the modular design possible in dendritic macromolecules, the chain ends lend solubility, and the trisarylamino groups act as efficient hole transport and recombination centers which direct energy, via the phenylacetylene segments, to a luminescent chromophore at the central core.

The choice of interior building blocks for the construction of dendritic macromolecules has not been limited to those based on "organic" units, or even covalent bonds. A number of groups have reported the synthesis of dendrimers containing silicon [43] and phosphorus [44]although however the more intriguing use of organometallic building blocks has involved the use of transition metal centers. This greatly expands the possible uses of dendrimers and opens up the development of novel transition metal-based dendritic catalysts. A range of

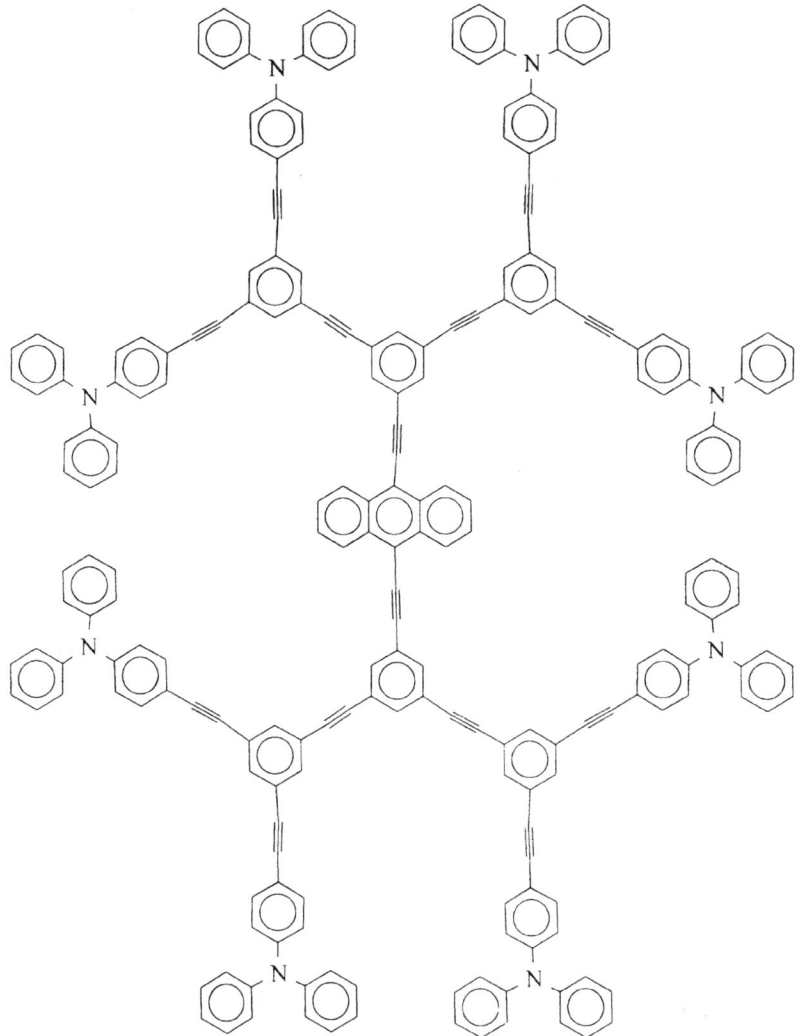

30

Fig. 5. Chromophore labeled dendritic poly(arylacetylenic), 30, for LED applications

transition metal-containing dendrimers have been prepared and the nature of the bonding in these systems range from "ferrocene-like" sandwich complexes [45, 46], to metal-carbon sigma-bonds [47, 48], and coordination complexes [49]. A prime example of the latter is shown in Fig. 6 where an internal layer of

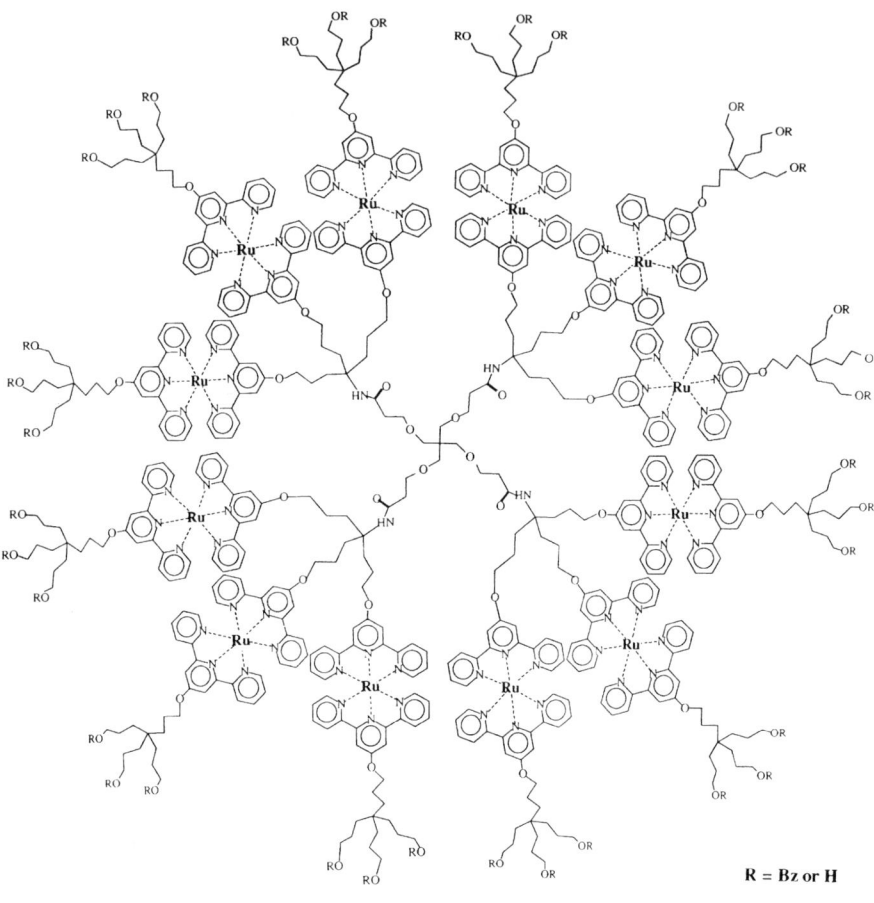

31

Fig. 6. Use of ruthenium-containing building blocks for dendrimer construction, **31**

building blocks comprises ruthenium terpyridyl units, **31**. Therefore the "organic" inner building blocks of the dendrimer are connected to the outermost branching layer, which contains 36 terminal groups, by 12 ruthenium co-ordination complexes. As a result of the design of this macromolecule the interior ruthenium atoms are surrounded by a "surface', the nature of which can be changed by changing the terminal group. This permits the solubility and glass transition temperature of dendrimers based on **31** to be varied according to the potential application [50].

It is also possible to construct dendritic macromolecules based entirely on co-ordination chemistry. Unlike the above example where only a single layer of interior building blocks relied on co-ordination chemistry, Balzani et al. have employed co-ordination complexes at every stage of the dendritic process and these organometallic macromolecules are the first examples of dendrimers where the core unit and all of the building blocks rely on non-covalent bonding for their branching [51]. The success of the synthetic strategy relies on the use of a predetermined combination of traditional bidentate ligands with multifunctional tetradentate ligands, which are capable of forming complexes with two metal centers. This permits the construction of dendritic macromolecules in which a central hexaco-ordinate ruthenium atom is connected to 2 branching layers of 3 and 6 ruthenium centers respectively and finally to a surface layer comprised of 12 ruthenium complexes. The structure therefore contains 22 metals atoms and the nature of the metal atom can be varied by replacing the ruthenium atoms at various stages with, for example, osmium centers to give a multi-transition metal dendrimer with precise arrangement of different transition metals. Such dendrimers have been postulated as harvesting molecules for solar energy and, in an important step towards achieving this goal, Balzani has shown that excitation of the ruthenium centers by light results in an extremely efficient energy transfer to the osmium centers [52, 53].

4.2
Central Core or Focal Point Group

As the name implies, the core unit, or focal point group, is located at the center of the dendritic macromolecule and, as such, constitutes only a small proportion of the overall dendrimer structure. However the role of functional groups at the central core can have a profound effect on a number of physical properties and be extremely useful in investigating the structure of dendrimers. The central role played by the focal point group in the success of the convergent growth approach has also been well documented in the above discussions.

As has been demonstrated by numerous authors, a variety of different functionalities can be introduced at the focal point of dendritic macromolecules. For the polyether dendrimers, such as **20**, at least 20 different focal point groups have been introduced and a number of these have been built into the structure for a specific purpose [54]. For example, a series of polyether dendrimers, such as **32**, containing a single unique solvatochromic probe, N-methyl-p-nitroaniline, at the focal point group have been prepared and studied to gain insight into the unique nanoenvironment at the "center" of dendritic macromolecules [55] (Scheme 11). The absorption maximum of the solvatochromic probe was observed to increase with the molecular weight of the dendrimer in low polarity solvents (Fig. 7). This unusual behavior is a consequence of the more effective shielding of the solvatochromic probe by the surrounding dendrimer as the generation number increases. A more significant observation was that a plot of absorption maximum vs generation number displays a marked discontinuity between generation three and four which has been correlated with a shape transition from an extended to a more globular shape as the steric requirements of the dendrimer increases. Additional evidence for this shape transition in dendritic

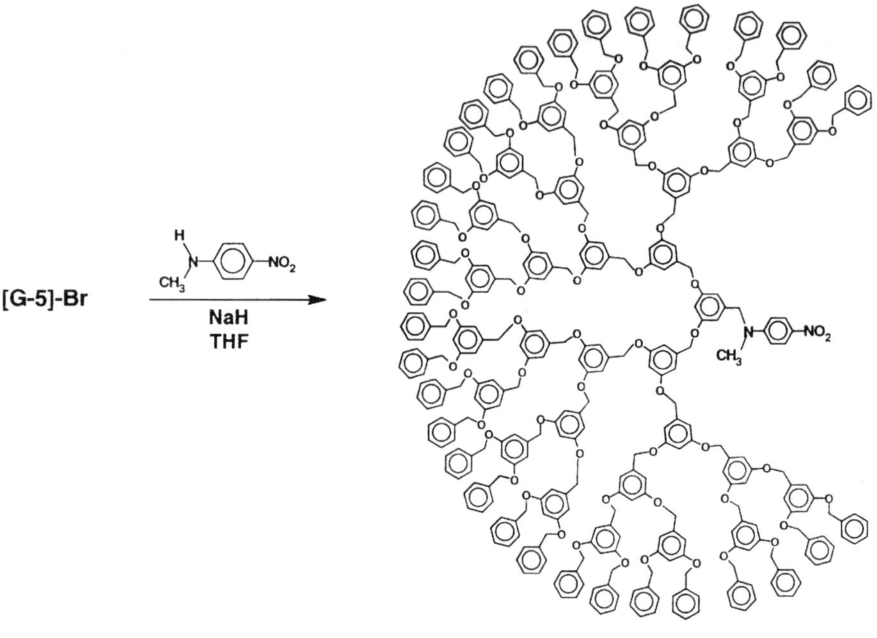

[G-5]-Br →(NaH, THF, with H-N(CH₃)-C₆H₄-NO₂) **32**

Scheme 11

macromolecules has been obtained from examination of intrinsic viscosity behavior for convergent polyether dendrimers [56], molecular modeling experiments of Goddard et al. [57], and photoinduced electron-transfer measurements on PAMAM dendrimers by Turro et al. [58] and Godipas et al. [59]

This ability of dendritic macromolecules to provide a unique nanoenvironment for functional groups at the central core, or focal point, coupled with the ability to introduce a wide variety of functional groups has been exploited by Aida et al. [60]. In an elegant series of experiments a series of dendritic macromolecules containing a central porphyrin core has been prepared and the fluorescence behavior of these systems with quenchers of different steric sizes studied. In both organic and aqueous systems, it was found that low generation dendrimers (e.g., [G-1]) displayed no selectivity to the nature of the quencher. In sharp contrast, higher generation dendrimers such as the third generation water soluble derivative, **33**, showed significant different quenching behavior depending on the size and charge of the quencher (Fig. 8). In effect the surrounding dendrimer acts as a size and charge selective barrier leading to selective isolation of the porphyrin core. The analogy of this behavior with many biological systems is intriguing and has recently been probed by Aida et al. [61–63]. In attempting to mimic the behavior of haemoproteins, a series of dendrimers containing an iron(II) porphyrin 1-methylimidazole core was prepared and the ox-

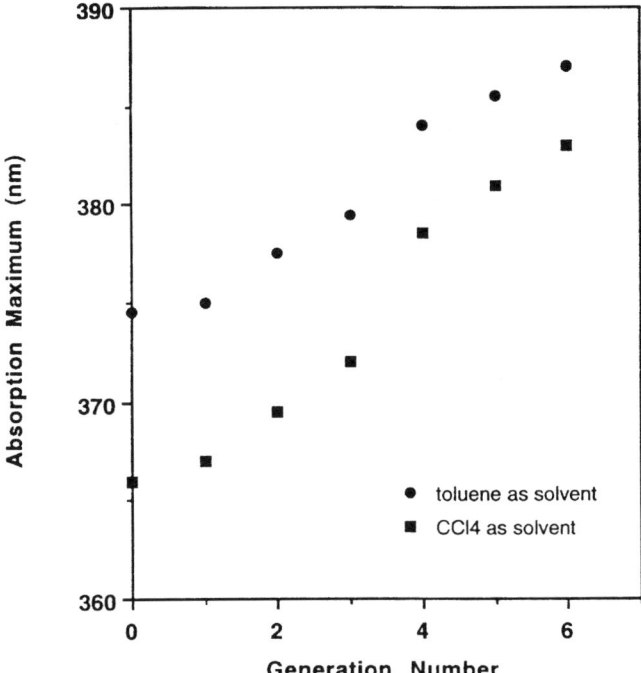

Fig. 7. Plot of absorption maximum vs generation number for polyether dendrimers labeled with a focal point solvatochromic group

ygen and carbon monoxide binding properties examined. Surprisingly the stability of the dioxygen species to dimerization and water promoted oxidation increased significantly on increasing the size of the surrounding dendrimer. In fact the fifth generation material displayed a level of stability not previously obtained for a synthetic mimic. The extraordinary stability of these dioxygen adducts is again due to the steric and hydrophobic protection of the active site by the surrounding dendrimer. These studies illustrate that dendrimers may have great potential for future catalytic or energy transfer applications.

The attachment of dendritic fragments to a central core can also be used to alter dramatically selected physical properties of the core moiety. Two dramatic examples of this feature are the synthesis of dendrimers which contain either buckminsterfullerene, C_{60} [64], or carboranes as the core unit [65]. In the former, polyether dendrimers, containing a single azide functional group at the focal point, were heated with C_{60} to give the dendrimer functionalized azafulleroid, **34** (Scheme 12). Unlike the starting C_{60}, which has limited solubility in most solvents, **34** was observed to have excellent solubility in a wide range of solvents. In fact, the solubility was similar to the starting dendrimer while the electrochemical and optical properties of the fullerene nucleus were maintained.

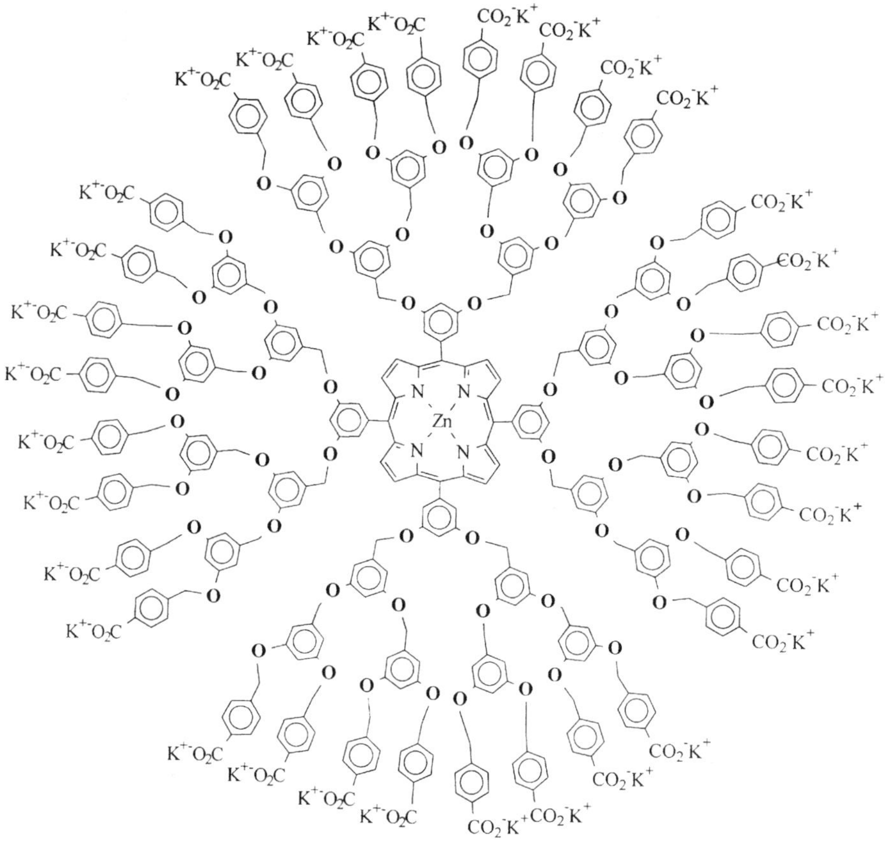

33

Fig. 8. Structure of dendritic micelle with a porphyrin core, 33

Similarly, a series of hydroxy-terminated poly(ether) dendrimers, **35**, with a single carborane nucleus at their core, were observed to have water solubilities comparable to that of chloroacetic acid, or D,L-valine. Again this is in direct contrast to the starting carborane nucleus which is insoluble in aqueous solutions and permitted the use of **35** in neutron capture therapy. Similar effects have also been observed with dendrimers containing calixarenes [66] and porphyrins [67] as the central units.

The functionality at the central core of the dendritic macromolecule can also be exploited as a unique reactive site for a variety of purposes. Perhaps the most dramatic example is the introduction of self-assembling units at the focal point to give dendritic fragments, **36**, which undergo spontaneous self-assembly to give larger dendrimers [68] (Fig. 9). In this case the central core of the dendrimer is a self-as-

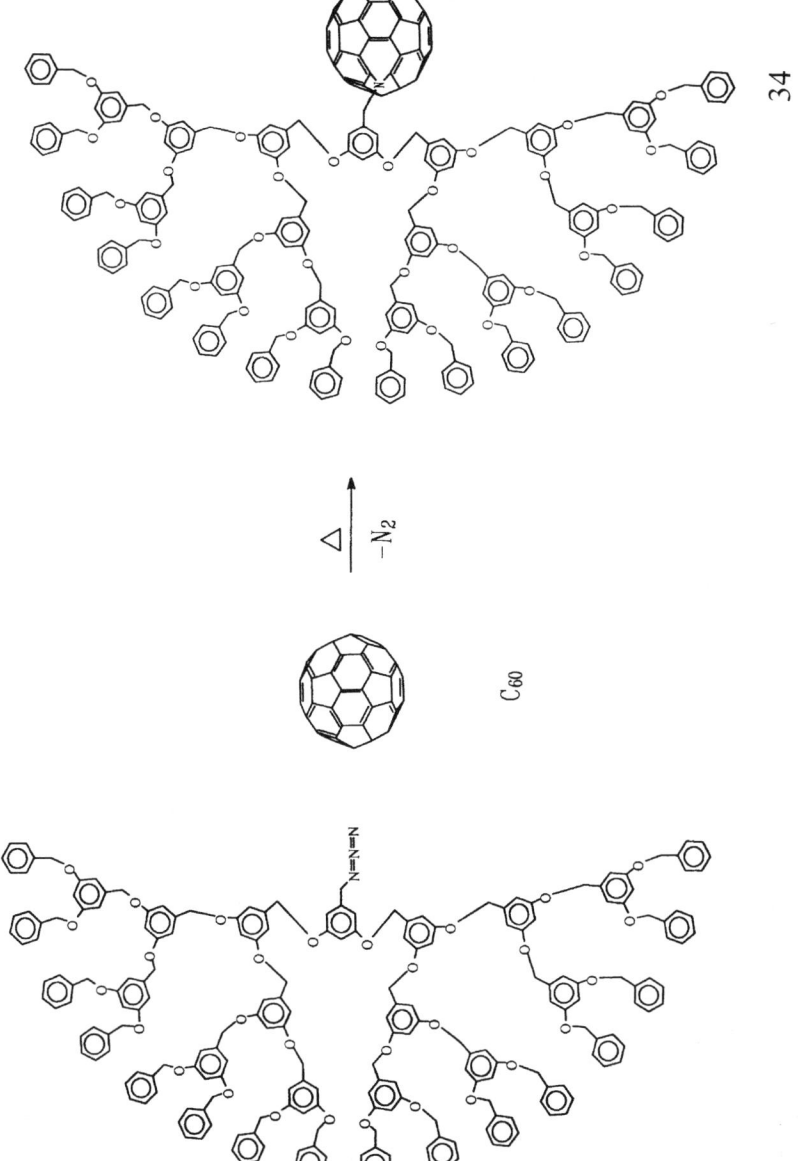

Scheme 12

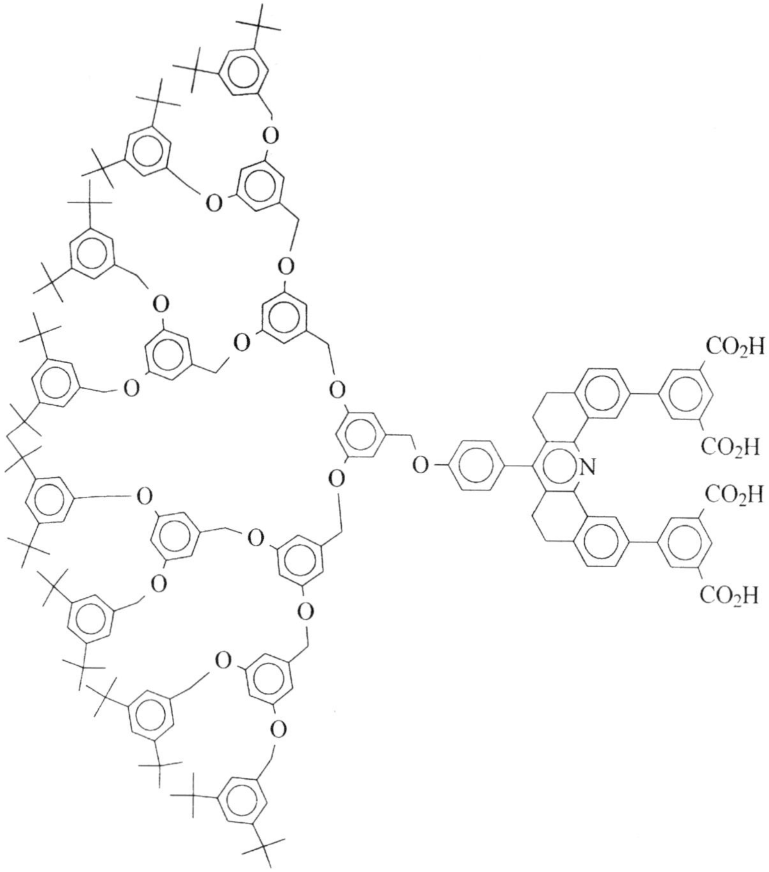

Fig. 9. Self assembling dendritic fragment, **36**, capable of forming hexameric complex

sembled hexameric unit which is connected by hydrogen bonds only (Fig. 7). The synthesis of **36** is an elegant demonstration of the wide range of possible dendritic structures that can be prepared by the accurate control of the chemistry and functionality of the core unit and illustrates that the discrete construction of very large three-dimensional macromolecules from simpler units is possible.

It is also possible to introduce polymerizable groups at the focal point of dendritic macromolecules and a number of authors have demonstrated the synthesis of novel hybrid dendritic-linear block copolymers by this methodology [69–73]. Initial experiments in the use of the dendritic macromonomers involved the preparation of a series of polyether dendrimers, such as the third generation derivative, **37**, which contains a single styrene at its focal point [74]. Copolymerization of **37** with styrene then affords a hybrid block copolymer, **38**, in which

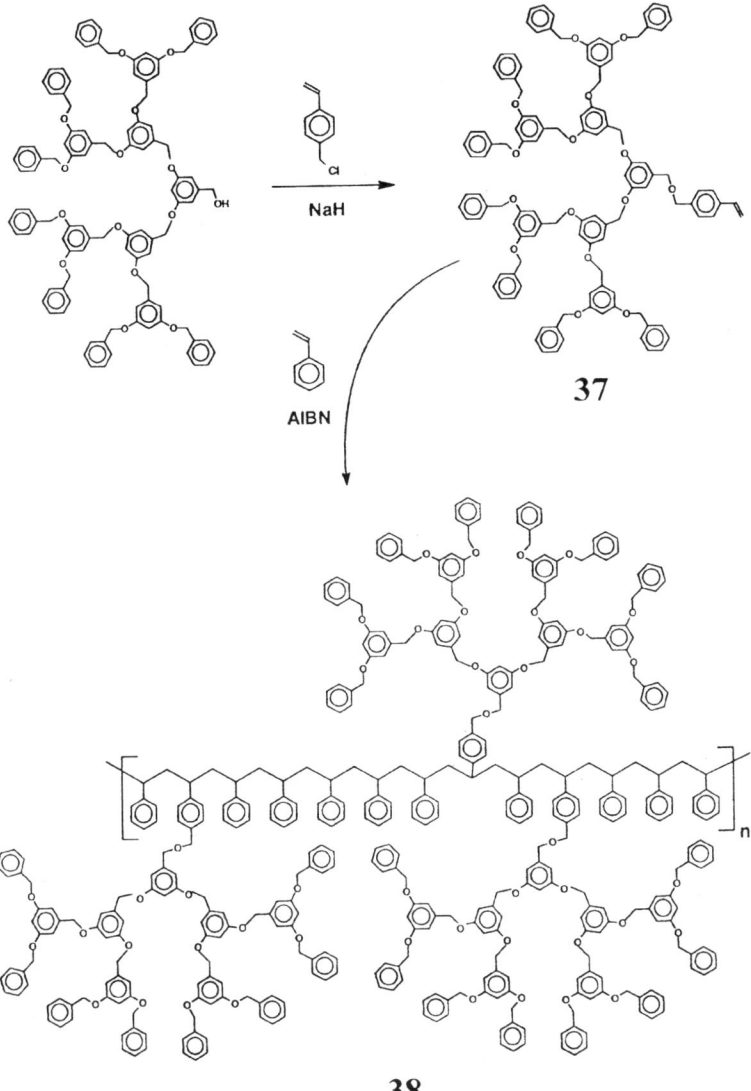

Scheme 13

globular dendritic polyether blocks are grafted to a linear polystyrene backbone (Scheme 13). Interestingly, a single glass transition temperature was observed for **38** and this lack of phase separation in dendritic block copolymers has also been observed for a variety of other structures. Because the focal point of convergent dendrimers can be constructed with almost any type of functional group, a wide variety of other hybrid dendritic-linear block copolymers has been prepared based on both addition and condensation chemistry [75–79]. A number of groups have also introduced initiator groups at the focal point of dendritic macromolecules. This permits the construction of a special type of hybrid dendritic-linear block copolymer in which a single linear chain is connected to the focal point of the dendrimer. For example, Fréchet and Gitsov have shown that the potassium alkoxide of a fourth generation alcohol, [G-4]-OH 18, is an extremely efficient initiator for the ring opening polymerization of caprolactone and, in contrast to other initiators, gives very high molecular material (M_n>300,000) with very low polydispersities (PD=1.07) [80]. The incorporation of initiating groups for "living" free radical polymerizations at the focal point of dendrimers has recently been accomplished and these dendritic initiators, such as **39**, provide rapid access to a variety of interesting hybrid dendritic-linear block copolymers (Fig. 10) [81].

4.3
Chain End Groups

Unlike traditional linear polymers that have only two chain end groups, irrespective of the degree of polymerization, numerous chain ends characterize dendritic macromolecules, the number of which increases with increasing generation number, or degree of polymerization. As was detailed earlier, for an AB_2 monomer connected to a trifunctional core unit the number of chain ends is equal to 3×2^g where g is the generation number. Alternatively, the number of chain ends is equal to the degree of polymerization+1 for convergently grown dendritic fragments based on AB_2 building blocks. In contrast to linear polymers the chain ends of dendritic macromolecules constitute a substantial portion of the molecule and since they may primarily reside at the periphery, or surface, of the dendrimer the chain ends have the greatest opportunity to interact with the surrounding media. Manipulating the nature of the functional groups at the chain ends can therefore easily control the physical and chemical properties of dendritic macromolecules.

In both the divergent and convergent growth approaches it is possible to change the nature of the chain end functional groups, though the divergent growth approach allows a wider variety and easier introduction of chain end functional groups. This is due to the fact that the convergent growth approach starts at the chain ends and the chain end functional groups are therefore present throughout the growth process. This dictates that the chain ends must be compatible with the chemistries used for growth while also permitting purification of the intermediate dendrimers. In contrast, the attachment of the chain ends is inherently the final step in the preparation of dendritic macromolecules by the divergent growth approach, which overcomes the above problems. This permits the introduction of reactive functionalities and also allows the easy synthesis of a wide variety of functionalized dendrimers from the same "base" dendrimer.

The stepwise nature of the convergent growth approach coupled with the limited number of steps required for generation growth does however permit accu-

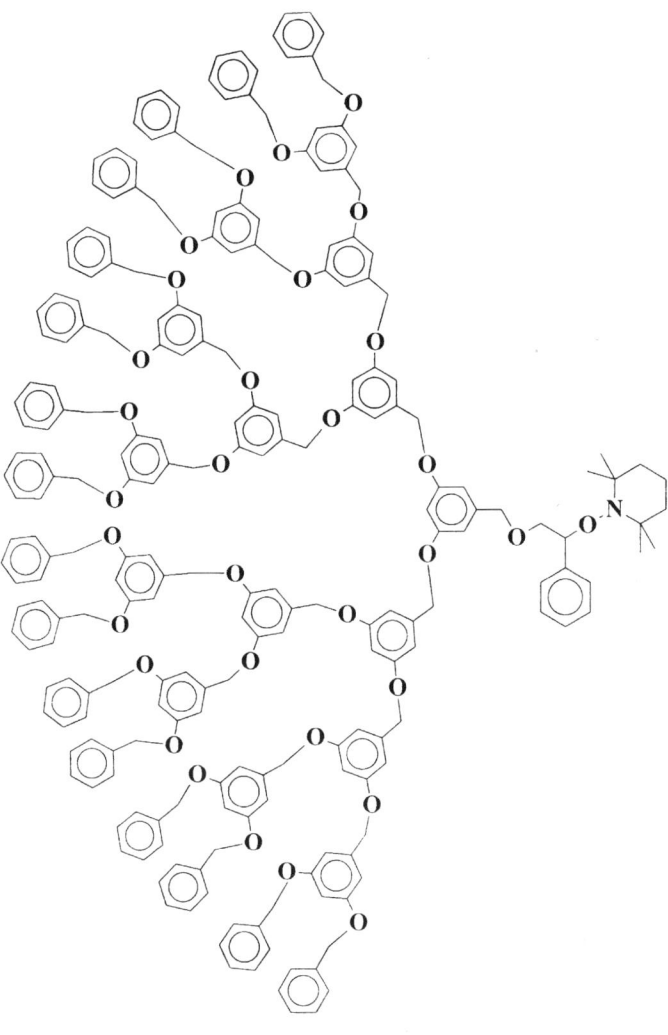

39

Fig. 10. Dendritic initiator, 39, for "living" free radical polymerizations

rate control over the number and placement of chain end functional groups. This degree of control is either not possible or extremely difficult by the divergent growth approach. To demonstrate this strategy for the control of chain end functional groups the synthesis of two extreme cases have been reported [17, 82]. In the first extreme case, a dendritic macromolecule containing 48 chain ends in which there is a single unique functional group was prepared. In the alternate example, dendritic macromolecules with 16 chain ends having one type of functional group and 32 chain ends with a different functionality were prepared. Significantly these functional groups were arranged into discrete blocks or "surface" areas and due to the branching sequence extensive mixing of these blocks is not possible.

The strategy used to prepare both extreme cases of chain end functionalized dendrimers is conceptually the same as the convergent growth approach and employs one or more stepwise alkylation steps as the cornerstone of the construction process. For example, the synthetic blueprint used for the synthesis of a "monofunctionalized" dendrimer is shown schematically in Scheme 14. In this case the stepwise alkylation step is employed at every stage of the convergent growth approach with the monomer unit, **40**, being initially alkylated with a "non-functionalized" dendritic fragment, **41**, followed by alkylation with a "mono-functionalized" dendritic fragment, **42**. The presence of a single functional group at the chain ends of the dendrimer is therefore maintained throughout the synthesis. The degree of versatility associated with control of chain ends by the convergent growth approach can be better appreciated when it is consider that slight variations in where these stepwise alkylation steps are introduced into the synthetic strategy permits the preparation of dendrimers with 1,2,3, or n functional groups located at the chain ends. A degree of control over their placement is also possible and an illustration of this can be found in the synthesis of the second extreme case, **43**. For **43** the stepwise alkylation step is not employed until the final core attachment reaction where a completely functionalized dendritic wedge is coupled with the core, and reaction of this purified monoaddition product with two equivalents of a fully functionalized dendrimer bearing different functional groups leads to the desired "surface-block" dendrimer (Scheme 15) [17].

This ability to control the placement of functional groups at the periphery of dendrimers has been utilized in the construction of two unique macromolecules, which may have technological relevance. In one case a polyether dendrimer, **44**, was prepared in which one half of the surface was functionalized with electron-withdrawing groups while the other half was functionalized with electron-donating groups. This partitioning of the surface was found to result in a considerable increase in the dipole moment of **44** when compared with similar dendrimers where no partitioning of the chain end groups was present [83]. Conceptually such dendrimers may find a use in the design of high speed switches or novel display devices. A similar partitioning approach has been utilized in the preparation of amphiphilic dendrimers, **45**, in which one hemisphere has terminal carboxylate groups while the other is functionalized with hydrophobic benzyl ether groups (Fig. 11) [84]. These materials were found to have high activity at interfaces and could be used to prepare novel membranes or ion transport agents.

One immediate effect of being able to construct dendritic macromolecules with a wide variety of different chain ends is that the solubility characteristics of the dendrimer can be finely tuned. For example, polyether dendrimers based on 3,5-dihydroxybenzyl alcohol as the repeat unit can be prepared with dodecyl

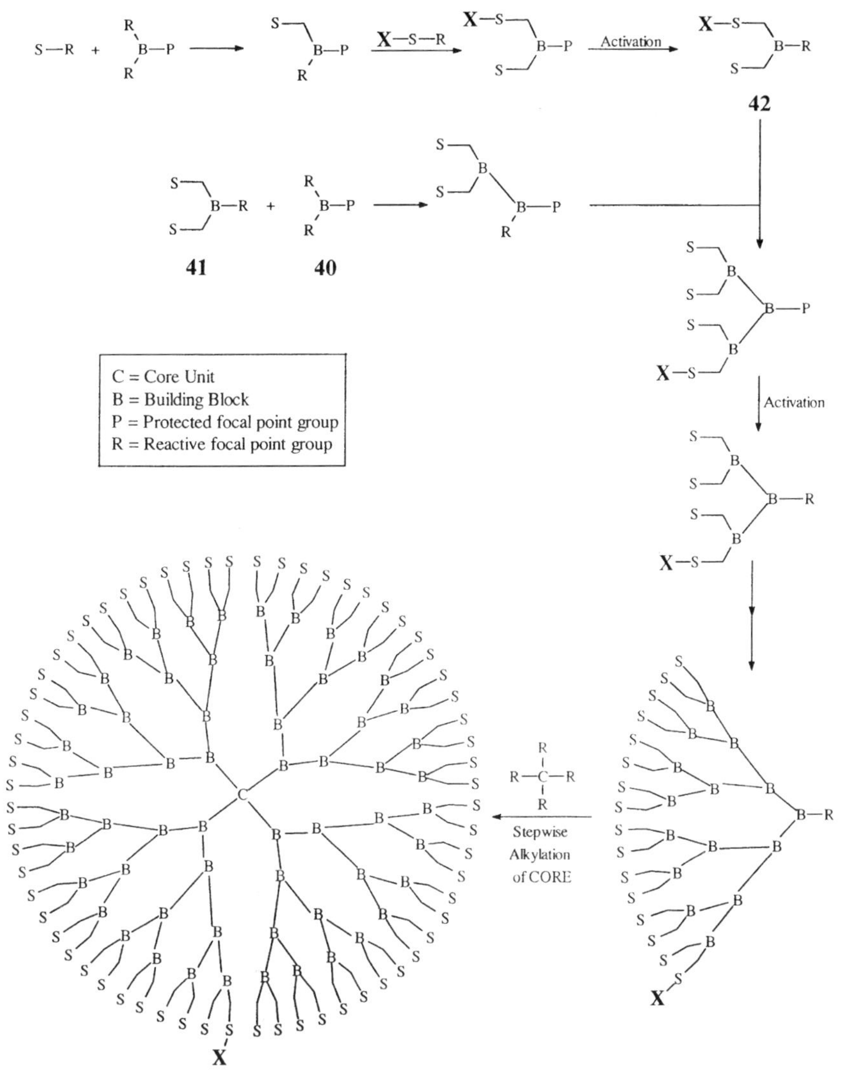

Scheme 14

chain ends which leads to solubility of the dendrimer in very non-polar solvents such as cyclohexane [85]. Conversely, the same dendrimer can be prepared with carboxylate end groups, **46**, which leads to solubility in only very polar solvents such as water. Significantly, the above two dendritic macromolecules have exactly the same internal structure, although their solubility is dramatically different and can be manipulated to encompass the whole solvent spectrum by changing the nature of the chain ends.

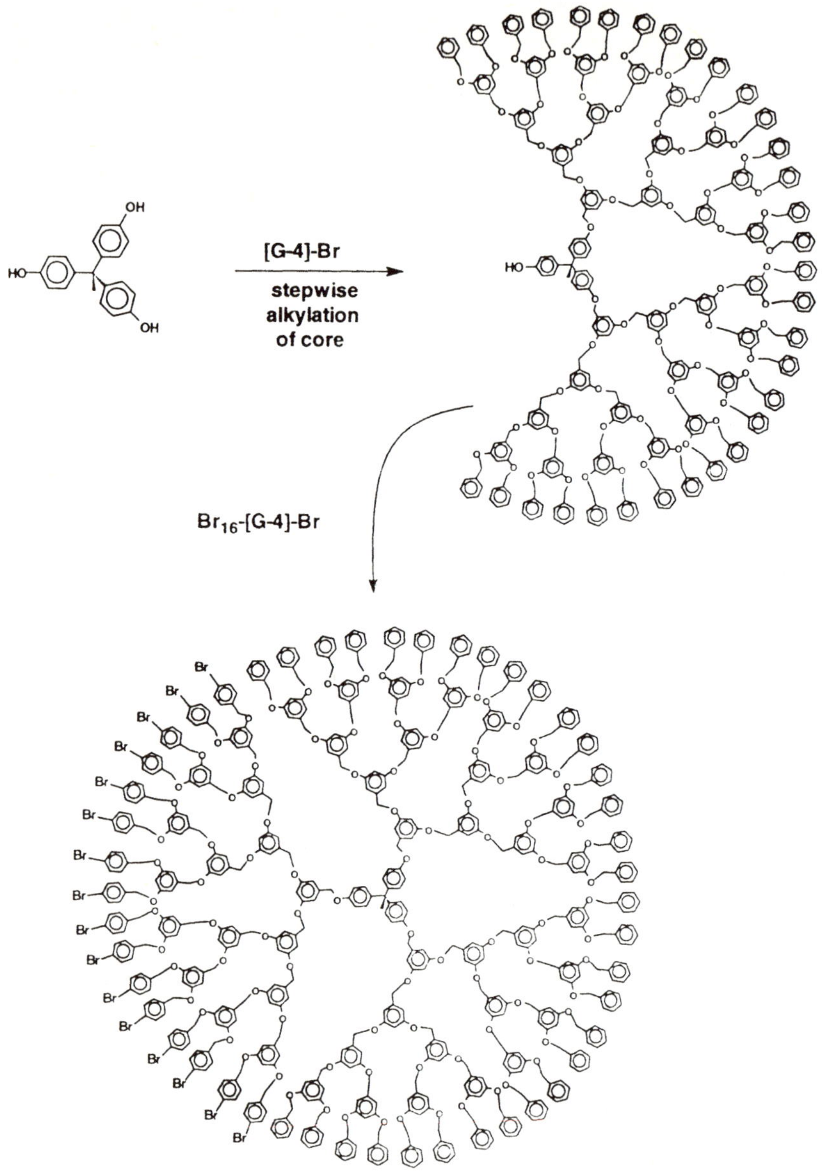

[G-4]-Br
stepwise alkylation of core

Br$_{16}$-[G-4]-Br

43

Scheme 15

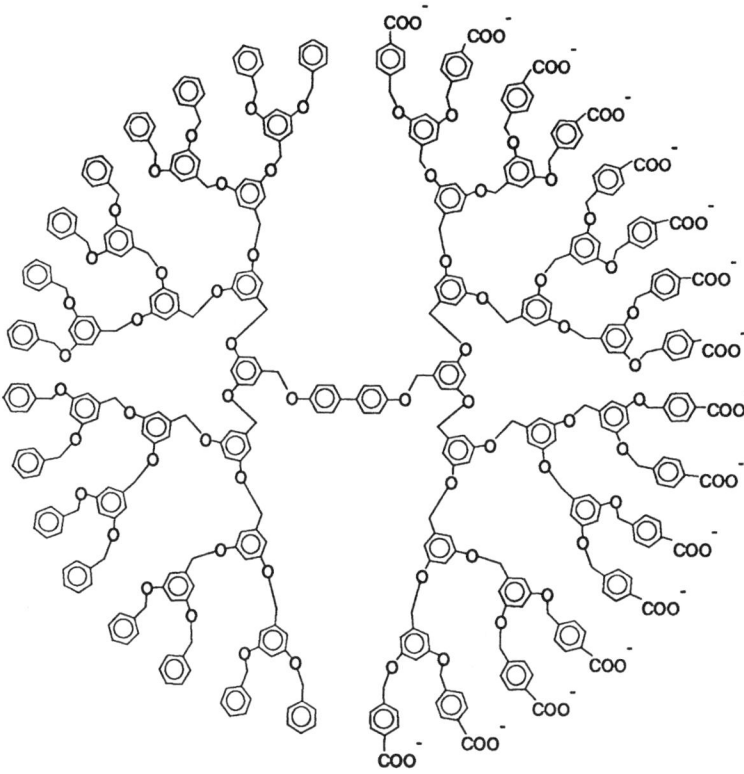

Fig. 11. Amphiphilic dendrimer, 45

The structure and properties of water soluble dendrimers, such as **46**, is, in itself, a very promising area of research due to their similarity with natural micellar systems. As can be seen from the two-dimensional representation of **46** the structure contains a hydrophobic inner core surrounded by a hydrophilic layer of carboxylate groups (Fig. 12). However these dendritic micelles differ from traditional micelles in that they are static, covalently bound structures instead of dynamic associations of individual molecules. A number of studies have exploited this unique feature of dendritic micelles in the design of novel recyclable solubilization and extraction systems that may find great application in the recovery of organic materials from aqueous solutions [84, 86–88]. These studies have also shown that dendritic micelles can solubilize hydrophobic molecules in aqueous solution to the same, if not greater, extent than traditional SDS micelles. The advantages of these dendritic micelles are that they do not suffer from a critical micelle concentration and therefore display solvation ability at nanomolar

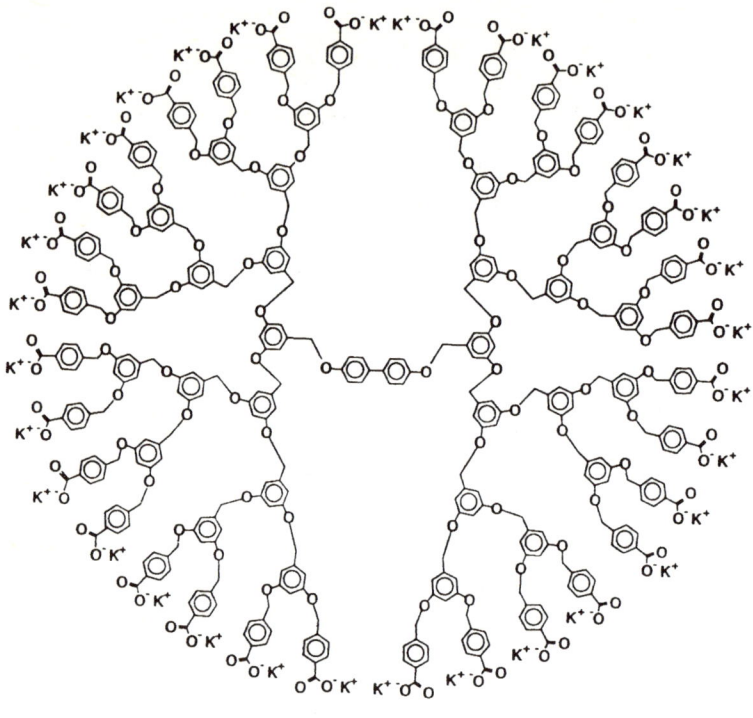

46

Fig. 12. Dendritic micelle, 46

concentration levels. It is also possible to construct dendritic micelles with specific binding sites in the hydrophobic interior, which permits selective solubilization and extraction. In an elegant series of experiments, Newkome et al. have also demonstrated that the molecular size of dendritic micelles is very sensitive to the degree of ionization of the terminal carboxylate groups [89]. A series of 2D-NMR (DOSY) experiments revealed that the radius of a dendritic micelle underwent a marked change with pH, with the hydrodynamic radius being the largest at neutral and smallest at low pH which correlates with protonation of the carboxylate end groups. Water soluble dendrimers having saccharide binding aminoboronic acid groups at the chain ends have also been prepared by Shinkai et al. [90] while lysine dendrimers containing numerous carborane units at the chain ends have been prepared as protein labels in electron microscopy [91].

The nature of the chain end groups have also been shown to have a dramatic effect on the glass transition temperature of dendritic macromolecules. A variety of workers have investigated this question by the preparation of chain end

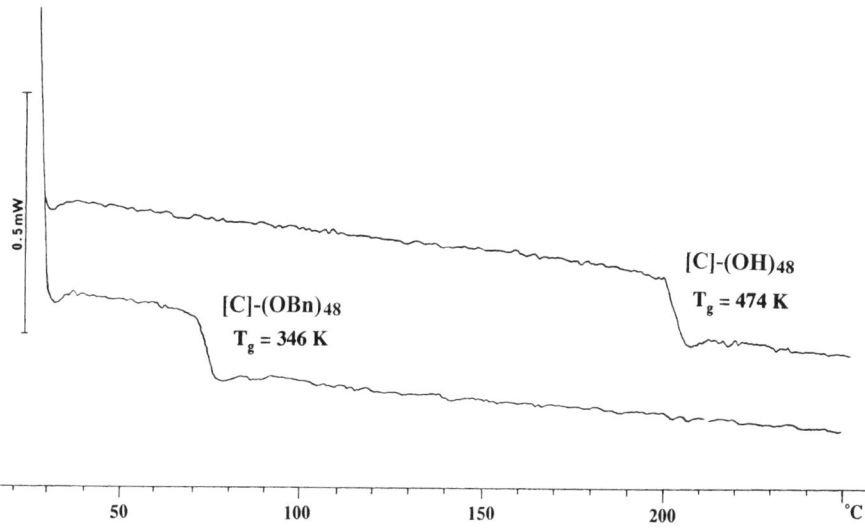

Fig. 13. Comparison of DSC traces for polyester dendrimers with phenolic and benzyloxy chain ends

functionalized dendritic macromolecules which have the same core unit and interior building blocks but differ in their chain end groups [92, 93]. From these results it was conclusively shown that the glass transition temperature of dendritic macromolecules can be manipulated by changing the chain end groups. For example, the glass transition temperature of a dendritic polyester based on 3,5-dihydroxybenzoic acid varies from 346 K to 474 K on going from benzyl ether chain ends to phenolic chain ends (Fig. 13) [94]. This dramatic increase of approximately 130 °C can only be attributed to changes in the chain end functional groups.

The specific construction of dendrimers with highly specialized surface groups for biological or redox activity has also been demonstrated using otherwise inactive dendritic macromolecules as building blocks. Roy et al. have introduced biological activity to PAMAM dendrimers by the introduction of sialinic acid units at the chain ends to give materials which show strong activity to the influenza A virus [95]. Similarly, a variety of redox active groups have been introduced at the chain ends of dendrimers. Starting from the basic PAMAM dendritic framework, Miller et al. have functionalized the terminal groups with perylene diimide electronophores with the result that the dendrimer is now electroactive and shows two reversible one-electron reductions per surface group [96]. Bryce et al. have also decorated otherwise inactive dendritic poly(arylesters) with tetrathiafulvalene units to give a series of dendrimers which exhibited multiple electron reversible oxidations and become conducting organic metals when complexed with 7,7,8,8-tetracyano-p-quinomethane [97].

Perhaps the most viable short-term use for dendritic macromolecules lies in their use as novel catalytic systems since it offers the possibility to combine the activity of small molecule catalysts with the isolation benefits of crosslinked polymeric systems. These potential advantages are intimately connected with the ability to control the number and nature of the surface functional groups. Unlike linear or crosslinked polymers where catalytic sites may be buried within the random coil structure, all the catalytic sites can be precisely located at the chain ends, or periphery, of the dendrimer. This maximizes the activity of each individual catalytic site and leads to activities approaching small molecule systems. However the well defined and monodisperse size of dendrimers permits their easy separation by ultrafiltration and leads to the recovery of catalyst-free products. The first examples of such dendrimer catalysts have recently been reported

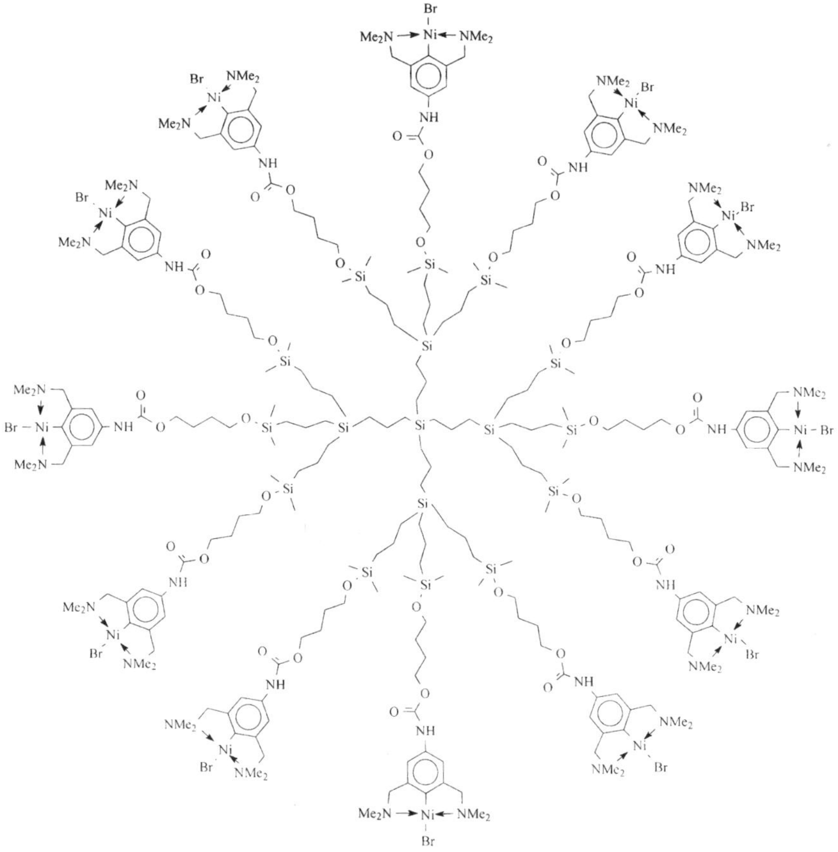

47

Fig. 14. Dendritic dodeca-nickel catalyst, 47, for Kharasch reaction

and involve the synthesis of a dodecanickel (II) species, **47**, in which the active arylnickel (II) complexes are covalently bound to the chain ends of a silane dendrimer (Fig. 14). Significantly, the dendrimer, **47**, displayed a catalytic ability comparable to that of the analogous non-dendrimer-bound small molecule for the regiospecific Kharasch addition of polyhaloalkanes to carbon-carbon double bonds [47].

5
Hyperbranched Macromolecules

As their name implies, hyperbranched macromolecules constitute a second class of highly branched globular synthetic macromolecules, which are closely related to dendrimers. Like dendrimers they have only been studied in detail recently (1989), though the concepts and principles underlying the preparation of hyperbranched macromolecules were first discussed by Flory more than 40 years ago [98]. Interestingly, hyperbranched and dendritic macromolecules share a number of common features. Both sets of macromolecules are prepared from AB_x monomers which results in a highly branched structure with an extremely large number of chain end functional groups. However the overall structure and method of construction for these novel materials differ substantially; hyperbranched macromolecules are prepared in a one-step synthetic strategy instead of the step-wise procedures used for dendritic macromolecules. This results in significant differences in structure and in some cases properties between hyperbranched and dendritic macromolecules.

The one-step procedure used for the preparation of hyperbranched macromolecule results in uncontrolled growth leading to a complex highly branched product, which contains both linear and dendritic sections. As shown in Scheme 16, polymerization of an AB_2 monomer, **48**, leads to a hyperbranched macromolecule, **49**, which may initially seem to have a complex structure. However further examination of **49** reveals only three different types of repeat units; a dendritic unit, **50**, in which both of the reactive B functionalities have formed polymeric linkages; a linear unit, **51**, in which only one of the two B functionalities have formed polymeric linkages; and a terminal unit, **52**, for which neither of the B functionalities have reacted. It is the presence of these linear units, or failure sequences, which leads to a non-ideal structure, which is intermediate between those of "perfect" dendrimers and linear macromolecules. To describe these hyperbranched macromolecules in more detail, a number of authors [87, 99] have developed the concept of degree of branching (DB) which is defined as (DB)=(no. of dendritic units+no. of terminal units)/((total no. of units) and since the number of dendritic units is essentially equal to the number of terminal units the relationship can be rewritten as (DB)=(2×no. of dendritic unit)/(2×no. of dendritic units+no. of terminal units).

Two different procedures have been developed for determining the relative percentage of the different sub-units in hyperbranched macromolecules. The most widespread method involves the identification of unique resonances for

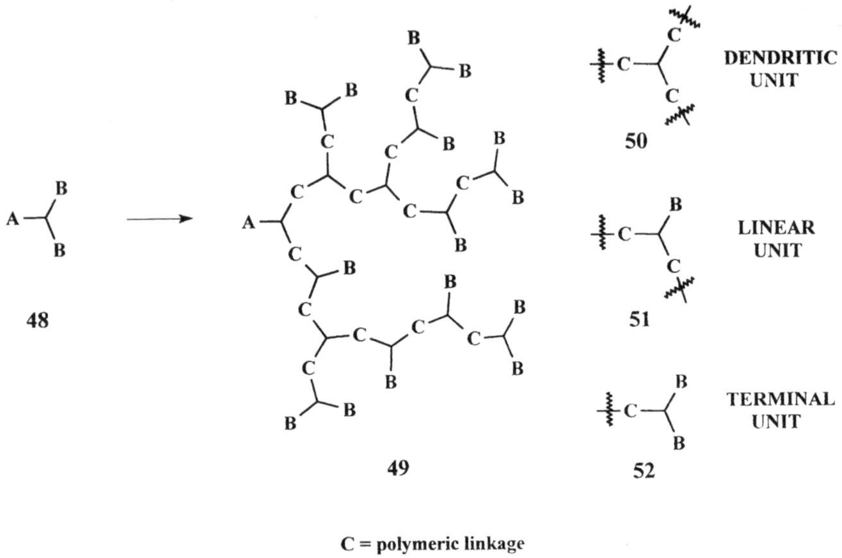

Scheme 16

C = polymeric linkage

the different sub-units by ^{1}H, ^{13}C, or ^{19}F NMR spectroscopy with the aid of model compounds [99–102]. For those systems where the NMR method is not applicable, an alternate methodology based on a functionalization/degradation procedure has been successfully employed [103]. One of the most widely studied hyperbranched systems is the phenol-terminated hyperbranched aromatic polyester, **53**, which is obtained from polymerization of 3,5-bis(trimethyloxysilyl)benzoyl chloride, **54** (Scheme 17) [104]. Analysis of the ^{1}H and ^{13}C NMR spectra of **53** and comparison with model compounds allows the relative percentage of the terminal, dendritic, and linear units to be determined. From this analysis a degree of branching of ca. 60% was calculated which correlates with a branch at approximately every second repeat unit. While this is not as highly branched as a dendritic macromolecule, **54** is still a very highly branched structure and would be expected to adopt a globular structure. The degree of branching has been determined for a number of other hyperbranched structures and values obtained for a DB range of 15–85% with typical values being in the 50–70% range [105, 106].

A wide range of other monomer units have been employed in the synthesis of hyperbranched macromolecules and the range of structures obtained is nearly as diverse as those for dendritic macromolecules. For example, hyperbranched polyphenylenes [93], polyesters [104, 107–109], polyethers [110–112], polyamides [113], polysilanes [114], polyetherketones [100], polycarbazoles [115], etc. [116–118] have been prepared. Interestingly, a number of groups have also used growth processes other than condensation chemistry to prepare hyper-

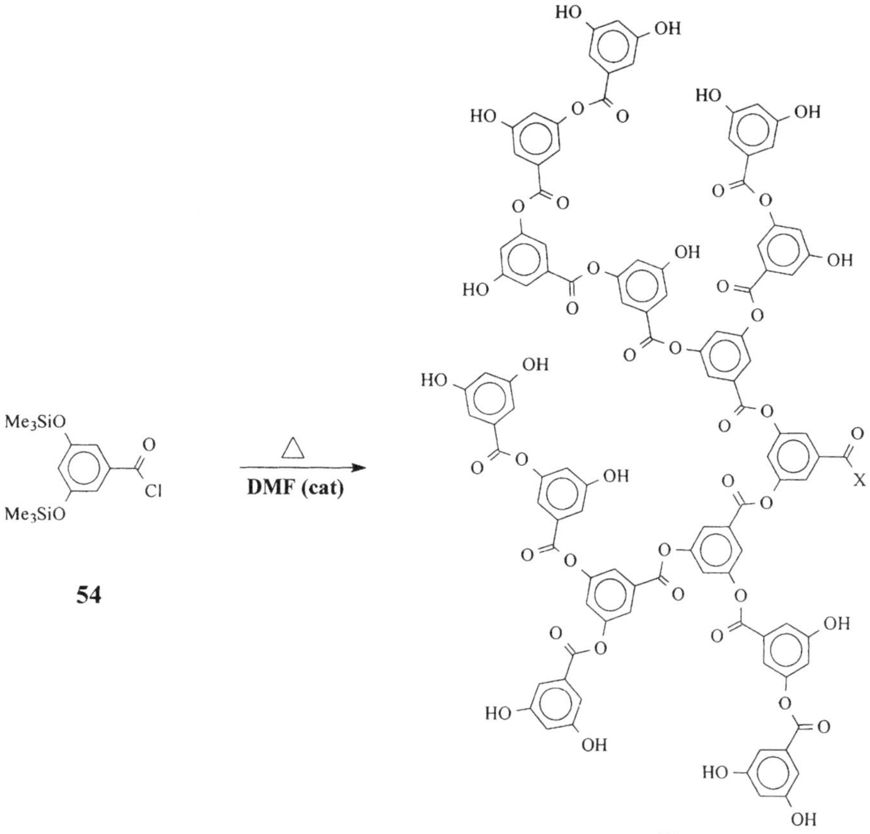

Scheme 17

branched macromolecules. Suzuki has reported an elegant application of ring-opening chemistry to the synthesis of hyperbranched polyamines with low polydispersities (<1.3) and controlled molecular weights [119]. A novel self condensing monomer approach, based on the cationic polymerization of chloromethylstyrene derivatives, has been developed by Fréchet et al. [120]. Meanwhile, Hawker has used a similar approach coupled with "living" free radical chemistry to prepare hyperbranched and branched polystyrene derivatives [121]. In this case a pseudo AB_2 monomer, **55**, is employed which contains both a polymerizable double bond as well as an initiating group. Polymerization of **55** under typical "living" free radical conditions leads to the hyperbranched polystyrene derivative, **56** (Scheme 18). It should be noted that the preparation of similar dendritic structures would be extremely complicated due to the requirements for controlled step-wise growth.

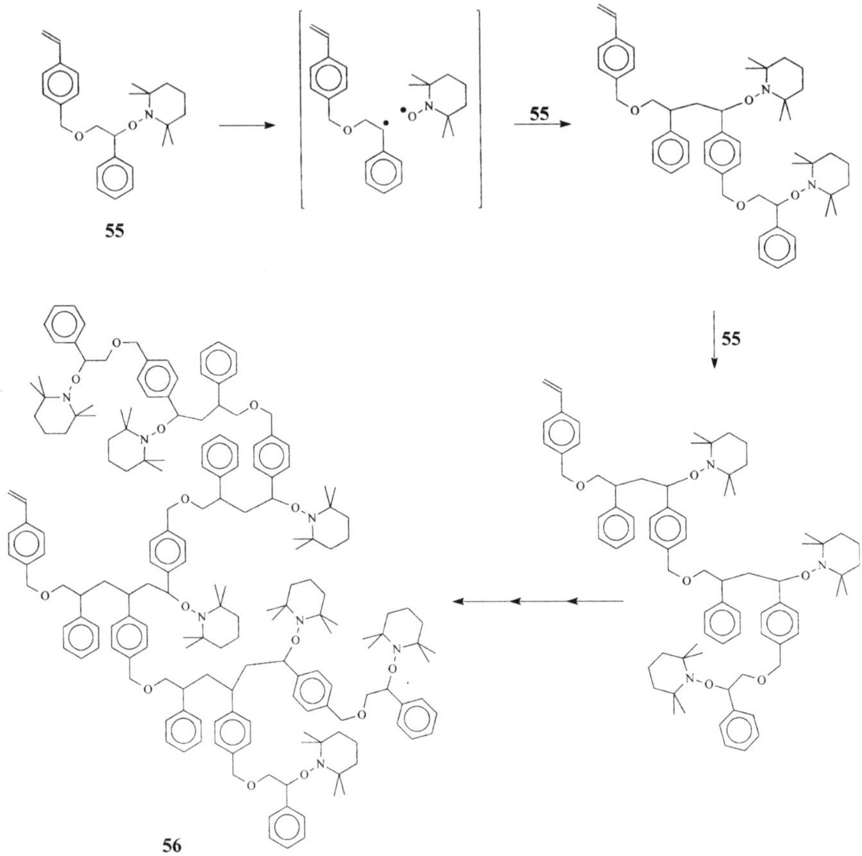

Scheme 18

A fascinating use of transition metal chemistry has also been employed by Reinhoudt et al. in the preparation of hyperbranched organopalladium species. In this case the monomer units, 57, are stabilized by chelated acetonitrile molecules at each of the two palladium centers. On gentle heating the acetonitrile molecules are lost and the single benzyl cyano group at the focal point of the monomer unit (Scheme 19) fills the co-ordination sphere of the palladium atoms. This results in a hyperbranched structure, 58, which is essentially formed by self assembly and leads to large organopalladium spheres with diameters of approximately 200 nm. Significantly the self assembly of 57 was shown to be completely reversible and addition of acetonitrile to 58 results in isolation of monomer units only [122].

The use of AB_x monomers for the preparation of both hyperbranched and dendritic macromolecules leads to the presence of an extremely large number of chain end functional groups for both systems. In analogy with dendrimers, the

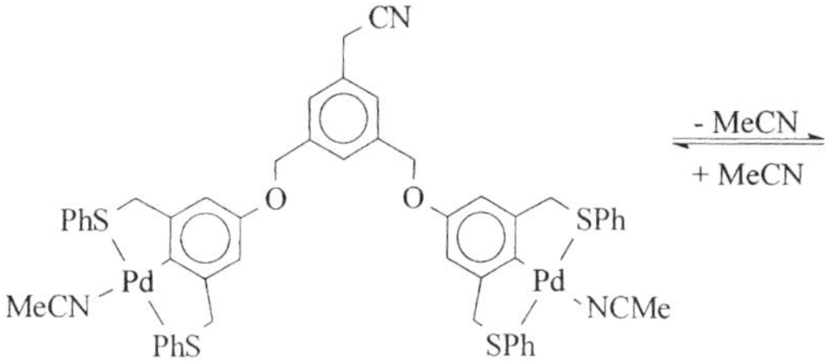

Scheme 19

57 ⇌ (−MeCN / +MeCN) Self-assembled hyperbranched organopalladium macromolecule 58

physical properties of hyperbranched macromolecules can also be manipulated by controlling the nature of the chain ends. For example, Kim has demonstrated that hyperbranched polyphenylene derivatives not only have increased solubility when compared with their linear analogs but this solubility can be controlled by changing the chain end groups [93]. In fact a carboxylate terminated polyphenylene has been shown to act as a unimolecular micelle leading to solubilization of hydrophobic guests in a similar manner to dendritic micelles [87]. Similarly the glass transition temperature of hyperbranched macromolecules has been shown by a number of authors to be dependent on the nature of the chain end functional groups, again in a manner reminiscent of dendrimers [123, 124].

While it can be expected that a number of physical properties of hyperbranched and dendritic macromolecules will be similar, it should not be assumed that all properties found for dendrimers will apply to hyperbranched macromolecules. This difference has clearly been observed in a number of different areas. As would be expected for a material intermediate between dendrimers and linear polymers, the reactivity of the chain ends is lower for hyperbranched macromolecules than for dendrimers [125]. Dendritic macromolecules would therefore possess a clear advantage in processes, which require maximum chain end reactivity such as novel catalysts. A dramatic difference is also observed when the intrinsic viscosity behavior of hyperbranched macromolecules is compared with regular dendrimers. While dendrimers are found to be the only materials that do not obey the Mark-Houwink-Sakurada relationship, hyperbranched macromolecules are found to follow this relationship, albeit with extremely low "a" values when compared to linear and branched polymers [126].

From the above discussion it can be seen that hyperbranched macromolecules are similar to dendrimers in a number of respects. A large degree of variation is possible in their structure and to a limited degree structural control is possible. Obviously the extremely high degree of control associated with dendritic macromolecules is not possible, although the readily availability of hyperbranched systems due to their one step synthesis does make these materials attractive from a commercial viewpoint. Exploitation of the globular, highly functionalized structure of hyperbranched macromolecules does lead to changes in their physical properties when compared to linear materials and a number of these property changes can be directly linked with observations from the field of dendrimers.

References

1. Fréchet JMJ (1994) Science 263:1710
2. Tomalia DA, Baker H, Dewald J, Hall G, Kallos G, Martin S, Ryder J, Smith P (1985) Polym J 17:117
3. Newkome GR, Yao Z, Baker GR, Gupta VK (1985) J Org Chem 50:2003
4. Buhleier E, Wehner W, Vogtle F (1978) Synthesis 155
5. Denkewalter RG, Kolc JF, Lukasavage WJ (1981) US Pat 4,289,872
6. Aharoni SM, Crosby CR, Walsh EK (1982) Macromolecules 15:1093
7. de Brabandr van den Berg EMM, Meijer EW (1993) Angew Chem Int Ed Engl 32:1308
8. Fréchet JMJ, Jiang Y, Hawker CJ, Philippides A (1989) Proc IUPAC Int Symp Functional Polymers 19
9. Miller TM, Neenan TX (1990) Chem Mater 2:346
10. Hawker CJ, Fréchet JMJ (1990) J Chem Soc Chem Commun 1010
11. Hawker CJ, Fréchet JMJ (1990) J Am Chem Soc 112:7638
12. Wooley KL, Hawker CJ, Fréchet JMJ, Wudl F, Srdanov G, Shi S, Kao J (1993) J Am Chem Soc 115:9836
13. Ferguson G, Callagher JF, McKervey AM, Madigan E (1996) J Chem Soc Perkin Trans 1 599
14. Wooley KL, Hawker CJ, Fréchet JMJ (1991) J Am Chem Soc 113:4252
15. Hawker CJ, Fréchet JMJ (1996) In: Ebdon JR, Eastmond GC (eds) New methods of polymer synthesis. Blackie, London, p 290
16. Walker KL, Kahr MS, Wilkins CL, Xu ZF, Moore JS (1994) J Am Soc Mass Spectrom 5:731
17. Leon JW, Fréchet JMJ (1995) Polym Bull 35:449
18. Wooley KL, Hawker CJ, Fréchet JMJ (1991) J Chem Soc Perkin Trans 1 1059
19. Hawker CJ, Fréchet JMJ (1992) J Am Chem Soc 114:8405
20. Miller TM, Kwock EW, Neenan TX (1992) Macromolecules 25:3143
21. Ihre H, Hult A, Soderlind E (1996) J Am Chem Soc 118:6388
22. Haddleton DM, Sahota HS, Taylor PC, Yeates SG (1996) J Chem Soc Perkin Trans 1 649
23. Miller TM, Neenan TX, Zayas R, Bair HE (1992) J Am Chem Soc 114:1018
24. Morikawa A, Kakimoto M, Imai Y (1991) Macromolecules 24:3469
25. Morikawa A, Kakimoto M, Imai Y (1992) Macromolecules 25:3247
26. Padias AB, Hall HK, Tomalia DA, McConnell JR (1987) J Org Chem 57:435
27. Moore JS, Xu Z (1991) Macromolecules 24:5893
28. Worner C, Mulhaupt R (1993) Angew Chem Int Ed Engl 32:1306
29. Moors R, Vogtle F (1993) Chem Ber 126:2133
30. Hall HK, Polis DW (1987) Polym Bull 17:409
31. Rengan K, Engel R (1991) J Chem Soc Chem Commun 987
32. Nagasaki T, Ukon M, Arimori S, Shinkai S (1992) J Chem Soc Chem Commun 608
33. Nagasaki T, Kimura O, Ukon M, Arimori S, Hamachi I, Shinkai S (1994) J Chem Soc Perkin Trans 1:75
34. Lorenz K, Holter D, Stuhn B, Mulhaupt R, Frey H (1996) Adv Mater 8:414
35. Chessa G, Scrivanti A (1996) J Chem Soc Perkin Trans 1:307
36. Newkome GR, Lin X, Weis CD (1991) Tet Assym 2:957
37. Kremers JA, Meijer EW (1994) J Org Chem 59:4262

38. McGrath DV, Wu M, Chaudhry U (1996) Tet Letts 37:6077
39. Chang HT, Chen C, Kondo T, Siuzdak G, Sharpless KB (1996) Angew Chem Int Ed Engl 35:182
40. Newkome GR (1994) Advances in dendritic macromolecules, vol 1. JAI Publ
41. Xu Z, Moore JS (1993) Angew Chem Int Ed Engl 32:1354
42. Wang P, Liu Y, Devandoss C, Bharathi P, Moore JS (1996) Adv Mater 8:237
43. Lorenz K, Holter D, Stuhn B, Mulhaupt R, Frey H (1996) Adv Mater 8:414
44. Launay N, Slany M, Caminade A, Majoral J (1996) J Org Chem 61:3799
45. Laio YH, Moss JR (1993) J Chem Soc Chem Commun 1774
46. Fillaut JL, Astruc D (1993) J Chem Soc Chem Commun 1320
47. Knapen JWJ, van der Made AW, de Wilde JC, van Leeuwen PWNM, Wijkens P, Grove DM, van Koten G (1994) Nature 372:659
48. Huck W, Veggel F, Kropman B, Keim E, Reinhoudt D (1995) J Am Chem Soc 117:8293
49. Lange P, Beruda H, Hiller W, Schmidbaur H (1994) Z Naturforsch 49:781
50. Newkome GR, Cardullo F, Constable EC, Moorefield CN, Thompson AMWC (1993) J Chem Soc Chem Commun 925
51. Denti G, Campagna S, Serroni S, Ciano M, Balzani V (1992) J Am Chem Soc 114:2944
52. Campagna S, Denti G, Serroni S, Ciano M, Juris A, Balzani V (1992) Inorg Chem 31:2982
53. Balzani V (1994) New Sci 12 November 32
54. Devonport W, Hawker CJ (1996) Polym News (in press)
55. Hawker CJ, Wooley KL, Fréchet JMJ (1993) J Am Chem Soc 115:4375
56. Mourey TH, Turner SR, Rubenstein M, Fréchet JMJ, Hawker CJ, Wooley KL (1992) Macromolecules 25:2401
57. Naylor AM, Goddard WA, Kiefer GE, Tomalia DA (1989) J Am Chem Soc 111:2339
58. Moreno-Bondi MC, Orellana G, Turro NJ, Tomalia DA (1990) Macromolecules 23:910
59. Godipas KR, Leheny AR, Caminati G, Turro NJ, Tomalia DA (1991) J Am Chem Soc 113:7335
60. Jin R, Aida T, Inoue S (1993) J Chem Soc Chem Commun 1260
61. Jiang D, Aida T (1996) J Chem Soc Chem Commun 1523
62. Tomoyose Y, Jiang D, Jin R, Aida T, Yamashita T, Horie K, Yashina E, Okamoto Y (1996) Macromolecules 29:5236
63. Sadamoto R, Tomioka N, Aida T (1996) J Am Chem Soc 118:3978
64. Hawker CJ, Wooley KL, Fréchet JMJ (1994) J Chem Soc Chem Commun 925
65. Nemoto H, Wilson JG, Nakamura H, Yamamoto Y (1992) J Org Chem 57:435
66. Ferguson G, Gallagher JF, McKervey A, Madigan E (1996) J Chem Soc Perkin Trans 1 599
67. Bhyrappa P, Young JK, Moore JS, Suslick KS (1996) J Am Chem Soc 118:5708
68. Zimmerman S, Zeng F, Reichert D, Kolotuchn S (1996) Science 271:1095
69. Gitsov I, Wooley KL, Hawker CJ, Fréchet JMJ (1991) Polym Prep 32:631
70. Gitsov I, Fréchet JMJ (1994) Macromolecules 27:7309
71. Gitsov I, Wooley KL, Fréchet JMJ (1992) Angew Chem Int Ed Engl 31:1200
72. Gitsov I, Wooley KL, Hawker CJ, Ivanova PT, Fréchet JMJ (1993) Macromolecules 26:5621
73. Gitsov I, Fréchet JMJ (1993) Macromolecules 26:6536
74. Hawker CJ, Fréchet JMJ (1992) Polymer 33:1507
75. Chapman TM, Hillyer GL, Mahan EJ, Schaffer (1994) J Am Chem Soc 116:11,195
76. van Hest JCM, Baars MWPL, Delnoye DAP, van Genderen MHP, Meijer EW (1995) Science 268:1592
77. Claussen W, Schulte N, Schluter AD (1995) Macromol Rapid Commun 16:89
78. Gitsov I, Fréchet JMJ (1996) J Am Chem Soc 118:3785
79. Chen Y, Chen C, Liu W, Xi F (1996) Macromol Rapid Commun 17:401
80. Gitsov I, Ivanova PT, Fréchet JMJ (1994) Macromol Rapid Commun 15:387
81. Leduc M, Hawker CJ, Dao J, Fréchet (1996) J Am Chem Soc 118 (in press)

82. Hawker CJ, Fréchet (1990) Macromolecules 23:4726
83. Wooley KL, Hawker CJ, Fréchet JMJ (1993) J Am Chem Soc 115:11,496
84. Hawker CJ, Wooley KL, Fréchet JMJ (1993) J Chem Soc Perkin Trans 1:1287
85. Atkinson NA, Hawker CJ (unpublished results)
86. Newkome GR, Moorefield CN, Baker GR, Saunders MJ, Grossman SH, (1991) Angew Chem Int Ed Engl 30:1178
87. Kim YH, Webster OW (1990) J Am Chem Soc 112:4592
88. Tomalia DA, Berry V, Hall M, Hedstrand DM (1987) Macromolecules 20:1164
89. Newkome GR, Young JK, Baker GR, Potter RL, Audoly L, Cooper D, Weis CD, Morris K, Johnson CS (1993) Macromolecules 26:2394
90. James TD, Shinmori H, Takeuchi M, Shinkai S (1996) J Chem Soc Chem Commun 705
91. Qualmann B, Kessels MM, Musiol HJ, Sierralta WD, Jungblut PW, Moroder L (1996) Angew Chem Int Ed Engl 35:909
92. Wooley KL, Hawker CJ, Pochan JM, Fréchet JMJ (1993) Macromolecules 26:1514
93. Kim YH, Webster OW (1992) Macromolecules 25:5561
94. Hawker CJ, Fréchet JMJ (1992) J Chem Soc Perkin Trans 1 2459
95. Roy R, Zanini D, Meunier SJ, Romanowska A (1993) J Chem Soc Chem Commun 1869
96. Miller LL, Hashimoto T, Tabokovoc I, Swanson DR, Tomalia DA (1995) Chem Mater 7:9
97. Bryce MR, Devonport W, Moore AJ (1994) Angew Chem Int Ed Engl 33:1761
98. Flory PJ (1952) J Am Chem Soc 74:2718
99. Hawker CJ, Lee R, Fréchet JMJ (1991) J Am Chem Soc 113:4583
100. Hawker CJ, Chu F (1996) Macromolecules 29:4370
101. Percec V, Chu P, Kawasumi M (1994) Macromolecules 27:4441
102. Voit BI (1995) Acta Polymer 46:87
103. Hawker CJ, Kambouris P (1993) J Chem Soc Perkin Trans 1:2717
104. Wooley KL, Hawker CJ, Lee R, Fréchet JMJ (1994) Polym J 26:187
105. Malmstrom E, Johannson M, Hult A (1995) Macromolecules 28:1698
106. Chu F, Hawker CJ (1993) Polym Bull 30:265
107. Turner SR, Voit BI, Mourey TM (1993) Macromolecules 26:4617
108. Turner SR, Walter F, Voit BI, Mourey TM (1994) Macromolecules 27:1611
109. Massa DJ, Shriner KA, Turner SR, Voit BI (1995) Macromolecules 28:3214
110. Uhrich KE, Hawker CJ, Fréchet JMJ, Turner SR (1992) Macromolecules 25:4583
111. Percec V, Kawasumi M (1992) Macromolecules 25:3843
112. Percec V, Chu P, Ungar G, Zhou J (1951) J Am Chem Soc 117:11,441
113. Kim YH (1992) J Am Chem Soc 114:4947
114. Mathias LJ, Carothers TW (1991) J Am Chem Soc 113:4043
115. Zhang Y, Wang L, Wada T, Sasabe H (1996) J Polym Sci Polym Chem 34:1359
116. Miller TM, Neenan TX, Kwock EW, Stein SM (1993) J Am Chem Soc 115:356
117. Spindler R, Fréchet JMJ (1993) Macromolecules 26:4809
118. Kumar A, Ramakrishnan S (1993) J Chem Soc Chem Commun 1453
119. Suzuki M, Ii A, Saegusa T (1992) Macromolecules 25:7071
120. Fréchet JMJ, Hemmi M, Gitsov I, Aoshima S, Leduc MR, Grubbs RB (1995) Science 269:1080
121. Hawker CJ, Fréchet JMJ, Grubbs RB, Dao J (1995) J Am Chem Soc 117:10,763
122. Huck WTS, van Veggel FCJM, Kropman BL, Blank DHA, Keim EG, Smithers MMA, Reinhoudt DN (1995) J Am Chem Soc 117:8293
123. Kim YH, Beckerbauer R (1994) Macromolecules 27:1968
124. Stutz H (1995) J Polym Sci Polym Phys 33:333
125. Fréchet JMJ, Hawker CJ, Wooley KL (1994) J M S Pure Appl Chem A31:1627
126. Fréchet JMJ, Hawker CJ (1996) Comprehensive polymer science (in press)

Received: August 1998

Macroporous Thermosets by Chemically Induced Phase Separation

Joachim Kiefer, James L. Hedrick*, Jöns G. Hilborn

Polymers Laboratory, Department of Material Sciences, Swiss Federal Institute of Technology, CH-1015 Lausanne, Switzerland
*IBM Research Division, Almaden Research Center, 650 Harry Road, San Jose California 95120–6099, USA
E-mail: joens.hilborn@epfl.ch

Macroporous polymers are treated in detail in this review, focusing on how to predict phase behavior and to prepare thermosetting materials. The formation of porous polymers is initially classified into methods of gas blowing, emulsion derived and phase separation. In chemically induced phase separation that involves a reactive polymer precursor and a solvent, phase separation is driven by the change in free energy of the system given by changes in enthalpy and entropy. Guidelines based on the components' molecular structure are given for the prediction and verification of phase behavior and hence the final porous morphology. The influence of the kinetics is followed by processing the two-phase polymer-liquid to give either isolated or co-continuous porosities. Finally the influence of dispersed liquid droplets or voids on polymer thermoset fracture properties is discussed with a closure of the article by a section on the lowering of dielectric constant through void incorporation.

Keywords. Porous polymer, Solubility parameter, Phase separation, Epoxy, Cyanurate, Toughness, Image analysis, Morphology

List of Symbols and Abbreviations 163

1   Introduction to Foams and Macroporous Polymers 164

1.1   Gas Derived Polymeric Foams 164
1.2   Emulsion Derived Polymeric Foams 166
1.3   Porous Polymers via Phase Separation Processes 167

2   Chemically Induced Phase Separation (CIPS) 168

2.1   Procedure for the Preparation of Solvent-Modified
      and Macroporous Epoxy Networks via Chemically Induced
      Phase Separation (CIPS) 169
2.2   Thermodynamic Aspects of Phase Separation Processes 170
2.2.1 Flory-Huggins Theory 172
2.3   Types of Phase Separation Processes and Nomenclature 173
2.3.1 Thermally Induced Phase Separation 174
2.3.2 Chemically Induced Phase Separation 175

2.4	Calculation of Schematic Phase Diagrams for the Preparation of Solvent-Modified and Macroporous Thermosets via CIPS	177
2.5	Guidelines for the Preparation of Solvent-Modified and Macroporous Thermosets via CIPS	181
3	**Selection Criteria for the Preparation of Solvent-Modified and Macroporous Epoxy Networks with Tailored Morphologies Prepared via CIPS**	**183**
3.1	Gradient Oven for the Determination of Phase Separation Lines	183
3.2	Influence of the Solubility Parameter on the Phase Separation Behavior	184
3.3	Reconstruction of Real Phase Diagrams Based on the Solubility Parameter Approach	189
4	**Morphology Development of Solvent-Modified and Macroporous Epoxy Networks Prepared via CIPS**	**194**
4.1	Image Analysis of Macroporous Epoxy Networks	197
4.2	Phase Separation Mechanism in Hexane-Epoxy Systems	203
4.3	Influence of Reaction Parameters on the Morphology of Cyclohexane-Modified Epoxy Networks Prepared via CIPS	206
4.3.1	Influence of Solvent Concentration	207
4.3.2	Influence of Curing Temperature	209
4.3.3	Influence of Drying Procedure	211
4.3.4	Influence of Reaction Rate	213
4.4	Characteristics of Epoxy Networks Prepared with Hexane and Cyclohexane via CIPS	216
5	**Toughness of Solvent-Modified and Macroporous Epoxies Prepared via CIPS**	**218**
5.1	Calculation of Stress Distribution in Macroporous Epoxies	223
5.2	Fracture Toughness of Solvent-Modified and Macroporous Epoxies Prepared via CIPS	227
6	**Macroporous Cyanurate Networks**	**235**
6.1	Materials Selection and Chemistry of Cyanate Ester Resins	236
6.2	Morphology, Thermal and Dielectric Properties	237
7	**Conclusions**	**242**
References		**243**

List of Symbols and Abbreviations

a	parameter introduced in the relationship between the Flory-Huggins interaction parameter and the absolute temperature
A	surface area used for image analysis
b	slope of the relationship between the Flory-Huggins interaction parameter and the inverse of absolute temperature
c	concentration of pores from image analysis
d	mean diameter of narrow size from image analysis
d_1, d_2	mean diameters of distinguished peaks of bimodal size distributions
D	diffusion constant
IPD	interparticle distance (surface to surface distance)
n	total number of domains used for image analysis
n_i	molar number of component i
N_0	number of nuclei at the start of phase separation
$\overline{M}_n$	number average molar mass
$\overline{M}_{n0}$	number average molar mass of precursor mixture
q	conversion, extent of reaction
q_{nucl}	critical extent of reaction for nucleation
r	radius
R	universal gas constant
t	surface roughness
T	absolute temperature
T_g^∞	ultimate glass transition temperature of thermoset network
V_i	molar volume of component i
V_r	molar volume of the unit cell serving as reference volume
W	number of possible arrangements
x	integer used in Flory's lattice model
x_1, x_2	number fractions of a bimodal size distribution
z_i	ratio of the molar volume of component i with respect to the reference volume
ΔE_v	molar energy of vaporization
ΔG	Gibbs free energy of mixing
ΔG^V	Gibbs free energy of mixing per unit volume
ΔG_n	nucleation energy
ΔH	enthalpy of mixing
ΔH^V	enthalpy of mixing per unit volume
ΔS	entropy of mixing
ΔS^V	entropy of mixing per unit volume
χ	Flory-Huggins interaction parameter
$\vec{\delta}_i$	vectorial representation of solubility parameter of component i
μ_i	chemical potential of component i
ϕ_0	initial volume fraction of solvent
ϕ_c	critical volume fraction of phase inversion given by the interception of binodal and spinodal line

ϕ_p	critical volume fraction for phase separation
ϕ_i	volume fraction of component i
v_i	volume of one polymer chain
v_{sol}	volume of one solvent molecule
ρ	density of neat, fully crosslinked thermoset
ρ_{por}	density of macroporous thermoset
CIPS	chemically induced phase separation
DGEBPA	diglycidyl ether of bisphenol-A
LCST	lower critical solution temperature
PACP	2,2'-bis(4-amino-cyclohexyl)propane
PMMA	poly(methyl methacrylate)
PS	poly(styrene)
PSD	pore size distribution
SEM	scanning electron microscopy
TIPS	thermally induced phase separation
UCST	upper critical solution temperature

1
Introduction to Foams and Macroporous Polymers

It is our intention to present strategies based on *chemically induced phase separation* (CIPS), which allow one to prepare porous thermosets with controlled size and distribution in the low μm-range. According to IUPAC nomenclature, porous materials with pore sizes greater than 50 nm should be termed *macroporous* [1]. Based on this terminology, porous materials with pore diameters lower than 2 nm are called *microporous*. The nomination *mesoporous* is reserved for materials with intermediate pore sizes. In this introductory section, we will classify and explain the different approaches to prepare porous polymers and to check their feasibility to achieve *macroporous thermosets*. A summary of the technologically most important techniques to prepare polymeric foams can be found in [2, 3].

Among the various techniques for the preparation of macroporous polymers, one can distinguish, in the main, three different routes: The first involves the use of gases as the void-forming medium. The second approach is based on the intermediate of emulsions, formed by tailor-made block-copolymers and the subsequent removal of the dispersed phase. The third category is classified by the use of a phase separation process to generate a two-phase morphology, finally resulting in a porous morphology.

1.1
Gas Derived Polymeric Foams

Gases can be generated during the processing of polymers by using blowing agents to yield polymeric foams widely used as insulating and packaging materials. Shutov distinguishes between physical and chemical blowing agents [4]. According to this classification, chemical blowing agents are compounds which deliver a gas as a result of the chemical interaction between several compounds or a thermal decomposition reaction. Physical blowing agents are substances which can easily undergo a transition from the liquid to the gas state either by a pressure reduction or by a temperature increase. The temperature increase can result from external sources or exothermic heat development during a crosslinking reaction. One important class of such physical blowing agents were the chlorofluorocarbons. Nowadays, they are known to destroy the ozone layer in the stratosphere. Since the world-wide banning of chlorofluorocarbons, the foam industry has been looking for new methods to avoid blowing agents and is using environmental friendly as well as economically viable gases, such as carbon dioxide, nitrogen, or simply air, to produce polymeric foams.

One alternative is to select precursors which form a gas as a reaction product in situ during the network formation of thermosets. However this approach is restricted to a very limited number of precursors reacting via a polycondensation mechanism to split off a gas. For example, flexible polyurethane foams are commercially produced using CO_2 that is liberated as a reaction product of the isocyanate monomer with water [5]. Very recently, Macosko and coworkers studied the macroscopic cell opening mechanism in polyurethane foams and unraveled a microphase separation occurring in the cell walls. This leads to nanosized domains, which are considered as hard segments and responsible for a rise in modulus after the cell opening [6].

A wide variety of microcellular thermoplastic foams, such as polypropylene, polystyrene, polycarbonate, polyvinylchloride, polyethyleneterephthalate, or acryl-butadiene-styrene with bubble densities up to 10^9 cells/cm^3 can be prepared via gas nucleation [7–9]. This method consists of a two-step procedure. In the first step the thermoplastic polymer is saturated with a non-reactive gas such as carbon dioxide or nitrogen at fairly elevated pressures. Upon saturation, the polymer is removed from the pressure reactor in order to produce a supersaturated sample. This supersaturated polymer is then heated to a temperature near T_g in the glassy state, thus inducing nucleation and growth of gas bubbles. The careful adjustment of foaming parameters such as temperature and time allows one to control the size of the pores within relatively narrow limits with average sizes typically 10–50 µm, and enables one to produce products with complex shapes [8]. Owing to the lower gas solubility and the fact that crystals prevent nucleation, semicrystalline polymers are difficult to foam by this procedure [7], and the gas nucleation technique is therefore limited to preparing amorphous thermoplastics.

When studying the plasticization effect of CO_2 in polymethylmethacrylate (PMMA) [10], Goel and Beckman explored a different strategy for using CO_2 to produce microcellular PMMA. They saturated the polymer with CO_2 at much higher pressures (25–35 MPa), thus placing the system in the supercritical region, where both components are miscible. By carrying out a rapid pressure quench, the two components become immiscible and start to phase separate, resulting in a highly opened porosity [11, 12]. Currently DeSimone and coworkers intend to prevent the agglomeration of pores by synthesizing tailored surfactants with CO_2-philic blocks such as fluoropolymers or silicones [13, 14].

Mainly due to the high diffusion rates of gases in polymers, the above techniques – apart from gas nucleation – provide foams with an open porosity and a very large pore size distribution, often ranging over several orders of magnitude. Furthermore, the in situ development of a gas during the processing reaction is very sensitive to the processing conditions. Consequently, it is difficult to control the final morphology. Even though some of the above techniques in principle also allow one to produce porous thermosets, gas derived foams cannot give the desired narrow size distribution in the low µm-range.

1.2
Emulsion Derived Polymeric Foams

Further techniques that also provide for polymeric foams are based on emulsion techniques. Depending on the ratios of the inner phase, the outer phase and the surfactant, either micellar, cylindrical, lamellar, or other structures can be formed. Such self-assembling systems are of substantial interest to design new materials with structural arrangements on a molecular level suitable for advanced applications [15]. In this strategy, the inner phase consists of a volatile liquid or polymeric substance, and the outer phase is a polymerizable monomer, thus leading to a rigid matrix upon polymerization. A foam is obtained upon removal of the inner phase, which can be realized by evaporation, thermal decomposition, or extraction with a convenient solvent. As the volume fraction of the inner phase exceed the highest density packaging of spheres (74%), the inner phase can no longer stay separated, resulting in the formation of interconnected, polyhedral cells. Ruckenstein and coworkers applied this concentrated emulsion pathway to synthesize porous polymers having highly interconnected pores [16]. The volume fraction of the continuous phase can become as low as 1%. However such extremely high porosities require that the continuous phase must be polymerized to ensure the mechanical stability of the membrane [17–19].

Emulsion derived foams prepared via the concentrated emulsion pathway are characterized by highly interconnected pores, thus offering density values as low as 0.02 g/cm^3 and a relatively narrow size distribution in the µm-range resulting from a thermodynamically stable system. This principle allows for the synthesis of organic as well as inorganic foams that offer a wide range of applications [20, 21]. Recently such technique has been applied to form injectable siloxane foams where the emulsified liquid was removed supercritically in order to avoid pore collapse [22].

In contrast to the highly interconnected pores mentioned previously, closed pores can also be obtained by microemulsion polymerization if the initial volume fraction of the dispersed phase is kept lower than 30%. Recently two systems have been reported where the polymerization of the continuous phase and the subsequent removal of the liquid dispersed phase resulted in the formation of porous thermoplastic materials such as PS [23] or PMMA [24–26]. Both groups observed independently that the initial morphology could not be retained as the final pore size was several µm, whereas the domain size in microemulsions is usually in the range 10–100 nm. As the final morphology results from the uncontrolled breakup of the microemulsion, the emulsion polymerization route does not seem to offer a satisfactory way to produce porous thermosets with a controlled closed cell morphology in the µm-range.

Very well defined porous structures having a closed cell morphology in the nm-range can also be obtained from the self-assembly of tailor-made block copolymers. Hedrick and coworkers have explored a general methodology, enabling the synthesis of "nanoporous" polyimides [27–35]. The well controlled porosity results from the self-assembly of triblock or graft copolymers consisting of a thermally labile and thermally stable block. Thermal degradation of the thermally labile inner phase leads to a polymeric matrix containing well dispersed voids. SAXS [35] and TEM [36] measurements revealed closed pores with sizes ranging from 5 to 20 nm, depending on the block length of the thermally labile block. This mesoporous structure led to materials with significantly lower density and dielectric constant, thus offering potential for applications in microelectronics [37, 38].

1.3
Porous Polymers via Phase Separation Processes

Porous polymers can also be derived from a phase separation process by carrying out a temperature quench. Porous PS foams with densities as low as 0.02–0.2 g/cm^3 are produced by phase separation starting from a PS-cyclohexane system [39, 40]. The phase diagram of this particular system is well known and exhibits a so-called upper critical solution temperature. Above the critical solution temperature the cyclohexane and polystyrene are miscible. A phase separation is initiated by cooling below the binodal or spinodal line, thus resulting in a two-phase morphology. This method has later been termed *thermally induced phase separation* (*TIPS*), as the phase separation is initiated by a thermal quench [41, 42]. Controlling the morphology requires a detailed knowledge of the phase diagram as well as the kinetics and thermodynamics of the different phase separation mechanisms. Depending on the quench rate, the phase separation proceeds either via nucleation and growth or via spinodal decomposition, resulting in different morphologies. The theoretical backgrounds of TIPS have much in common with CIPS, and these aspects will be further outlined in the next section. After phase separation, a porous morphology is achieved by evaporation of the solvent. Thus the pore size can be varied from less than 1 µm up to around 100 µm. The low volume fraction of the polymeric phase leads to a highly interconnected, porous structure. This technique allows only for the preparation of films less than 1 mm in thickness that can be used as membranes.

Polymeric membranes with highly interconnected pores have been produced commercially for more than 30 years based on a phase inversion process. These phase inversion membranes were first used for the desalination of sea water [43]. Phase inversion membranes have later entered diversified applications ranging from oil recovery to dialysis [44, 45]. In the strategy of the phase inversion process a polymer-solvent mixture is immersed into a non-solvent bath. Precipitation leads to the formation of a polymer-rich and a solvent-rich phase. The phase separation results from the solvent interchange and the simultaneous temperature quench caused by the immersion into the non-solvent bath. Despite the very large industrial use and economic importance of these products, the pore formation mechanisms are only moderately understood, and explanations of the structure development are rather qualitatively discussed than theoretically confirmed. Nevertheless the influence of processing parameters on the structure development is quite well understood and the morphology of the membranes can be varied from a sponge-like structure to a finger-like structure [46, 47].

Porous materials used for chromatography result from a chemically induced phase separation using chain-wise polymerization of vinyl-containing monomers crosslinked with a portion of divinyl functional monomers. Fréchet has improved this technique for the preparation of porous PS beads [48]. In this approach the inner phase consists of a mixture containing the reactive styrene and divinylbenzene monomers as well as an unreactive polymeric porogen. After polymerization, the soluble polymeric porogen is removed, leaving behind macroporous beads with pore sizes of around 100–500 nm.

The above summary shows that there is gap in existing techniques and none can provide thermosets having closed pores with a narrow size distribution in the µm-range. However, we will combine the knowledge of the above techniques to develop a new, general strategy to give macroporous thermosets. The basic ideas and theoretical backgrounds of this new technology are explained in Sect. 2.

2
Chemically Induced Phase Separation (CIPS)

Thermoset materials with a controlled morphology in the low µm-range are found in toughened thermosets where the second polymeric phase has been generated via a phase separation process. Indeed the methodology we sought to

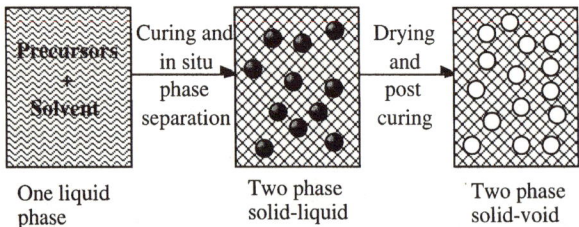

Fig. 1. General strategy of the *Chemically Induced Phase Separation* (CIPS) technique to prepare macroporous thermosets having a closed cell morphology

use is very similar, as we intend to use a non-reactive low molecular weight liquid instead of using a reactive, oligomeric moiety to phase separate. The choice of the liquid is crucial, as it must be a moderately good solvent for the precursor monomers to allow the components to be miscible in the uncured state, thus giving initially a homogenous mixture. During the curing reaction, the low molecular weight liquid should turn into a non-solvent and start to phase separate into discrete liquid domains. At this stage the system consists of two phases – a solid, highly crosslinked matrix containing dispersed liquid droplets. This liquid, being volatile, can subsequently be removed by a diffusion process by holding the system in the rubbery state of the crosslinked network and above the boiling point of the liquid. This should ideally result in the formation of pores with sizes and distributions which result from the phase separation process. Hence neither ripening nor collapse should occur upon removal of the low molecular weight liquid. The principle idea of this so called *Chemically Induced Phase Separation* (CIPS) technique is shown in Fig. 1.

2.1
Procedure for the Preparation of Solvent-Modified and Macroporous Epoxy Networks via Chemically Induced Phase Separation (CIPS)

The precursor monomers used to build up a highly crosslinked network are shown in Scheme 1 and consist of diglycidyl ether of bisphenol-A (DGEBPA) and 2,2'-bis(4-amino-cyclohexyl)propane (PACP).

In addition to the general procedure of the CIPS technique described in Fig. 1, the principle experimental procedure for the preparation of solvent-modified and macroporous epoxy networks via the CIPS technique is given below [49, 50].

For fundamental studies on the preparation of macroporous epoxies via CIPS an initial amount of around 2–3 g of *DGEBPA* was sufficient to realize reproducible samples suitable for further characterizations. The curing agent *PACP* was then added in a stoichiometric amount considering that the diamine can react with four epoxy groups.

The desired volume of low molecular weight liquid was added and the three components – epoxy, curing agent, and low molecular weight liquid – were mixed together under gentle stirring at room temperature, until a clear and ho-

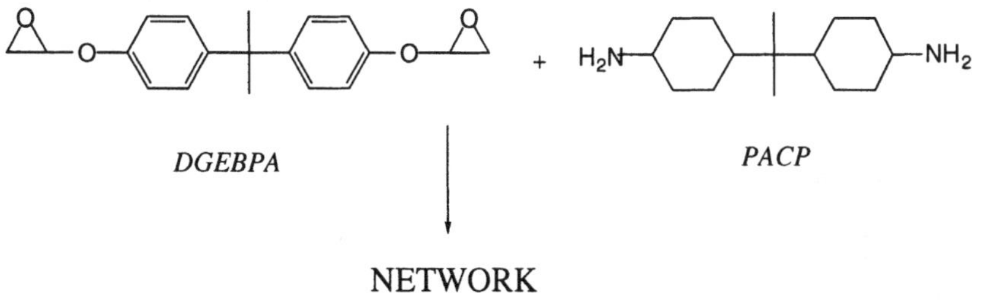

DGEBPA + PACP → NETWORK

Scheme 1

mogeneous solution was obtained. This one-phase system gave the starting system. Part of this clear solution was then transferred into a glass tube sealed at one end. This glass tube was then placed in liquid nitrogen in order to cool the one-phase system below the melting point of the pure low molecular weight liquid and attached to a vacuum line. The glass tube was then sealed in the upper part under an applied vacuum. Hence one obtained a closed system thus allowing simulation of real processing conditions. The sealed glass tube was then placed in a pre-heated sandbath for curing. Alternatively, the well controlled conditions using a glass tube could very well be exchanged for a standard closed mold with the only difference being that the curing temperature may not exceed the boiling point of the liquid (be careful with exotherms!). Hence, any shape and size of object could be envisaged using standard thermoset processing technologies. If the appropriate thermodynamic conditions were fulfilled, phase separation took place leading to a two-phase system consisting of a rigid, highly crosslinked network forming the matrix and separated domains, which were liquid. Such a system will be denoted as *solvent-modified* in the following.

After curing, the sample was removed from the glass tube or mold and heated to above the ultimate glass transition temperature, T_g^∞, in a vacuum oven. The liquid was simply removed by holding the system at this temperature until constant weight was reached. A temperature above T_g^∞ has been selected to ensure that the material remained in the rubbery state, thus allowing for sufficient diffusion rates. However the temperature should not be too high to avoid thermal degradation and char formation. The final materials consisting of a highly crosslinked matrix and voids will be termed macroporous networks.

This experimental procedure to synthesize macroporous thermosets via *CIPS* is quite straightforward and similar to the well established method to yield toughened polymers prepared via a phase separation process. This process differs from the standard polymer-polymer phase separation in the ease of solvent diffusion resulting in droplet formation and in the subsequent pore formation step. The difficult experimental part is to find the correct reaction parameters to achieve the envisaged phase separation for a given precursor system. The first criterion is to find a suitable solvent, which will turn into a non-solvent during the curing reaction, and to investigate the influence of reaction parameters on the morphology and the reproducibility of the results. The experimental search for polymer-solvent systems undergoing phase separation is a time-consuming process. In the next sections we will present theoretical guidelines which should simplify the solvent selection while simultaneously allowing one to tailor the morphology.

2.2
Thermodynamic Aspects of Phase Separation Processes

This technique is called *Chemically Induced Phase Separation* (CIPS) as the thermodynamic origin of the phase separation can be regarded as a chemical quench, as will be explained below.

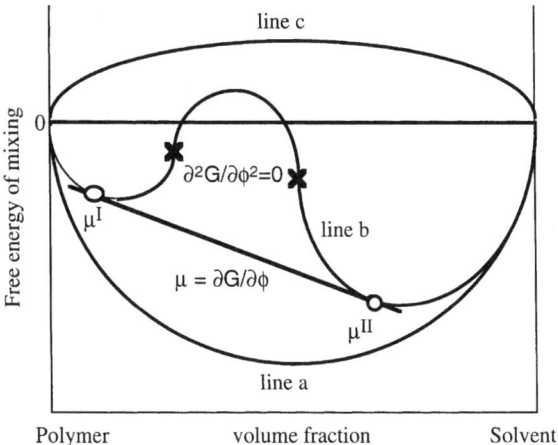

Fig. 2. Free energy curves corresponding to miscibility (*line a*), phase separation (*line b*), and immiscibility (*line c*)

From a thermodynamic viewpoint, any type of phase separation is the result of a change in the free energy of the system, ΔG [51–53]. ΔG can be separated into an enthalpy term, ΔH, resulting from the changes in intermolecular interactions, and an entropic contribution associated with configurational changes, ΔS.

The change in the free energy is generally expressed by the Gibbs equation:

$$\Delta G = \Delta H - T \cdot \Delta S \tag{1}$$

In polymeric systems exhibiting a high degree of flexibility, the change in the entropy often becomes dominating for polymer miscibility and phase separation processes. Thus it is possible that mixing takes place spontaneously even though the enthalpy of mixing is slightly endothermic.

A phase separation can be described just with regard to the shape of free energy curves for several states without the detailed knowledge of the expressions for ΔH and ΔS. Let us consider that two components are completely miscible. This will be the case if ΔG is negative over the entire composition range. This situation corresponds to line a in Fig. 2. The inverse case of complete immiscibility is thermodynamically given by positive ΔG values as shown by line c. A phase separation is equivalent to the transition from the miscible to the immiscible state. Mathematically this implies a change in the curvature of the ΔG-function. In this intermediate state (line b), the ΔG-function is characterized by a curve displaying two minima, one maximum and two inflection points. The inflection points are given by the points where the second derivation vanishes, thus:

$$\frac{\partial^2 \Delta G}{\partial \phi^2} = 0 \tag{2}$$

This is the thermodynamic condition for spinodal decomposition. For any composition in between the two inflection points $\partial^2 \Delta G/\partial \phi^2 < 0$ and the initially homogeneous mixture will split up into two phases with compositions given by the inflection points, thus minimizing the total free energy.

Let us now consider a system composed of a polymer and a solvent. For compositions in between the inflection points, solvent molecules will diffuse into the solvent-rich phase, and the polymer molecules diffuse in the polymer-rich phase. Thus diffusion occurs against a concentration gradient. Therefore, this type of phase separation is known as up-hill diffusion. The up-hill diffusion leads to a spontaneous decomposition and it is therefore also named spinodal decomposition. The formation of two phases via spinodal decomposition occurs immediately upon reaching the spinodal decomposition region and does not require any activation energy.

In the intermediate state (line b) the free energy curve also delivers the points that satisfy the thermodynamic conditions for equilibrium [51–54]. Thus the chemical potentials of each component (polymer or solvent) must be the same in both phases (I and II). For a polymer-solvent system, this is mathematically expressed by

$$\mu_{Pol}^{I} = \mu_{Pol}^{II}$$
$$\mu_{Sol}^{I} = \mu_{Sol}^{II} \qquad (3)$$

The chemical potential, μ_i, of phase i is defined as

$$\mu_i = \partial \Delta G / \partial n_i = \frac{\partial \Delta G}{\partial \phi_i} \cdot \frac{\partial \phi_i}{\partial n_i} \qquad (4)$$

Graphically, the conditions for thermodynamic equilibrium are equal to two points which have a common tangent. These points give the composition of a polymer-rich phase *(I)* and a solvent-rich phase *(II)* that can coexist in thermodynamic equilibrium. The summation of such points is also called the coexistence curve or *binodal line*.

2.2.1
Flory-Huggins Theory

Any parameter that influences either the enthalpy or the entropy of the system might change the free energy in such a manner that phase separation occurs. For a system like those studied here, being composed of a polymer and a solvent, the entropy of mixing, ΔS, has been calculated independently by Flory [53, 55, 56] and Huggins [57, 58] based on the lattice model. It is assumed that each solvent molecule occupies one cell of the lattice. The volume of the smallest unit serves as reference volume. It is stated that the macromolecule can be composed of adjacent segments with volumes identical to the solvent molecule. Thus the volume occupied by the macromolecule is given by

$$v_{Pol} = x \cdot v_{Sol} \tag{5}$$

where x is an integer and v_{Pol} and v_{Sol} represent the volume of one polymer chain and one solvent molecule, respectively. Several assumptions were necessary to calculate the number of possible arrangements. Thus, it was assumed that each site can be occupied only by either the solvent or the macromolecule. For the statistical calculations, the lattice was first filled with the polymer chains, and the solvent molecules were then placed in the remaining free sites after the complete filling process. This treatment requires that the size of the second component is very small like a solvent molecule, thus v_{Pol} must be much higher than v_{Sol}. This assumption is fulfilled for polymer-solvent systems used in the CIPS technique to prepare macroporous thermosets. Furthermore, all polymer molecules were assumed to have the same size and the placement of adjacent macromolecules should not lead to concentration differences in the lattice. These calculations finally lead to

$$\Delta S = -R \cdot \{n_{Pol} \cdot \ln \phi_{Pol} + n_{Sol} \cdot \ln \phi_{Sol}\} \tag{6}$$

where n_{Pol}, n_{Sol} represent the molar numbers of polymer and solvent respectively and ϕ_{Pol}, ϕ_{Sol} their volume fractions. R is the gas constant.

The volume fractions of polymer and solvent are defined as

$$\phi_{Pol} = \frac{n_{Pol} \cdot V_{Pol}}{V_T}; \quad \phi_{Sol} = \frac{n_{Sol} \cdot V_{Sol}}{V_T} \tag{7}$$

Therein V_{Pol} and V_{Sol} represent the molar volumes of polymer and solvent respectively, and V_T the total volume which is given by

$$V_T = n_{Pol} \cdot V_{Pol} + n_{Sol} \cdot V_{Sol} \tag{8}$$

In addition to the calculations of changes in the free entropy mixing, Flory introduced the interaction parameter, χ, to account for the intermolecular interactions between polymer and solvent molecules, thus giving [53]

$$\Delta H = R \cdot T \cdot \chi \cdot n_{Sol} \cdot \phi_{Pol} \tag{9}$$

The combination of entropy and enthalpy changes finally lead to the well-known Flory-Huggins equation [53]:

$$\Delta G = R \cdot T \cdot \{n_{Pol} \cdot \ln \phi_{Pol} + n_{Sol} \cdot \ln \phi_{Sol} + \chi \cdot \phi_{Pol} \cdot n_{Sol}\} \tag{10}$$

2.3
Types of Phase Separation Processes and Nomenclature

The Flory-Huggins equation (Eq. 10) allows one to reconstruct schematic phase diagrams to express the phase separation behavior as discussed below.

According to the Flory-Huggins Equation (Eq. 10) the free energy of mixing is a function of the temperature, the interaction parameter, and the volume fraction of polymer and solvent. A phase separation can result if ΔG changes its curvature as a consequence of the variation of one or several parameters contributing to the free energy. This has led to several possibilities of initiating phase separation processes depending on the variable that causes the change in the free energy, as explained below.

2.3.1
Thermally Induced Phase Separation

One variable in the Flory-Huggins equation is the temperature. Thermally induced phase separation is generally used as the expression to characterize all techniques where phase separation results from temperature changes. The occurrence of phase separation and the shape of the phase diagram depend strongly on the interaction parameter, which is specific for each system. The interaction parameter does not have a constant value, but it depends on temperature [59], composition [60], and also pressure [61]. Several experimental methods exist to determine the interaction parameter. These are based on techniques like either light- or neutron scattering and osmometry for high dilutions, or inverse gas chromatography near the polymer melt [59, 62].

Two types of phase diagrams, which are of great technological importance, derive from the simplification that the interaction parameter varies lineary with the inverse of temperature:

$$\chi = a + \frac{b}{T} \tag{11}$$

This relationship, which is most widely used to take into account the temperature dependence of the interaction parameter, was first confirmed experimentally by Huggins [58].

An upper critical solution temperature (*UCST*) behavior is obtained from Eqs. (10) and (11) if the value of χ is positive and increases linearly with *1/T* [63]. The resulting phase diagram is schematically represented in Fig. 3a. If χ is negative and decreases linearly with *1/T*, the system displays a *lower critical solution temperature* (*LCST*) behavior, as shown in Fig. 3b. The schematic phase diagrams shown in Fig. 3 contain two lines and several regions: The inner line is called the *spinodal line* and the outer line the *binodal line*. The binodal line results from the free energy curve by interconnecting all the points having a common tangent as a function of the temperature. Hence this represents the equilibrium or coexistence curve. Entering the *metastable region*, which is limited by the spinodal and binodal line, will initiate phase separation which will proceed via a *nucleation and growth mechanism*. Similar to the construction of the binodal line, the spinodal line results from the summation of inflection points of free energy curves as a function of temperature. If the area enclosed by the spinodal line is entered, the phase separation will take place via *spinodal decomposi-*

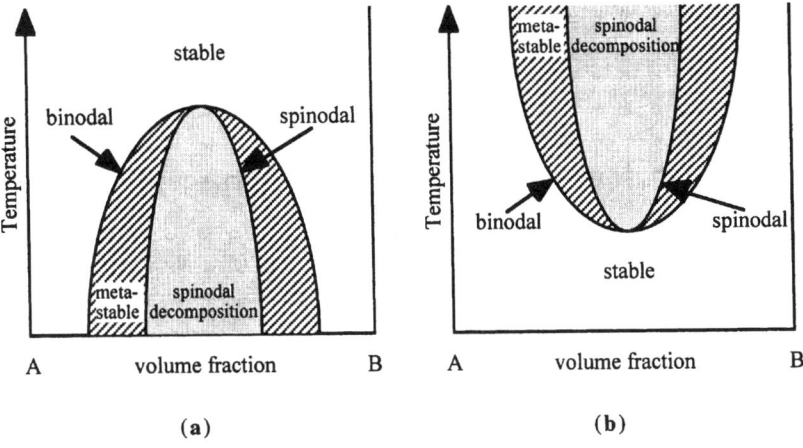

Fig. 3a,b. Schematic phase diagrams displaying: **a** upper critical solution temperature (*UCST*) behavior; **b** lower critical solution temperature (*LCST*) behavior

tion. Such schematic phase diagrams derived from the Flory-Huggins equation are very useful to explain the morphology development upon phase separation.

The area outside the binodal line is called the *stable region*. Here the two components are completely miscible. Starting from the stable region a phase separation is achieved, if the temperature is lowered (*UCST*) or raised (*LCST*) in such a manner that the metastable region or the domain for spinodal decomposition are entered. The transition from one area of a phase diagram into another is called quench. Any technique where phase separation results from a temperature quench is therefore known as *thermally induced phase separation* (*TIPS*). The phase separation mechanism and the final morphology depend on the region that is entered during the temperature quench. However, it is difficult to reconstruct these phase diagrams, as this requires an accurate description of the interaction parameter as a function of temperature, composition, and molecular weight of the polymer [64].

The *TIPS* technique is largely applied to yield a wide variety of polymer blends deriving from an *UCST* or *LCST* behavior [65] and has also proven to be useful for the preparation of porous thermoplastic polymers by the use of a phase separating solvent [41, 42].

2.3.2
Chemically Induced Phase Separation

A change in the free energy resulting in phase separation can also be achieved by continuously changing the molecular weight of the polymer, thus contributing to changes in entropy. Such a change in miscibility is largely applied to the synthesis of thermoplastic polymers, where the polymerization is carried out in a solvent. The polymerization product, having a high molecular weight and being non-soluble, does precipitate and can be separated from the low molecular weight products. During the curing of thermosetting polymers a highly crosslinked network structure is generated. The curing process is accompanied by an increase in molecular weight and can be used to initiate a phase separation, thus giving a two-phase morphology. This procedure is, for instance, applied to produce toughened thermosets or polymer dispersed liquid crystals. If it is desired to reconstruct phase diagrams, it becomes necessary to express changes in entropy and enthalpy as a function of conversion.

Let us first consider the changes in entropy and calculate the free entropy of mixing per unit volume, ΔS^V, instead of ΔS [66–68].

Combination of Eqs. (6)–(8) then gives

$$\Delta S^V = \frac{\Delta S}{V_T} = -R \cdot \left\{ \frac{\phi_{Pol}}{V_{Pol}} \cdot \ln \phi_{Pol} + \frac{\phi_{Sol}}{V_{Sol}} \cdot \ln \phi_{Sol} \right\} \qquad (12)$$

Let us now introduce

$$z_{Pol} = \frac{V_{Pol}}{V_0} \quad \text{and} \quad z_{Sol} = \frac{V_{Sol}}{V_0} \qquad (13)$$

where V_0 is the molar volume of precursor monomers used to build up a thermosetting network. It is then assumed that V_0 is equivalent to the reference volume used in the original lattice model. Under this condition the molar volumes are linked to the number average molecular weight by [67–69]

$$z_{Pol} = \frac{V_{Pol}}{V_0} = \frac{\overline{M}_n}{\overline{M}_{n0}} \qquad (14)$$

The number average molecular weight of the precursor monomer mixture, $\overline{M}_{n0}$, is determined by the molecular weights and ratio of the precursor monomers. The number average molecular weight distribution, $\overline{M}_n$, of a thermosetting polymer reacting by polycondensation can be derived from statistical calculations as a function of the extent of reaction. The resulting equations depend strongly on the type of reaction and the functionality of the precursor monomers [70]. In our case, the curing reaction of bifunctional epoxy resins with tetrafunctional diamines, the molecular weight distribution is linked to the extent of reaction or conversion, q, by [67, 68, 71]

$$\overline{M}_n = \overline{M}_{n0} / \left(1 - \frac{4}{3} q \right) \qquad (15)$$

It is important to note that this statistical calculation is only valid as long as the kinetics of network formation is totally controlled by the reactivity between the precursor monomers. With the formation of an infinite network at gelation and corresponding increase in viscosity, the reaction is slowed down considerably. Consequently, Eq. (15) is only valid prior to gelation.

Combination of Eqs. (12)–(15) leads to

$$\Delta S^v = \frac{R}{V_0} \cdot \left\{ \left(1 + \frac{4}{3}q\right) \cdot \phi_{Pol} \cdot \ln\phi_{Pol} + \frac{\phi_{Sol}}{z_{Sol}} \cdot \ln\phi_{Sol} \right\} \qquad (16)$$

Similar to the entropic contribution, one can also express the free enthalpy of mixing per unit volume.

Disposing the Flory-Huggins modified equation, including the free entropy of mixing per total volume, ΔS^v, as a function of conversion and the enthalpy term expressed with the interaction parameter [66–68, 72]:

$$\Delta G^V = \frac{\Delta G}{V_T} = \frac{RT}{V_0} \cdot \left\{ \left(1 - \frac{4}{3}q\right) \cdot \phi_{Pol} \cdot \ln\phi_{Pol} + \frac{\phi_{Sol}}{z_{Sol}} \cdot \ln\phi_{Sol} + \chi \cdot \phi_{Pol} \cdot \phi_{Sol} \right\} \qquad (17)$$

the complete phase diagrams allowing for prediction of the phase behavior could be constructed. Williams et al. have widely applied this strategy to calculate ΔG^V-curves and explain the phase separation behavior of epoxy and cyanurate networks modified with rubbery and thermoplastic polymers [66–69, 71, 73, 74]. Investigating the equation to calculate ΔG^V as a function of conversion, one finds that only the interaction parameter χ is missing. However, the experimental determination of χ requires experimental techniques such as inverse gas chromatography [75, 76] or neutron scattering [59, 60, 77] and it is well known, that χ depends on numerous parameters, such as temperature [53, 58, 59, 63, 77], composition [60, 64, 78], and pressure [61].

2.4
Calculation of Schematic Phase Diagrams for the Preparation of Solvent-Modified and Macroporous Thermosets via CIPS

As an alternative to the use of Flory's interaction parameter, we intend to express ΔH based on the solubility parameter, δ, being slightly limited in accuracy but superior in simplicity, needing only pen and paper. According to its definition, δ is equal to the square root of the molar energy of vaporization, ΔE_v, per unit molar volume, thus [79]

$$\delta = \sqrt{\frac{\Delta E_v}{V}} \qquad (18)$$

ΔE_v can be determined from calorimetry for organic liquids. Even though it cannot be determined experimentally for macromolecules, the value for the solubility parameter can be calculated based on individual group contributions for nearly any type of macromolecules [80].

According to the Hildebrand concept, the free enthalpy of mixing ΔH is given by [79]

$$\Delta H = V_T \cdot \left(\delta_{Pol} - \delta_{sol}\right)^2 \cdot \phi_{Pol} \cdot \phi_{Sol} \qquad (19)$$

According to this definition ΔH cannot take negative values and miscibility is favored if the differences between the solubility parameters of polymer, δ_{Pol}, and solvent, δ_{Sol}, are small. This concept has been largely applied to check for the miscibility between polymer-solvent pairs. An initial miscibility is crucial in the strategy of *CIPS*, where the low molecular weight liquid should convert from a solvent into a non-solvent during the curing reaction and the question arises, whether the phase separation behavior can be simply related to the solubility parameters of the polymer and solvent. Thus the solubility parameter concept might help to select solvents suitable for phase separation.

For our calculations of DH it is taken into account that the solubility parameter $\vec{\delta}$ is a vectorial quantity [81, 80]. It can be divided into different contributions, resulting from polar (δ_p), hydrogen bonding (δ_h), or dispersive forces (δ_d) (see Fig. 4). These can be calculated from the molecular structure of the macromolecule by applying theories proposed by Small, Hoy, Fedors or Hoftyzer and van Krevelen [80]. Among these theories we have chosen the method of Hoftyzer and van Krevelen for the calculation of group contributions and $\vec{\delta}$. The direction of the solubility parameter depends on the chemical structure, whereas the vector length is a result of the sum of contributions from individual groups by taking into account the molar volume. Graphically this difference between the solubility parameter of the polymer, $\vec{\delta}_{Pol}$, and the solvent, $\vec{\delta}_{Sol}$, is given by vector $\vec{d}$ (Fig. 4). Mathematically $\vec{d}$ is expressed by

$$|\vec{d}| = \sqrt{\left\{\left(\delta_d^{Pol} - \delta_d^{Sol}\right)^2 + \left(\delta_p^{Pol} - \delta_p^{Sol}\right)^2 + \left(\delta_h^{Pol} - \delta_h^{Sol}\right)^2\right\}} \qquad (20)$$

and the Hildebrand equation becomes

$$\Delta H^V = \frac{\Delta H}{V_T} = \phi_{Pol} \cdot \phi_{Sol} \cdot |\vec{d}|^2 \qquad (21)$$

Equation (21) reduces to Eq. (19) if polar and hydrogen bonding interactions are neglected. Hence, not the absolute value of the solubility parameter, as suggested by the simplified Hildebrand concept, but its individual contributions must be considered in order to express the solution and phase separation behavior. Thus, even though the norm of the solubility parameter of various solvents is constant, the partial contributions can be very different, resulting in a totally different solution and phase separation behavior respectively.

It is schematically represented in Fig. 4, that the solubility parameter of the polymer changes its direction as well as its length during the crosslinking reaction. For our system, this is a consequence of the formation of hydroxyl groups during the epoxy-amine reaction [82, 83]. This results in an increase in polar and hydrogen bonding interactions. These changes might be very different if other precursor monomers are used. In the uncured state ($q=0$), the length of the vector $\vec{d}$ ($q=0$), pointing from the solubility parameter vector of the precursor polymer mixture to the solubility parameter vector of the solvent, is initially

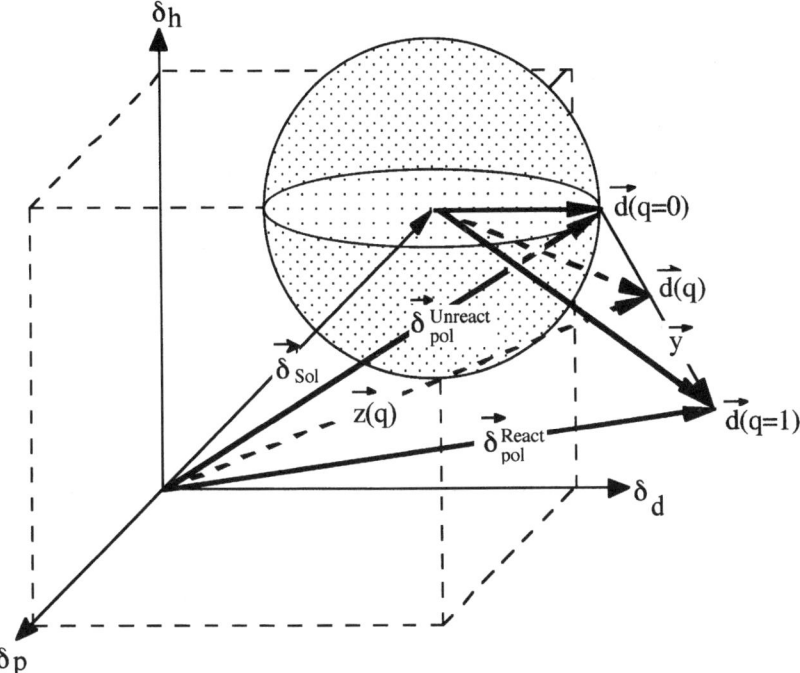

Fig. 4. Vectorial representation of solubility parameters used to calculate the free enthalpy of mixing as a function of conversion, q

short (Fig. 4). Thus the free enthalpy of mixing is low, and miscibility is favored. Upon reaction the length of $\vec{d}(q)$ increases, thus increasing the enthalpic contribution. It should be kept in mind, that according to the Hildebrand equation (Eq. 19), the free enthalpy of mixing is proportional to the square of $|\vec{d}(q)|$. However, more than the direction of $\vec{d}(q)$, its absolute value is of utmost importance. Thus not only changes in entropy but also enthalpy determine the phase separation behavior. The mathematical solution of the enthalpy of mixing is required to express $\vec{d}(q)$. Assuming that the change in solubility parameter of the polymer occurs linearly with conversion (see Fig. 4), $\vec{d}(q)$ is given by

$$\vec{d}(q) = \vec{z}(q) - \vec{\delta}_{Sol} \tag{22}$$

With

$$\vec{z}(q) = \vec{\delta}_{Pol}^{unreact} + q \cdot \vec{y} \tag{23}$$

and

$$\vec{y} = \vec{\delta}_{Pol}^{react} - \vec{\delta}_{Pol}^{unreact} \tag{24}$$

$\vec{d}(q)$ can be formulated as

$$|\vec{d}(q)| = \begin{pmatrix} \delta_d^{unreact} + q \cdot (\delta_d^{react} - \delta_d^{unreact}) - \delta_d^{Sol} \\ \delta_p^{unreact} + q \cdot (\delta_p^{react} - \delta_p^{unreact}) - \delta_p^{Sol} \\ \delta_h^{unreact} + q \cdot (\delta_h^{react} - \delta_h^{unreact}) - \delta_h^{Sol} \end{pmatrix} = \begin{pmatrix} a \\ b \\ c \end{pmatrix} \quad (25)$$

where the indices d, p, and h stand for the dispersive, polar, and hydrogen bonding contributions respectively.

The length of $\vec{d}(q)$ can then be calculated as a function of conversion according to

$$|\vec{d}(q)| = \sqrt{a^2 + b^2 + c^2} \quad (26)$$

Combination of Eqs. (16), (21), and (25) then yield the desired expression for ΔG^V as a function of conversion:

$$\Delta G^V = \frac{\Delta G}{V_T} = \frac{RT}{V_0} \cdot \left\{ \left(1 - \frac{4}{3}q\right) \cdot \phi_{Pol} \cdot \ln\phi_{Pol} + \frac{\phi_{Sol}}{z_{Sol}} \ln\phi_{Sol} \right\} + |\vec{d}(q)|^2 \cdot \phi_{Pol} \cdot \phi_{Sol} \quad (27)$$

Similar to the derivation of schematic phase diagrams showing an *UCST* or *LCST* behavior, as shown in the previous section, it is now possible to plot a schematic phase diagram as a function of conversion, q, such as shown in Fig. 5. The binodal and spinodal curves are obtained by interconnecting all the coexisting points and inflection points of the ΔG^V-curves as a function of q similar to the procedure described in [67–69].

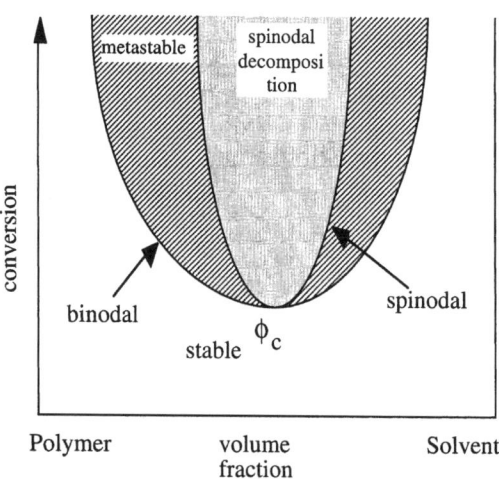

Fig. 5. Schematic phase diagram for Chemically Induced Phase Separation (CIPS)

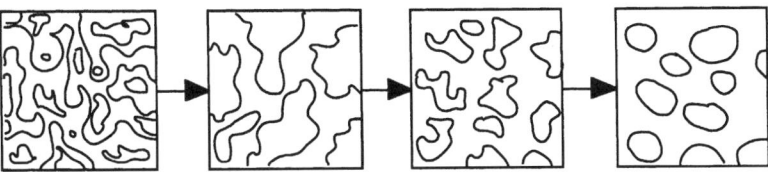

Fig. 6. Morphology development upon spinodal decomposition

This schematic phase diagram is now used to explain the generation of a second phase during the curing of thermosetting polymers. Upon the curing reaction, the conversion increases continuously. Depending on the concentration and on the reaction rate, a system which is completely miscible in the uncured state ($q=0$) can enter the metastable or the spinodal decomposition region, thus leading to phase separation. In the metastable region the phase separation proceeds via a nucleation and growth mechanism, whereas a spinodal decomposition occurs if the area enclosed by the spinodal line is entered. This transition from the stable to the metastable or alternatively spinodal region can be regarded as a chemical quench. Consequently, this type of phase separation is named *chemically induced phase separation* (CIPS) [49, 50, 84–91] because it refers to its thermodynamic origins. Alternatively the term *reaction induced phase separation* has been widely used to describe thermosets which were toughened by the generation of rubber or thermoplastic particles [68].

The morphologies that can be generated via the *CIPS* technique will depend on the phase separation mechanism. The energy balance of a system that phase separates via a nucleation and growth mechanism is governed by the energy required to enlarge the new surface and by the energy gain that results from the formation of a higher volume. Hence the second phase will take the shape that offers the best volume to surface ratio. Therefore the domains formed via nucleation and growth will be spherical. If the metastable region is entered during the curing reaction, the final morphology will then be the result of the competing effects between the nucleation and growth of the separated domains and the continuous advancement in crosslinking which is accompanied by an increase in viscosity and thus lowers the growth rate.

In polymeric materials, the morphology development upon spinodal decomposition proceeds through various stages [92, 93]. In the early stage of decomposition a co-continuous structure develops. A dispersed two-phase structure results only in the late stage of phase separation and the shape of the domains is not uniform. The morphology development upon spinodal decomposition is presented in Fig. 6.

2.5
Guidelines for the Preparation of Solvent-Modified and Macroporous Thermosets via CIPS

It has been shown above that the knowledge of the thermodynamic origins of phase separation processes allows one to reconstruct schematic phase diagrams.

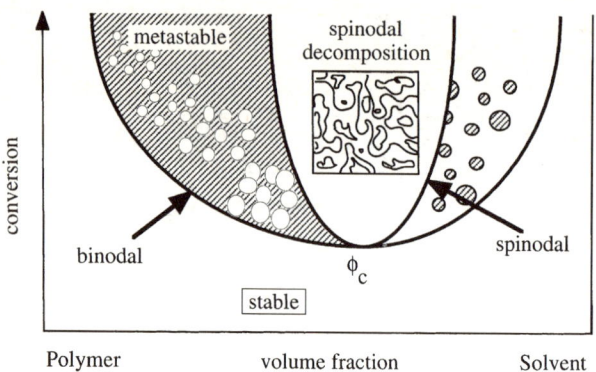

Fig. 7. Various strategies of the Chemically Induced Phase Separation technique to generate different types of morphologies

This also provides useful strategies to achieve porous thermosets having tailored morphologies such as interconnected or closed pores. These guidelines are summarized in Fig. 7 and discussed below.

If a closed cell morphology is desired the volume fraction must be kept low to avoid interconnection or even phase inversion and the phase separation should proceed via a nucleation and growth process. The nucleation and growth mechanism is favored if the transition from the stable into the metastable region occurs slowly, thus allowing the system to reach the equilibrium concentration. Therefore the reaction rate should be kept low. By this route the final size of the liquid droplets will be the result of the competing effects of nucleation and growth and the continuous advancement in crosslink density. The amount of solvent should be limited by the phase inversion concentration that is given by the interception of the binodal and spinodal line. This point is called the critical concentration ϕ_c.

At concentrations above ϕ_c, a phase separation would lead to the formation of particles dispersed in a liquid matrix. The composition of such particles should be given by the binodal line. Thus such particles will still contain enough solvent to undergo a phase separation. Indeed such an internal phase separation can be used to prepare porous polymeric particles with potential for application as chromatography beads [48].

An open porosity might be realized by interconnection of droplets generated via a nucleation and growth mechanism using a higher solvent concentration. Alternatively very high reaction rates might be used to enable for spinodal decomposition and suppress the coalescence of domains occurring in its late stage. Both strategies should allow for the preparation of materials having interconnected pores which might become useful for membrane applications.

3
Selection Criteria for the Preparation of Solvent-Modified and Macroporous Epoxy Networks with Tailored Morphologies Prepared via CIPS

Imagine now that you want to prepare a macroporous thermoset using a predetermined set of precursor monomers. The first thing you would need to do is to select a suitable solvent, initially dissolving the precursors but upon crosslinking turning to a non-solvent giving the desired phase separation.

Here we want to discuss whether the solubility parameter concept which has been explained above can be used to simplify the solvent selection. Hence we will recalculate free energy curves, reconstruct phase diagrams, and discuss the advantages and limitations of the solubility parameter approach in comparison with experimental findings. Our aim is to show how this concept can be used for the initial selection of a suitable solvent without performing time-consuming experiments, thus allowing for the synthesis of a wide variety of solvent-modified and macroporous thermosets with tailored morphologies via the *CIPS* technique.

3.1
Gradient Oven for the Determination of Phase Separation Lines

In order to get experimental information about the concentration and temperature range where phase separation occurs, it is required to perform a series of isothermal curing experiments varying these parameters [49]. By interconnecting individual phase separation points, say 10% solvent at 40 °C is needed while 17% solvent has to be employed at 100 °C, a phase separation line is constructed. In order to obtain the temperature and concentration ranges where phase separation occurs with a very few experiments, a gradient oven made out of aluminum, schematically shown in Fig. 8, was constructed.

The temperature gradient is achieved by cooling one end with water, while the opposite end is placed on a heating plate. A sealed glass tube filled with the homogeneous mixture, containing the unreacted precursor monomers together with the desired amount of solvent, is then inserted into the gradient oven and allowed to cure. The temperature along the tube shows a temperature distribution as given in Fig. 9. After curing, the glass tube is opened and the sample is heated to 200 °C in a vacuum oven for 120 h in order to complete the network formation and to evaporate fully the low molecular weight liquid. The resulting sample now represents a linear gradient in temperature and may be sectioned at various positions to investigate any morphology and density in the temperature range 20–150 °C for this given solvent composition. The phase separation lines were recorded within relatively narrow temperature limits. The lower limit is given by the lowest curing temperature, identical to the temperature at the cold end of the gradient oven, thus predetermined by the cooling media. In our studies we used water as the cooling fluid, thus limiting the lower temperature to 20 °C. The maximum temperature, identical to the temperature on the heating

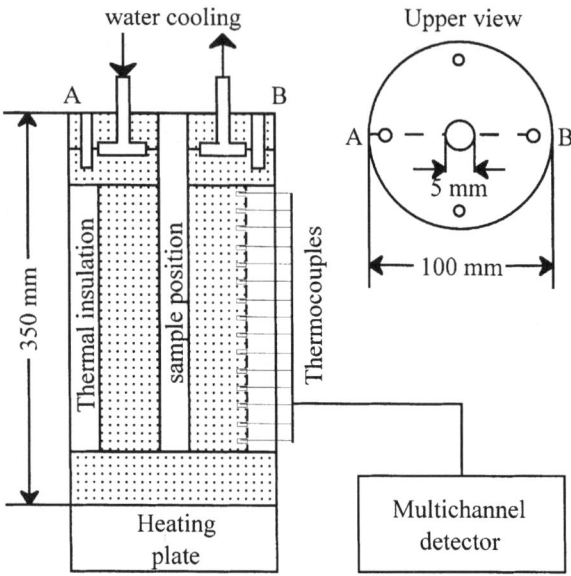

Fig. 8. Schematic representation of the gradient oven

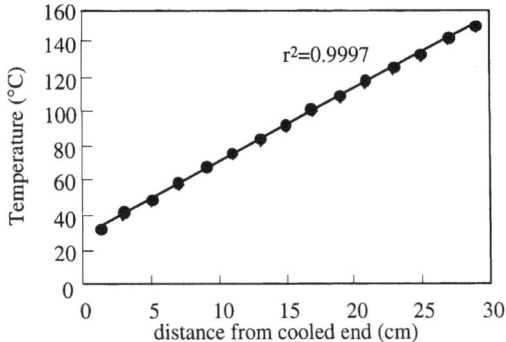

Fig. 9. Typical temperature distribution achieved with the gradient oven

plate, was restricted to 150 °C, owing to the risk of explosion due to the pressure buildup, by operating in a closed system at temperatures above the boiling point of the liquids used for our experiments.

3.2
Influence of the Solubility Parameter on the Phase Separation Behavior

To illustrate the influence of the solubility parameter on the phase separation behavior, a series of solvents has been selected which have either different values

Table 1. Solubility parameters and group contributions of polymer and solvents used for the synthesis of macroporous epoxies via CIPS

	Chemical structure	δ	δ_d	δ_p	δ_h
			$(J/cm^3)^{1/2}$		
DGEBPA		18.2	16.9	1.4	6.4
PACP		20.05	18.3	0	8.2
Epoxy fully cured	NETWORK	20.95	17.6	2.1	11.2
Hexane		14.7	14.7	0	0
Octane		15.1	15.1	0	0
Decane		15.4	15.4	0	0
Cyclohexane		15	15	0	0
Methylcyclohexane		16	16	0	0
2,6-Dimethyl-4-heptanone		16.1	15.2	4.4	3.4
Dibutylether		15.8	15	2.3	4.2

of $|\vec{\delta}_{Sol}|$, but similar chemical structures, or similar values of $|\vec{\delta}_{Sol}|$ but different chemical structures and phase separation lines were recorded with the gradient oven.

Consequently, we have chosen solvents in order to change separately either the norm of the solubility parameter or its direction (see Fig. 4). These solvents are listed in Table 1. It can be clearly seen that the polar and hydrogen bonding interactions are zero for all of the aliphatic and cycloaliphatic alkanes. This allows one to change only the value of $|\vec{\delta}_{Sol}|$ without changing its direction. For a second series of experiments, we compare 2,6-dimethyl-4-heptanone, dibutylether and methyl-cyclohexane which have nearly identical lengths, but different vector directions.

Also shown in Table 1 are the solubility parameter values of the precursor monomers and the fully cured epoxy network, which were calculated from the

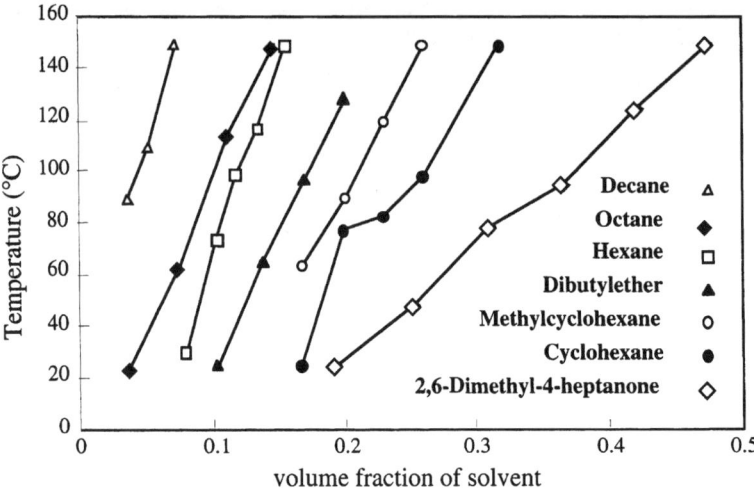

Fig. 10. Phase separation lines for the synthesis of macroporous epoxies via CIPS

group contributions following the Hoftyzer-van Krevelen approach. As can be seen from Table 1 and Fig. 4, the solubility parameter of the polymer changes its overall length as well as its direction with the curing reaction, thus leading to a change in the enthalpic contribution. Consequently the variation of entropy and enthalpy during the curing reaction are both responsible for a phase separation. Hence the quantitative description of the free energy of mixing which considers the changes of $|\vec{\delta}_{Pol}|$ and $\vec{d}$ with conversion (Eq. 27) should allow one to reconstruct real phase diagrams.

Plotted in Fig. 10 are phase separation lines, which were obtained with the gradient oven for epoxies cured in the presence of the above solvents [88]. At the left side of these lines, no phase separation occurs and the materials stay transparent. The right side of these phase separation lines gives the temperature and composition ranges where phase separation occurs. The onset of the phase separation, which gives one single point on the phase separation line, can easily be detected for each concentration, as the samples become opaque as a consequence of the formation of liquid domains in the μm-range.

One common feature in all the systems is the shift in the onset of phase separation towards higher concentrations as the curing temperature is raised. This is typical for *UCST* behavior. This phenomena is due to a higher miscibility of the epoxy precursors and the solvent at higher temperatures. The onset of phase separation represents in a first approximation the amount of solvent which remains dissolved in the completely cured network. All the alkanes which show large differences in the polar and hydrogen bonding contributions compared with the crosslinked epoxy network demonstrate low miscibility with the neat resin and the phase separation occurs at fairly low concentrations (see Fig. 4 and Eq. 27). It can also be observed that increasing the size of solvent molecule lowers the miscibility, which is only partially due to a larger value of $|\vec{d}|$. A second criterion for the phase separation is the conformational freedom of the solvent molecule, that is not considered in the Flory-Huggins expression for the entropy. Upon the formation of a highly crosslinked network the conformational freedom of long, flexible solvent molecules is greatly reduced, equivalent with a considerable loss in entropy. Therefore the decane undergoes a phase separation at very low solvent concentrations. Hence the phase separation lines shift to lower concentrations starting from hexane over octane to decane.

The same trend is also observed when methylcyclohexane is compared to cyclohexane. The addition of only one methylene group in the structural unit causes a considerable influence in the phase separation behavior. Even though, these cycloaliphatic solvents display very similar group contributions to the linear alkanes, the onset of phase separation starts at higher concentrations. It is experimentally observed, that the cyclohexane shows a higher miscibility than the methylcyclohexane. This behavior is not expected from the differences in the values of solubility parameter, but it can be explained with regard to the conformational freedom of the solvent. The cyclohexane has only two configurations – the boat and the chair form – whereas the linear alkanes have four (hexane) to eight (decane) flexible points. The addition of one methylene group to cyclohexane introduces a new flexible point, thus allowing for a higher change in entropy, leading to phase separation starting at lower concentrations.

The influence of the conformational freedom becomes even more evident, when the phase separation behavior of the more polar solvents, like dibutylether and 2,6-dimethyl-4-heptanone, are compared to each other and to methylcyclohexane, which all have nearly identical values for $|\vec{\delta}_{Sol}|$, but various degrees of conformational freedom. According to Eq. (27) it is expected that both polar solvents should display a higher miscibility than the methylcyclohexane, and that the 2,6-dimethyl-4-heptanone should undergo a phase separation at lower concentrations than the dibutylether. The discrepancy between these theoretical expectations and the experimental results are explained with regard to molecular flexibility of the different solvent molecules. Owing to the highly flexible ether linkage together with the attached flexible linear chains, the dibutylether displays a high level of conformational freedom, whereas the structural units of the 2,6-dimethyl-4-heptanone and methylcyclohexane (shown in Table 3) are examples of molecules with a lower conformational freedom. Therefore the loss in entropy becomes more important in the ether-based system than for the ketone, and the phase separation starts at lower concentrations for the dibutylether than for the methylcyclohexane and 2,6-dimethyl-4-heptanone. These examples demonstrate the importance of the conformational freedom of the solvent molecule and the solubility parameter for the phase separation behavior.

One particularly interesting system is the epoxy 2,6-dimethyl-4-heptanone as up to 40 wt % of this solvent can be easily mixed together with the epoxy precursors to generate a phase separation process. This allows one to verify experimentally the possible morphologies which were predicted based on the schematic phase diagram at concentrations below the phase inversion (see Fig. 7). Shown

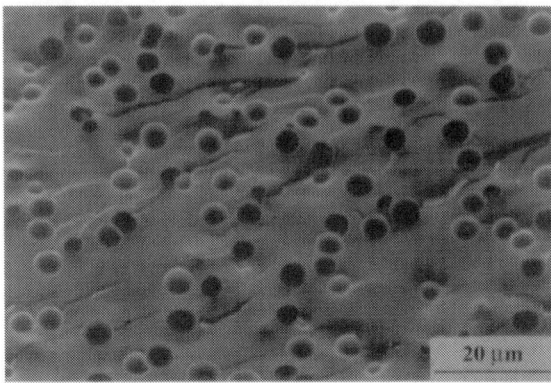
a) Narrow size distribution obtained with 25 wt% 2,6-dimethyl-4-heptanone

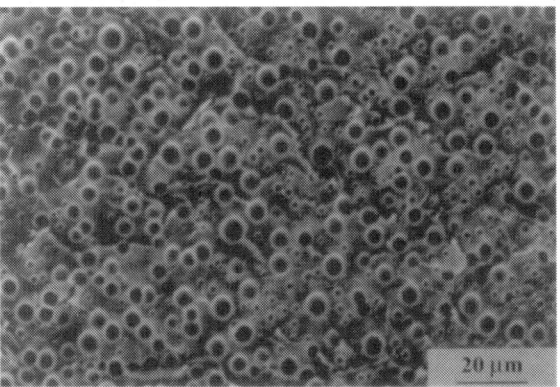
b) Bimodal size distribution obtained with 30 wt% 2,6-dimethyl-4-heptanone

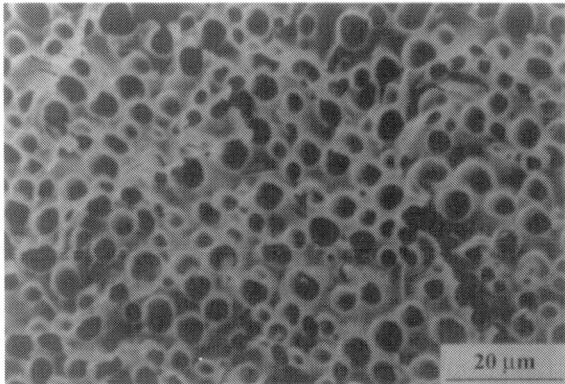
c) Co-continuous strutcture obtained with 35 wt% 2,6-dimethyl-4-heptanone

Fig. 11. Scanning electron micrographs of macroporous epoxies prepared via the *CIPS* technique with various amounts of 2,6-dimethyl-4-heptanone at constant curing temperature, T=40 °C

in Fig. 11a–c are the scanning electron microscopy (SEM) micrographs of macroporous epoxies, prepared via *CIPS* with various amounts of 2,6-dimethyl-4-heptanone, sampled from the low temperature end of the gradient oven, thus corresponding to a cure at room temperature. Experimental techniques for sample preparation and SEM measurements are described in [49, 50, 89]. It was concluded from these measurements that the morphology is not changed during the drying procedure, and thus the pore size is equivalent to the domain size after the phase separation. By exceeding the critical concentration for the phase separation, a narrow sized porosity is generated, as it is seen from the SEM micrograph taken from a sample prepared with 25 wt % 2,6-dimethyl-4-heptanone (Fig. 11a). Upon a further increase in solvent concentration to 30 wt % of 2,6-dimethyl-4-heptanone, a bimodal distribution appears (Fig. 11b). Bimodal distributions are also observed in all the alkane based systems [49, 88, 89]. The explanation for the development of such bimodal distributions will be given in the next section.

As the solvent content is raised further to 35 wt % of 2,6-dimethyl-4-heptanone, the pores become interconnected (Fig. 11c). The size distribution is very narrow compared to other foaming procedures, where a gas is used as the pore forming agent. Therefore the *CIPS* technique might become an alternative route for the preparation of polymeric membranes with a narrow size distribution in the µm-range. The generation of such a morphology has been claimed in the previous section based on the schematic phase diagram (Fig. 7). However it must be mentioned that such an open porosity has only been observed in a small concentration and temperature range. As the concentration is raised further to 40 wt % 2,6-dimethyl-4-heptanone, an inhomogeneous structure containing domains with sizes larger than 100 µm was found. Such a morphology is typically observed when the solubility limit of the solvent in the unreacted precursor mixture is exceeded. Thus the generation of an open porosity via the CIPS technique seems to become possible only if the phase separation gap is relatively large. The gradient oven can hence greatly facilitate the selection of suitable solvents, which show such large phase separation gaps. Furthermore it accounts for an eventual small offset in calculated solubility parameters.

3.3
Reconstruction of Real Phase Diagrams Based on the Solubility Parameter Approach

The above discussion demonstrates that the solubility parameter concept in combination with the gradient oven is a useful tool to select a convenient solvent, which could undergo a phase separation during the crosslinking reaction.

Based on Eq. (27) it is now possible to reconstruct free energy curves as a function of conversion for any polymer-solvent system discussed above. Plotted in Fig. 12 are the results of ΔG^v as a function of conversion for the 2,6-dimethyl-4-heptanone based system. These results clearly show the theoretically predicted change in curvature which is accompanied with the development of a curve having two minima and two inflection points [52, 53, 63, 68].

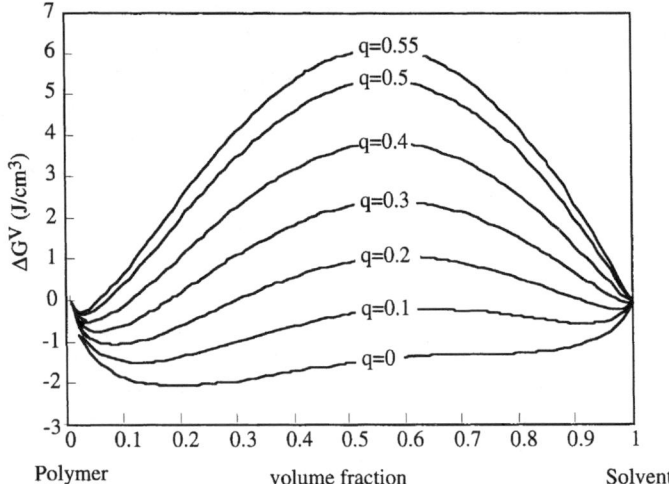

Fig. 12. Free energy curves for 2,6-dimethyl-4-heptanone modified epoxies calculated with the solubility parameter approach at a constant temperature (315 K) as a function of conversion

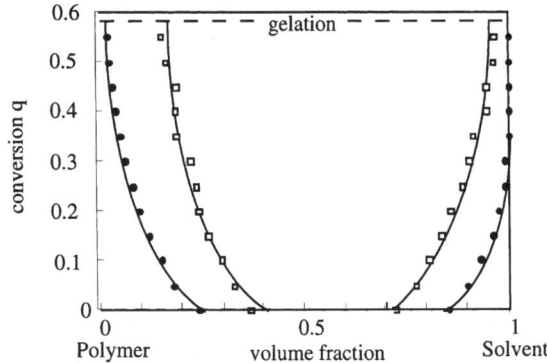

Fig. 13. Phase diagram for the 2,6-dimethyl-4-heptanone modified epoxy system derived graphically from Fig. 12

Based on these results the points for thermodynamic equilibrium are determined graphically from the two points, which have a common tangent, thus leading to the construction of the binodal line. The interconnection of inflection points as a function of conversion yields the spinodal line. Thus one obtains the phase diagram shown in Fig. 13. As Eq. (15) is only valid prior to gelation, phase diagrams should not be extrapolated beyond this limit. According to Flory's theory of gelation [53], the extent of reaction at gelation takes a value of $q_{gel}=0.577$ for the reaction of a stoichiometric amount of a bifunctional epoxy and a tetrafunctional amine.

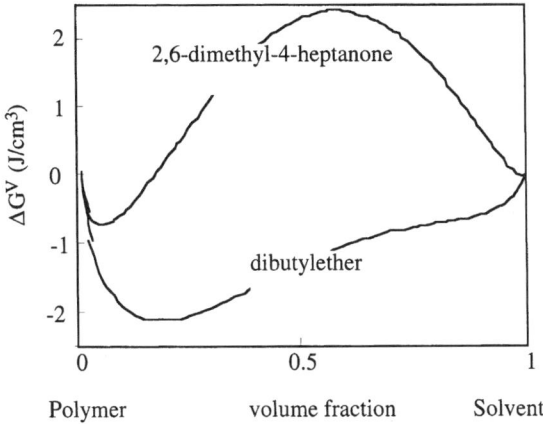

Fig. 14. Free energy curves calculated with the solubility parameter approach for the 2,6-dimethyl-4-heptanone- and dibutylether-modified epoxy systems at a constant temperature (T=315 K and conversion (q=0.3)

As stated earlier, the solubility parameter approach is an approximation. This becomes obvious by comparing the 2,6-dimethyl-4-heptanone to dibutylether. Experimentally it has been found that 2,6-dimethyl-4-heptanone displays a higher miscibility than dibutylether. The calculations show that for any temperature and conversion dibutylether should display a higher miscibility. This is reflected in Fig. 14, where free energy curves of 2,6-dimethyl-4-heptanone-modified systems and dibutylether-modified systems are compared at a constant temperature (T=315 K) and a conversion of 30%. One can clearly see, that the phase separation should just have started for the dibutylether-modified system with an equilibrium concentration of around 25 vol. % solvent, whereas the change in curvature is much more advanced in the 2,6-dimethyl-4-heptanone-modified system exhibiting an equilibrium concentration of around 5–10 vol. % 2,6-dimethyl-4-heptanone in the polymeric matrix. The reason for this discrepancy between experimental observations and calculated data is two-fold.

The first limitation arises from the approximation introduced by the calculation of individual contributions for the solubility parameters. Based on the group contributions proposed by Hoftyzer and van Krevelen which is used for our calculations, the polar contribution of the ether group is only half the value of ketone. Additionally, the chemical environment that influences the local charge is not taken into account in this approach. For the calculation of group contributions, it is not important whether an electron-withdrawing, a neutral or an electron-donating group is adjacent to the functional group. The complex interactions in polymer-solvent systems exhibiting polar and hydrogen bonding interaction, encountered in the ketone and ether based systems, cannot be easily quantified or predicted. This error becomes less important in the alkane-modified systems, where only weak van-der-Waals interactions are active. Moreover

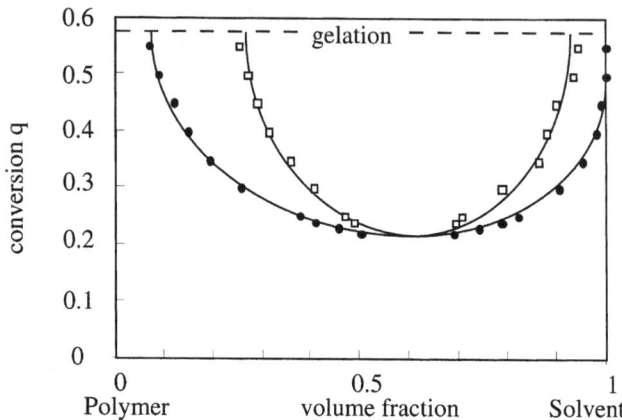

Fig. 15. Phase diagram for the dibutylether-modified epoxy system derived graphically from free energy curves

the calculations (Eq. 27) do not take into account the influence of the molecular freedom of the solvent molecule on the changes in entropy. Thus the solubility parameter approach provides a qualitative selection criteria, but it cannot yield quantitatively correct results.

Analogous to the construction of phase diagrams for the 2,6-dimethyl-4-heptanone-modified system, the phase diagram for dibutylether is plotted in Fig. 15. Again one does recognize that the shape of this phase diagram ideally resembles the one predicted by the Flory-Huggins theory as shown in Fig. 7. Based on this diagram one can conclude that the metastable region is entered at a higher extent of reaction. This might explain the fact that no bimodal distribution has been observed in the dibutylether-modified system, although the system has been tested to volume fractions up to 25%. Furthermore this phase diagram is in agreement with the observation that no phase separation occurs for volume fractions below 10%. We believe that in such a case a critical extent of reaction, q_{nucl}, is reached where the viscosity or crosslinked density become so high that nucleation is not possible. If q_{nucl} is reached before the metastable region is entered, nucleation cannot take place and the samples remain transparent. Our preliminary investigations gave no exact answers as to whether this imaginary critical conversion for nucleation, q_{nucl}, can be related to gelation and how it depends on the volume fraction of solvent [85].

Solving the modified Flory-Huggins-equation (Eq. 27), we find that the phase separation behavior of alkanes is fairly well predicted [85]. As an example, the free energy curves of epoxy networks modified with either hexane, octane, or decane at equal temperature and conversion are shown in Fig. 16. According to the calculated free energy curves, the miscibility is lowered as the chain length increases, which agrees with the experimental observations. One characteristic feature of all the alkanes is the fact that they possess a large conformational freedom, thus leading to a phase separation at low solvent concentrations. These free ener-

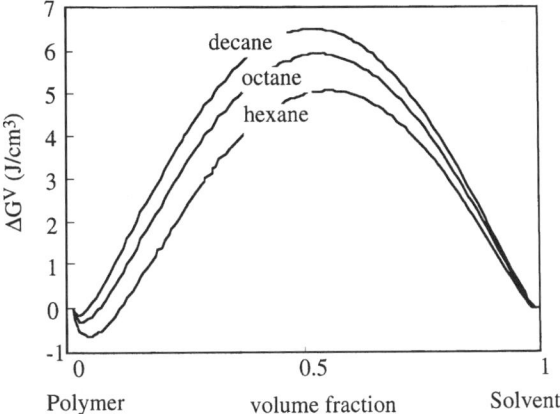

Fig. 16. Free energy curves for hexane-, octane-, and decane-modified epoxies at a constant temperature (T=315 K) and conversion ($q=0$)

gy curves allowed then to reconstruct phase diagrams and gave results similar to the one shown in Fig. 13 [85]. All these solvents display a very low miscibility even in the uncured state. This solubility limit is given by the intercept of the binodal line at zero conversion. A second boundary is given by the intercept of the binodal line at the imaginary value of q_{nucl}. Samples with concentrations between these two boundaries demonstrate phase separation and this concentration range is called the phase separation gap. However a small difference of around 5 vol. % exists between the theoretical prediction and experimental result. The isothermal curing experiments [49, 50] as well as the experiments performed with the gradient oven (Fig. 10) gave nearly identical phase separation gaps.

The above results demonstrate that the solubility parameter concept (Eq. 27) is very useful for reconstructing phase diagrams. Even though these phase diagrams enable one to explain the phase separation behavior of solvent-modified thermosets prepared via *CIPS*, the calculations are simplifications with limitations on predicting exactly the phase separation gap. These include the influence of conformational freedom of the solvent molecule on the entropic contribution and the influence of temperature on changes in intermolecular interactions. The understanding of these parameters on the phase separation behavior, as presented here, is crucial to selecting a suitable solvent to induce a phase separation during the curing reaction.

To overcome these problems a gradient oven was presented which allows one to find rapidly the real phase separation gap for a given set of polymer and solvent. These results may serve as general guidelines for the preparation of a wide variety of solvent-modified and macroporous thermosets with tailored morphologies via *CIPS*.

Within the following section it is intended to identify the phase separation mechanism and morphology development and to investigate the effect of reaction parameters on pore size and distribution.

4
Morphology Development of Solvent-Modified and Macroporous Epoxy Networks Prepared via CIPS

If hexane is used as the low molecular weight liquid, the desired phase separation is observed when precursor mixtures containing 6–15 wt % hexane are cured isothermally at 40 °C. Further discussion of the phase separation behavior requires more detailed consideration of the schematic phase diagram, as presented in Fig. 17, which resembles the real phase diagram shown in Fig. 13. Experimentally it is found, that no phase separation occurs with hexane concentrations equal to or lower than 5 wt %. Hence the critical amount for phase separation, ϕ_p, is given by the intercept of the binodal line and the imaginary value of q_{nucl}. Hence no phase separation occurs if q_{nucl} is reached before the metastable region is entered.

If hexane is chosen as the solvent, it has been found experimentally that the curing temperature is limited to temperatures below 50 °C to maintain homogeneous samples. It is believed that the exothermic heat development during the curing reaction locally can create temperatures exceeding the boiling point of hexane (68 °C at ambient pressure), when the isothermal curing temperature approaches the boiling point of the solvent. Upon isothermal curing at 40 °C, samples with concentrations of 5 wt % hexane or lower stay transparent and no sign of separated domains can be detected even with transmission electron microscopy. This corresponds to situation A in Fig. 17. Phase separation resulting in the formation of white, opaque samples is only observed at concentrations

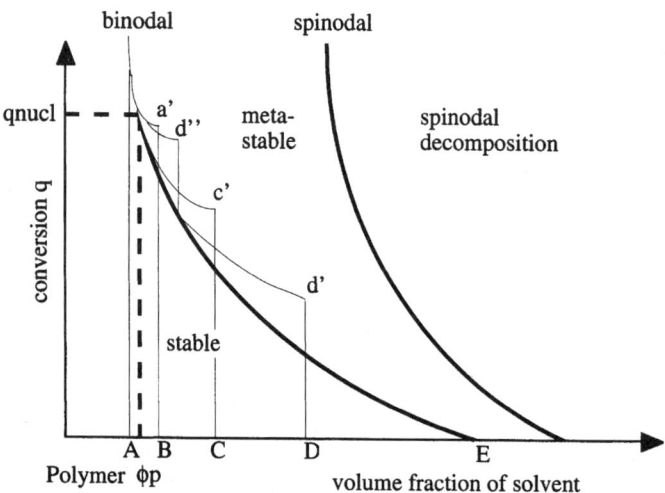

Fig. 17. Schematic phase diagram explaining the phase separation behavior observed during the curing of epoxies in the presence of hexane

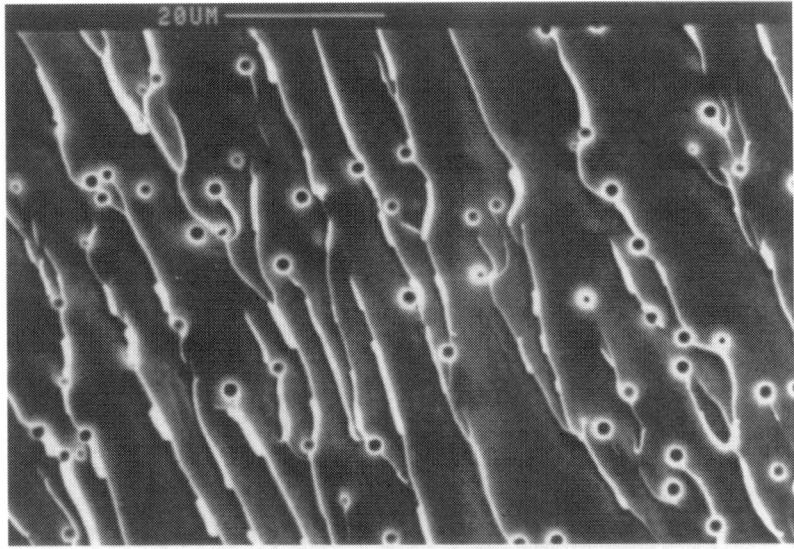

(a) 6 wt% hexane

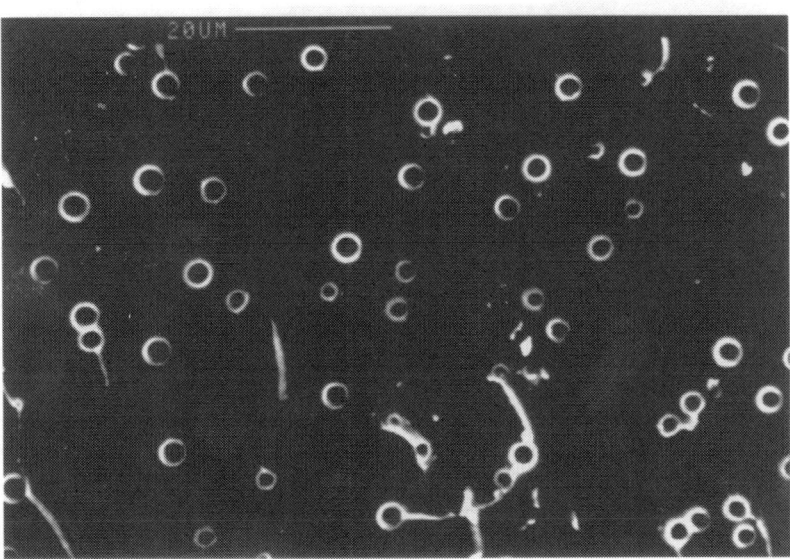

(a) 7,5 wt% hexane

Fig. 18a–b. Scanning electron micrographs on cryo fractured surfaces of: **a** macroporous epoxy prepared with 6 wt % hexane via the CIPS technique showing a narrow size distribution; **b** macroporous epoxy prepared with 7.5 wt % hexane via the CIPS technique showing a narrow size distribution. Reprinted from Polymer, 37(25). J. Kiefer, J.G. Hilborn and J.L. Hedrick, "Chemically induced phase separation: a new technique for the synthesis of macroporous epoxy networks" p 5719, Copyright (1996), with permission from Elsevier Science

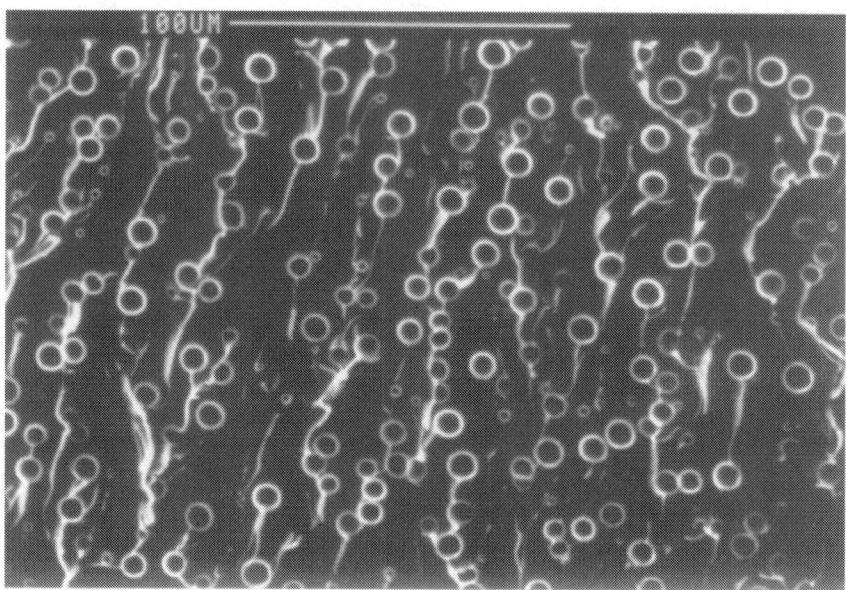

(c) 10 wt% hexane

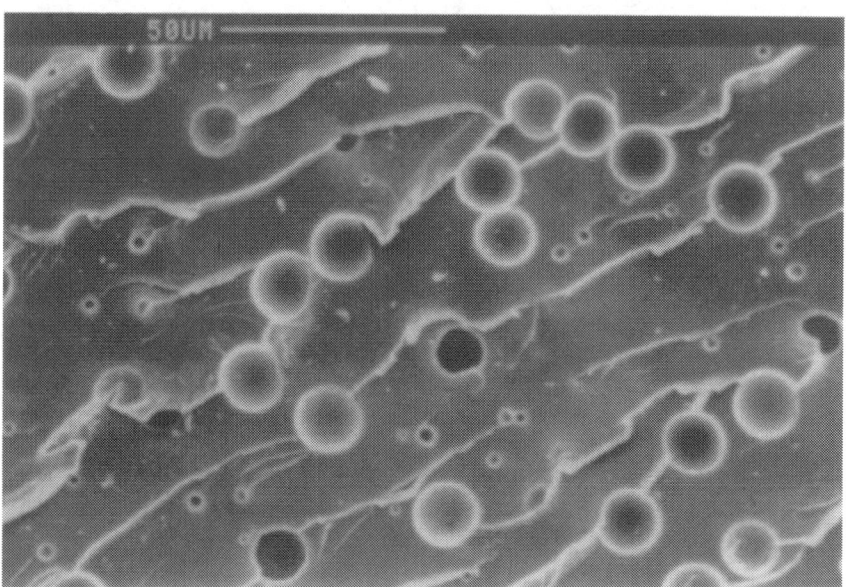

(d) 15 wt% hexane

Fig. 18c–d. c epoxy prepared with 10 wt % hexane via the CIPS technique showing a bimodal size distribution; **d** epoxy prepared with 15 wt % hexane via the CIPS technique showing a bimodal size distribution

above 6 wt % hexane. Hence, ϕ_p (hexane, 40 °C) is found experimentally from isothermal curing experiments to be situated between 5 and 6 wt % hexane.

The morphology of a fractured surface of a solvent-modified and macroporous epoxy networks was investigated by scanning electron microscopy (SEM). The fracture is influenced by the presence of voids leading to a rough fracture surface not representing a planar cut through the sample. Furthermore it can be argued, that the voids may have been fractured at the center, the bottom, or the top. Hence the domains seen with *SEM* on the fracture surface would deviate from the true size of the voids in such a manner that the void sizes observed are slightly smaller than the true sizes. However, it can be assumed that the fracture path passes through the meridian of the voids. This conclusion has been drawn from finite element analysis results on a circular hole in a polycarbonate plate, which predict the highest stresses to occur at the equator of the void, where the formation of shear bands starts [94]. Thus at this stage image analysis performed on fracture surfaces with *SEM* may provide a resonably good representation of the real size and distribution, at least allowing us to compare different systems and conditions.

SEM micrographs of macroporous epoxy networks prepared with concentrations above ϕ_c, such as 6 and 7.5 wt % hexane, are shown in Fig. 18a,b. It can clearly be seen that a closed cell morphology, including a narrow pore size distribution, has been achieved. These SEM micrographs indicate an increase in pore size, pore size distribution, and volume fraction with increasing amount of hexane once the critical concentration necessary for phase separation has been passed. These situations, resulting in the formation of a narrow pore size distribution corresponding to lines B and C, are represented in the schematic phase diagram (Fig. 17).

A further increase in the amount of solvent leads to the development of a bimodal pore size distribution, as observed with SEM on samples prepared with concentrations of 10–15 wt % hexane (Fig. 18c,d). Similar bimodal distributions have also been reported with the octane and decane based systems [88, 89] as well as in in rubber-modified epoxies prepared via phase separation [67, 95–98].

4.1
Image Analysis of Macroporous Epoxy Networks

Image analysis allows one to determine the size distributions of macroporous epoxies most correctly. This analysis include the number of pores, n, medium pore diameter assuming spherical domains, d, and the inter pore distance, *IPD*, which is an artificial value taking into account the average distance between the domains. In the following, it is intended to discuss further the validity of image analysis, which is usually performed analogously on fracture surfaces of toughened thermosets.

The particle diameters, d_i, of each particle are calculated from the detected particle surface, S_i, assuming that the particles are spherical, and that the observed surface represent the maximum surface resulting from a cut, thus being

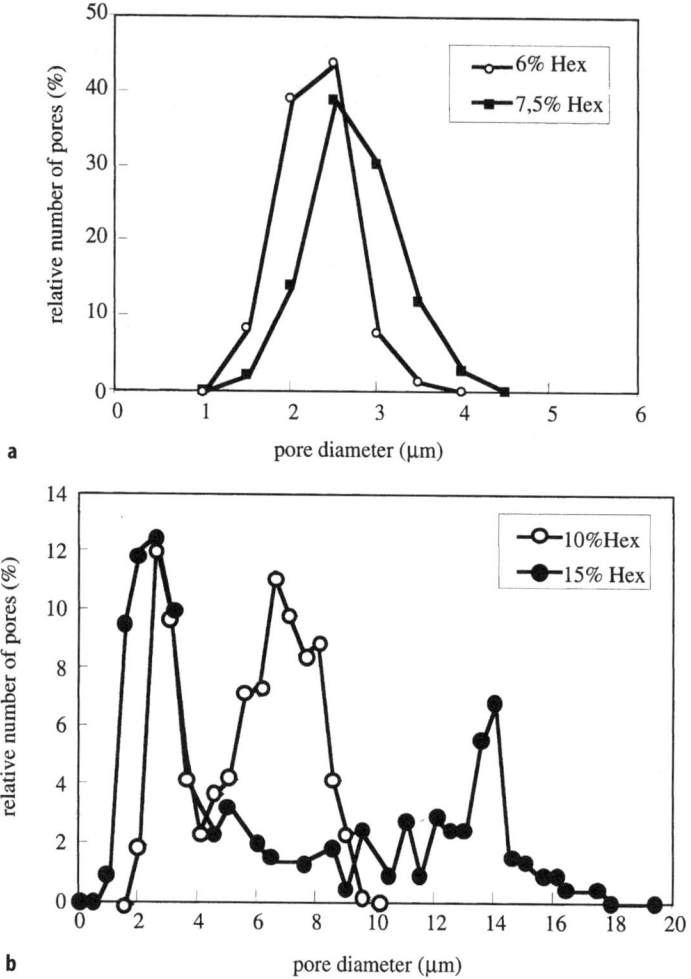

Fig. 19a,b. Pore size distributions obtained from image analysis on SEM micrographs showing: **a** narrow size distributions; **b** bimodal size distributions

equal to πr^2. An average particle diameter, d, is then automatically calculated according to:

$$d = \frac{\sum_i d_i}{n} = \frac{\sum_i 2 \cdot \sqrt{\frac{S_i}{\pi}}}{n} \qquad (28)$$

The pore size distributions calculated from image analysis of the sample with 6 and 7.5 wt % hexane are plotted in Fig. 19a, clearly showing a narrow size distribution and an increase of pore size with increasing hexane concentration.

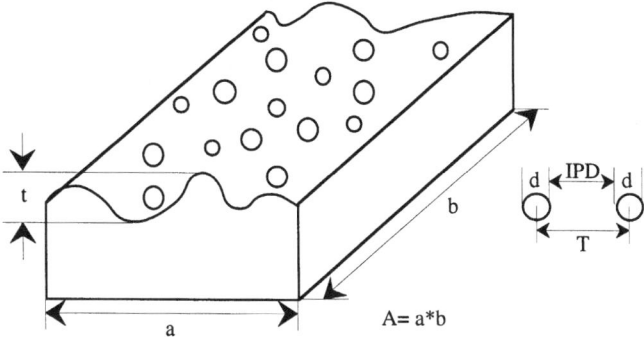

Fig. 20. Simplified model to calculate morphological characteristics with the mean diameter obtained from image analysis

Shown on the ordinate is the relative number of pores calculated as the percentage of the pore number for each class divided by the total number of pores. It can be seen that the maximum of the histogram decreases with increasing hexane concentration. This indicates that the pore size distribution becomes larger with increasing amount of solvent. This behavior can be explained regarding Fig. 17. For higher concentrations, the metastable region is entered at lower conversions. Therefore the domains have a longer time to grow and coalesce.

The image analysis clearly reveals the existence of a bimodal pore size distribution as seen from Fig. 19b, which has been suggested from the SEM micrographs. It can be seen that the domain size and size distribution of the large domains increase with the initial solvent concentration, whereas the size of the small domains remains unchanged. It should also be mentioned that the size distribution is not truly bimodal although consisting of two separated peaks. One detects pores with sizes in between the two peaks. Even though, the percentage of these domain sizes become very small it does not reach zero. Discussion of the kinetics of the phase separation will yield an explanation for this observation, as will be shown in the next section.

Digital image analysis directly gives the size distributions and a mean diameter, d. All other morphological characteristics are calculated with this mean diameter and the model shown in Fig. 20.

According to Fig. 20, the volume fraction should be calculated according to

$$\phi = \frac{\sum_i V_i}{V_T} = \frac{\sum_i \frac{4}{3} \cdot \pi \cdot r_i^3}{A \cdot t} \approx \frac{n \cdot \frac{\pi}{6} \cdot d^3}{A \cdot t} \tag{29}$$

Therein A is the total surface area, which has been scanned. The symbol t represents a measure for the surface depth or roughness, where the information is obtained from. The exact value of this surface roughness cannot be obtained from the *SEM* micrograph. The classical method for the calculation of further characteristics, such as the inter-pore distance, *IPD*, and volume fraction, ϕ, is based on the simplification that all the pores that are measured result from a depth that equals the diameter. This assumption greatly simplifies the calculations, but no estimations on the error are found in the literature.

With $t=d$, the volume fraction ϕ reduces to

$$\phi \approx \frac{n}{A} \cdot \frac{\pi}{6} \cdot d^2 \tag{30}$$

Image analysis also allows one to calculate the pore concentration, c, in millions per cubic mm ($10^6/\text{mm}^3$) according to

$$c \approx \frac{n}{A \cdot t} \cdot 1000 \tag{31}$$

Equation (31) is only valid if the surface area, A, is given in μm^2 and t in μm.

The calculation of the inter-pore distance derives from the assumption that the total volume is equal to the product of the number of pores and the cube of T, being the distance between the centers of two domains, as represented in Fig. 20.

With

$$n \cdot T^3 = V = A \cdot t = \frac{\pi}{6} \cdot \frac{n}{\phi} \cdot d^3 \tag{32}$$

and

$$T = IPD + d \tag{33}$$

the inter-pore distance, *IPD*, which is also called surface-to-surface distance, becomes

$$IPD = d \cdot \left\{ \left(\frac{\pi}{6 \cdot \phi} \right)^{1/3} - 1 \right\} \tag{34}$$

It should be emphasized that a correct calculation of the *IPD* requires the knowledge of the volume fraction of the dispersed phase. In the classical method, the calculation of the volume fraction is always based on the simplification that the surface roughness, t, equals the pore diameter (Eq. 30). This simplification can introduce considerable errors into the calculated value of the *IPD*. This might have drastic consequences for the validity of Wu's theory, which describes the *IPD* as the main morphological parameter being responsible for the improvement of toughness [99]. SEM can be used to get a better indication of the surface roughness. Therefore the samples were tilted in such manner that the angle between the incident electron beam and the surface becomes very small.

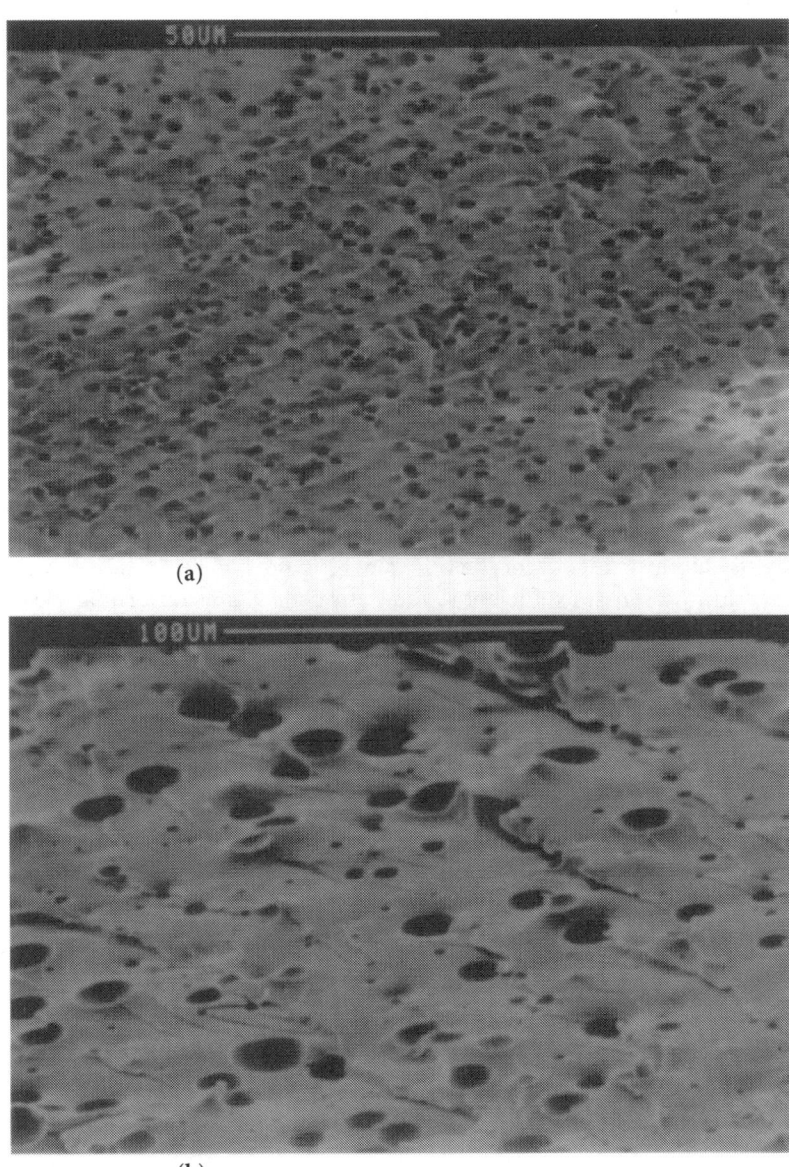

Fig. 21a,b. SEM micrographs showing the surface roughness of macroporous epoxy networks prepared with hexane via CIPS displaying: **a** narrow size distribution (7.5 wt % hexane); **b** bimodal size distribution (15 %wt hexane)

Such tilted photos are shown in Fig. 21 for a sample showing (a) narrow and (b) bimodal size distribution. However such micrographs do not allow one to determine t precisely.

Macroporous epoxies prepared via the CIPS technique represent an alternative for determining the volume fraction, ϕ, very accurately from the densities of the porous material, ρ_{por}, and the neat, fully crosslinked matrix, ρ, according to

$$\phi = 1 - \frac{\rho_{por}}{\rho} \qquad (35)$$

Together with the value of d determined with image analysis, this then allows one to calculate the *IPD* directly from Eq. (34) without the need for any simpflication and to evaluate the roughness t from Eq. (32) as

$$t = \frac{n}{A \cdot \phi} \cdot \frac{4}{3} \pi \cdot r^3 \qquad (36)$$

Interestingly, the values for volume fraction and *IPD* calculated via the classical method or via the density values differ significantly. These data are compared in Table 2 for the macroporous epoxies prepared with 6 and 7.5 wt % hexane displaying a narrow size distribution. A reasonable agreement for the density values exists, but the tendency for the IPD values is quite different. These two single values are, however, not sufficient to draw any conclusion concerning the error introduced by the simplification that $t=d$.

Image analysis for bimodal distributions is more complicated. With the classical method, the determination of the penetration depth, t, is critical, as one does not know whether to take the value of the larger or the smaller pores, or an intermediate one.

Table 2. Comparison of volume fraction and inter pore distance, *IPD*, for macroporous epoxies prepared with 6 and 7.5 wt % hexane via the *CIPS* technique calculated either with the classical method from *SEM* micrographs or based on density measurements

wt % hexane	$\phi_{density}$	ϕ_{SEM}	$IPD_{density}$	IPD_{SEM}
6	2.05	1.71	3.91	4.28
7.5	1.96	2.72	4.91	4.15

Table 3. Morphological characteristics of macroporous epoxies prepared via CIPS with various amounts of hexane showing either a monomodal or bimodal distribution

wt% hexane	Type of distribution	A (µm²)	n	d (µm)	ϕ (%)	IPD (µm)	t (µm)	c (10⁶/mm³)
6	Narrow	20,9	169	2.01	2.05	3.9	1.7	4.82
7.5	Narrow	21,7	185	2.47	1.96	4.9	3.4	2.48
10	Bimodal	75,6	452	5.3	9.3	4.1	6.6	0.01
12.5	Bimodal	36,5	312	13.2	11.2	8.6	12.8	0.003
15	Bimodal	95,1	206	6.8	8.9	5.1	10.4	0.54

The calculation of the *IPD* in bimodal size distributions requires the value of the volume fraction as well as the number fractions x_1 and x_2 and the mean diameters of the smaller and the larger domains d_1 and d_2. The *IPD* of a bimodal distribution can be calculated [100] by iteration from

$$\frac{6\phi}{\pi} = \frac{x_1 \cdot d_1^3}{(d_1 + IPD)^3} + \frac{x_2 \cdot d_2^3}{(d_2 + IPD)^3} \qquad (37)$$

with

$$x_1 = \frac{n_1}{n_1 + n_2} \text{ and } x_2 = \frac{n_2}{n_1 + n_2} \qquad (38)$$

wherein n_1 and n_2 are the number of particles taken into account for one peak of the bimodal distribution.

For macroporous epoxies the *IPD* can be determined from Eq. (37) by iteration using the volume fraction from density measurements. This also allows one to determine the total volume, V, and consequently the pore concentrations, $C_{bimodal}$, according to the following set of equations:

$$V = \frac{4}{3} \cdot \frac{\pi}{\phi} \cdot \left(n_1 \cdot r_1^3 + n_2 \cdot r_2^3\right) \qquad (39)$$

$$C_{bimodal} = \frac{\left(x_1 \cdot r_1^3 + n_2 + r_2^3\right)}{V} \qquad (40)$$

The mean diameter of a bimodal distribution is calculated from

$$d_{bimodal} = x_1 \cdot d_1 + x_2 \cdot d_2 \qquad (41)$$

The results of image analysis of macroporous epoxies showing a narrow and bimodal pore size distribution are summarized in Table 3. The volume fraction, ϕ, is always calculated from density measurements. The validity of the data obtained with digital image analysis is of utmost importance in order to draw correct conclusions concerning the structure-property relationships.

4.2
Phase Separation Mechanism in Hexane-Epoxy Systems

The previous discussion has shown that the CIPS technique allows one to produce macroporous epoxy networks with either a narrow or bimodal size distribution. However, no indication has been given on the type of phase separation mechanism to yield these morphologies. As discussed earlier, the formation of a closed cell morphology can result either from a nucleation and growth mechanism or from spinodal decomposition.

The observation of a bimodal size distribution is the key to unravel the phase separation mechanism with respect to the kinetics of the two types of phase separation processes.

Let us surmise a homogeneous nucleation. In that case the nucleation rate is given by [101]

$$\frac{dN}{dt} = N_0 \cdot D \cdot e^{-\frac{\Delta G_n}{RT}} \qquad (42)$$

Therein N_0 represents the number of nuclei at the start of phase separation, ΔG_n is the activation energy for the creation of a nuclei, and R and T are the gas constant and absolute temperature, respectively. In the case of a thermosetting material, the diffusion constant, D, is a complex function that depends on the temperature and the viscosity, which itself changes with the continuous advancement in crosslinking reaction and therefore with time. The integration gives rise to an exponential decrease in the number of nuclei with time after the start of phase separation [67].

The growth rate, characterized by the change of the radius with time, is proportional to the driving force for the phase separation, given by the differences between ϕ_2^c, the chemical composition of the second phase in the continuous phase at any time, and ϕ_2^{eq}, its equilibrium composition given by the binodal line. The proportionality factor, given by the quotient of the diffusion constant, D, and the radius, r, is called mass transfer coefficient. Furthermore the difference between the initial amount of solvent, ϕ_0, and ϕ_2^{eq} must be considered. The growth rate is mathematically expressed by [101]

$$\frac{dr}{dt} = \frac{D}{r} \cdot \frac{\left(\phi_2^c - \phi_2^{eq}\right)}{\left(\phi_0 - \phi_2^{eq}\right)} \qquad (43)$$

As the growth rate is inversely proportional to the domain radius, smaller domains are able to grow faster than larger ones, thus giving rise to a narrowing of the size distribution with time.

Based on Eqs. (42) and (43), the development of a narrow or bimodal size distribution can be qualitatively explained without the detailed knowledge of the real phase diagram nor the exact dependency of the diffusion constant as a function of time. The final morphology depends mainly on the extent of reaction at which the metastable region is entered and the difference between ϕ_p and ϕ_0, as discussed below.

For concentrations slightly above ϕ_p, the metastable region is entered at high conversions (lines starting from points B and C in Fig. 17). The composition remains unchanged until the supersaturation becomes so high that the nucleation energy is exceeded. The nucleation then starts at points a' and c' in a highly viscous surrounding. Even though the diffusion constant is low as a consequence of the high viscosity, the initially formed small domains grow fast ow-

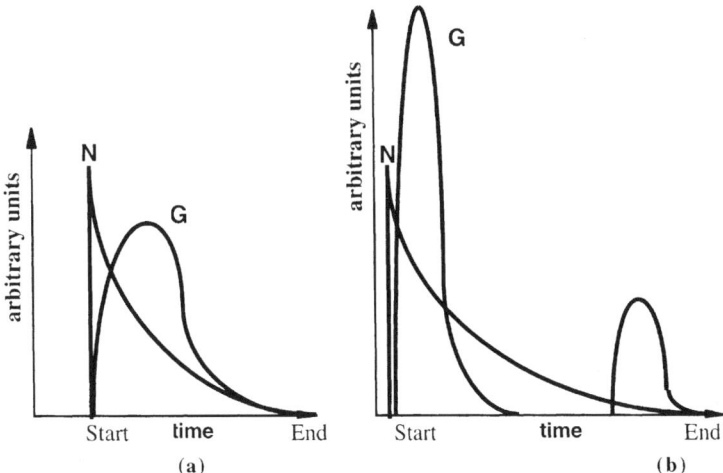

Fig. 22a,b. Nucleation rate (N) and growth rate (G) resulting in: **a** narrow size distribution; **b** bimodal size distribution

ing to the proportionality of the growth rate with $1/r$ (Eq. 43). After a certain period, the growth rate slows down close to zero due to two different contributions: first, as the composition approaches the equilibrium concentration, the driving force for phase separation, given by $\phi_2^c - \phi_2^{eq}$, tends to approach zero and, second, as gelation, accompanied by a dramatic increase in viscosity, is reached. This situation is responsible for the generation of narrow size distributions at concentrations slightly above ϕ_p. The time evolution of the nucleation and growth rates corresponding to a narrow pore size distribution are schematically shown in Fig. 22a. The trajectories shown in Fig. 17, which represent the composition change with increasing conversion, assume that a certain degree of supersaturation must be reached in order to overcome ΔG_n. Once the nucleation starts – at points a' and c' – the composition tends towards the equilibrium concentration with the simultaneous increase in conversion. Such trajectories have been calculated by Moschiar et al. [72] and Rozenberg et al. [97] in dependence of the interfacial tension between modifier and matrix material, but should occur much more easily here where the modifier is a solvent.

At higher concentrations of hexane, the metastable region is entered at lower conversions, hence lower viscosity. As the nucleation starts at the point d', the diffusion constant is high, leading to a very fast growth of the separated domains immediately after the start of the phase separation. This allows the system to reach the equilibrium concentration after a rather short period. Consequently, the driving force for the phase separation, given by $\phi^c_2 - \phi^{eq}_2$, becomes nearly zero, and the growth rate is considerably slowed down well before gelation. At this time, however, nucleation is still active, as shown in Fig. 22b, representing the corresponding nucleation and growth rates. After reaching the equilibrium concentration, the system can again become unstable, thus creating a second generation of nuclei at the imaginary composition d''. These nuclei are still able to grow relatively fast (owing to the proportionality of the growth rate with $1/r$ (Eq. 43), but their final size will remain smaller than for the droplets formed in the early stage of phase separation. This finally leads to a bimodal distribution. This behavior is schematically represented by the trajectory starting from point D in Fig. 17. The ongoing nucleation, which never stops between the two nucleation events, leads to the observation that the percentage of pores before the secondary nucleation becomes small but not zero. A very similar phase separation model has been proposed by Vazquez and coworkers [67]. Based on this phase separation model we conclude that the phase separation resulting in a bimodal distribution proceeds via a nucleation and growth mechanism. The development of such a bimodal distribution cannot be explained with regard to the kinetics of spinodal decomposition, which has been investigated by Inoue [93] and Char et al. [102].

Another characteristic concentration in Fig. 17 is point E, which gives the solubility limit of the solvent in the unreacted precursor mixture. If the solvent concentration exceeds this solubility limit, demixing occurs in the initial state.

To summarize, we conclude that a chemical quench at low curing temperatures leads to a continuous and smooth transition from the stable to the metastable region and therefore favors a nucleation and growth mechanism. In such a case the composition approaches the binodal line in order to achieve thermodynamic equilibrium and it automatically turns away from the spinodal decomposition region. Therefore the *CIPS* technique allows for the synthesis of macroporous thermosets with controlled morphologies providing either narrow or bimodal size distributions. In contrast, spinodal decomposition phenomena have been reported very often as a consequence of a temperature quench, especially in systems exhibiting a *UCST* behavior, where the exact phase diagram is not known in advance and the spinodal region is often entered.

4.3
Influence of Reaction Parameters on the Morphology of Cyclohexane-Modified Epoxy Networks Prepared via CIPS

Phase separation has first been observed by using hexane or cyclohexane as solvents [49, 50]. If cyclohexane is used as the solvent, the desired phase separation is observed at concentrations ranging from 14 wt % to around 25 wt % cyclohexane at a curing temperature of 40 °C. It is obvious, that the nature of the solvent has a great influence on the onset of phase separation, ϕ_p. The influence of the nature of the solvent, expressed by the solubility parameter, has already been discussed in Sect. 3. This section will be limited to discussing the influence of solvent concentration, curing temperature, drying procedure, and reaction rate on the morphology.

4.3.1
Influence of Solvent Concentration

Experimentally ϕ_p (cyclohexane, 40 °C) was found to be 13–14 wt % cyclohexane based on isothermal curing experiments [49, 50]. Regardless of the large difference in ϕ_p, the morphologies that are obtained by using either cyclohexane or hexane as solvent are very similar at concentrations slightly above ϕ_p. Figure 23a,b shows the *SEM* micrographs of samples cured isothermally at $T=$ 40 °C with 15 wt % and 20 wt % cyclohexane respectively. It can be seen that a closed cell morphology, including a narrow size distribution, is achieved.

The *SEM* micrographs reveal that the pore size and the volume fraction increases with increasing amount of solvent. This qualitative result is also confirmed by image analysis performed on an average of around 150–250 pores, clearly showing the expected increase of pore size with increasing amount of cyclohexane (Fig. 24). This phenomenon has been observed in any polymer-solvent system studied here.

The plot of the diameter vs the relative number of pores also provides useful information on the size distribution. A broadening of the size distribution will then give a reduction of the peak height. In contrast to the hexane system, no bimodal distribution is detected up to the solubility limit of cyclohexane at 25 wt %.

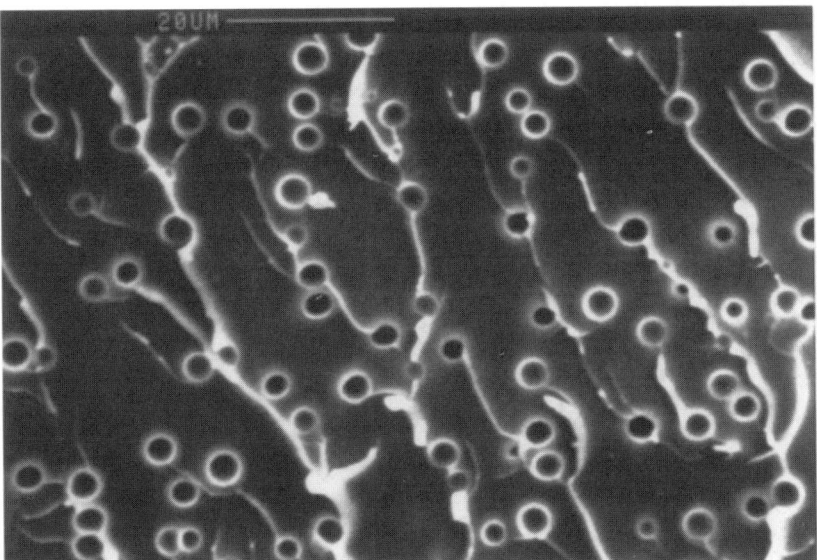

Fig. 23a. SEM micrographs of macroporous epoxy networks prepared via CIPS with 15 wt % cyclohexane at a curing temperature of 40 °C. Reprinted from Polymer, 37(25). J. Kiefer, J.G. Hilborn and J.L. Hedrick, "Chemically induced phase separation: a new technique for the synthesis of macroporous epoxy networks" p 5721, Copyright (1996), with permission from Elsevier Science

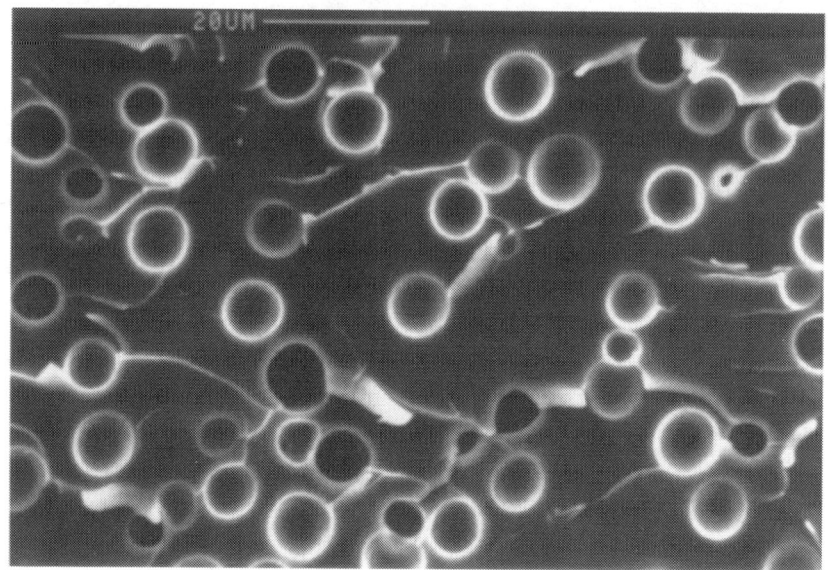

Fig. 23b. SEM micrographs of macroporous epoxy networks prepared via CIPS with 20 wt % cyclohexane at a curing temperature of 40 °C

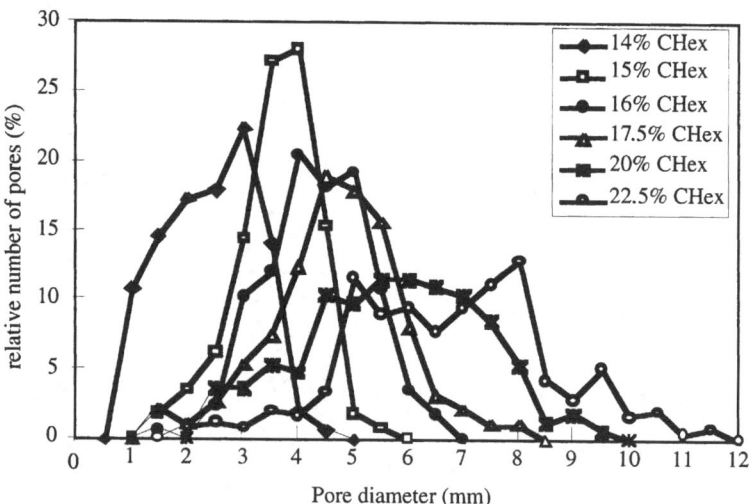

Fig. 24. Pore size distributions of macroporous epoxies prepared via CIPS with cyclohexane obtained from image analysis

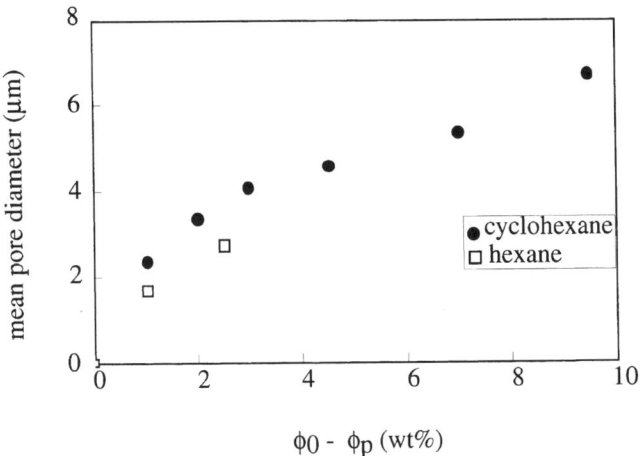

Fig. 25. Mean pore diameter, d, of macroporous epoxies prepared via *CIPS* with different solvents. $\phi_0 - \phi_p$ gives the difference between initial solvent concentration and critical solvent concentration for phase separation

In comparison to the results obtained for the samples prepared with hexane, it can be concluded that the mean pore size and volume fraction do not depend on the initial concentration of the solvent, ϕ_0, but mainly on the difference between ϕ_0 and ϕ_p (Fig. 25). Similar qualitative results are also reported for rubber-modified epoxies prepared via reaction induced phase separation [103].

4.3.2
Influence of Curing Temperature

The influence of curing temperature and concentration of cyclohexane on the phase separation behavior resulting from isothermal curing experiments is summarized in Fig. 26. It is seen that the critical amount for phase separation, ϕ_p, is lowered by decreasing the curing temperature which is characteristic for *UCST* behavior. The ϕ_p values give a phase separation line which forms a boundary between the regions of opaque and transparent samples. By comparing Fig. 26 and Fig. 10, one finds that nearly identical results are obtained using either the gradient oven or isothermal experiments to construct the phase separation line.

With respect to this temperature dependence, it is then intended to find out whether it is possible to lower the pore size by approaching the phase separation line in Fig. 26. In one series of experiments a constant cyclohexane concentration of 20 wt % is chosen and samples of identical composition are cured isothermally at temperatures ranging from room temperature to 120 °C. Phase separation occurs at curing temperatures equal to or below 104 °C, and transparent

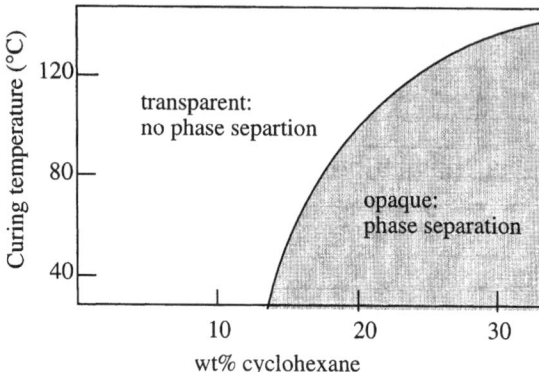

Fig. 26. Influence of isothermal curing temperature on the phase separation behavior of macroporous epoxies prepared via CIPS with cyclohexane

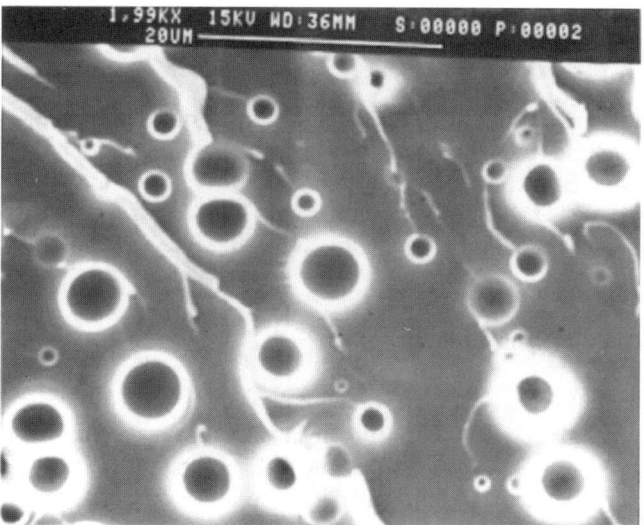

Fig. 27. SEM micrograph of macroporous epoxy prepared with 20 wt % cyclohexane cured at $T=104\,°C$. Reprinted from Polymer, 37(25). J. Kiefer, J.G. Hilborn and J.L. Hedrick, "Chemically induced phase separation: a new technique for the synthesis of macroporous epoxy networks" p 5723, Copyright (1996), with permission from Elsevier Science

samples are obtained by curing at temperatures above 105 °C. Thus the boundary is determined very precisely for this composition. Figure 27 shows the *SEM* micrograph of the sample prepared at 104 °C, exactly at the point where the start of phase separation should occur quasi simultaneously at the moment where the nucleation is suppressed, q_{nucl} (see Fig. 17). Pores with diameters ranging from

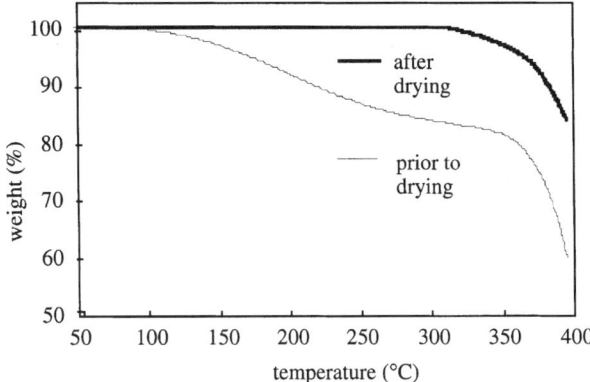

Fig. 28. Weight loss of epoxies prepared with 20 wt % cyclohexane before and after the drying procedure measured with TGA

around 1 µm to 6 µm appear. It seems that the smallest domains are inhibited from growing further. In a second series of experiments, samples with different cyclohexane concentrations are prepared and cured isothermally at 40, 80, or 120 °C and the critical amount for phase separation is determined with an accuracy of 1 wt %. Both attempts to approach the phase separation line by either increasing the curing temperature for a constant concentration or raising the concentration for a given curing temperature did not allow for the preparation of macroporous epoxies having pore diameters substantially less than 1 µm [85], although it is generally assumed that the critical size for nucleation is typically less than 10 nm.

It is still questionable whether the end of phase separation or q_{nucl} is linked to gelation or vitrification, and if a limited growth of the separated domains in the gelled or glassy state is still possible. Within their investigations on the phase separation behavior in rubber-modified epoxies, several research groups have come up with experimental evidence for the assumption that phase separation is still possible after gelation [66, 104–106]. Based on the above morphological observations, we conclude that, for our particular system, a time delay between the limit for nucleation, q_{nucl}, and the end of phase separation must exist, thus allowing the nuclei to grow to around 1 µm.

4.3.3
Influence of Drying Procedure

After phase separation, the creation of an isolated porous morphology is achieved by holding the sample at a temperature in the rubbery state, thus enabling for high diffusion rates well above the boiling point of the solvent. On the other hand the temperature must kept below the decomposition temperature of the network. Thus removal of the low molecular weight liquid is achieved by

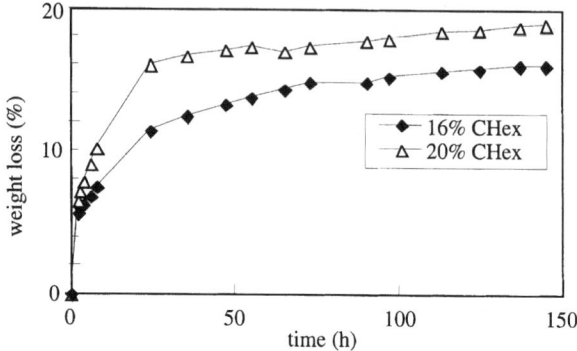

Fig. 29. Weight loss of cyclohexane modified epoxies by holding at $T=200$ °C

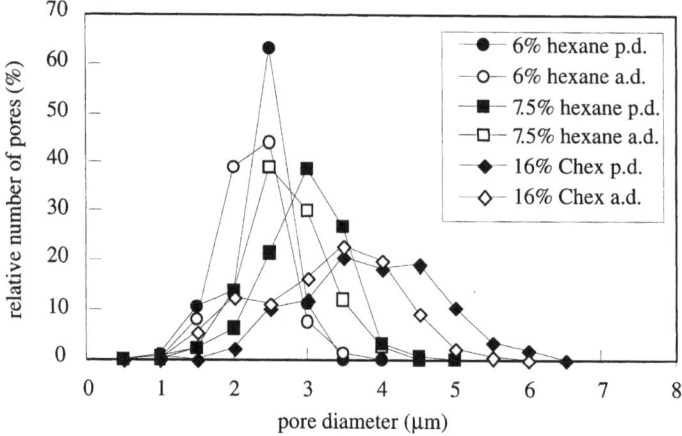

Fig. 30. Size distribution of solvent modified and macroporous epoxies prepared via CIPS with different types and amounts of solvent before and after the drying procedure

heating the samples above T_g^∞ of the neat matrix ($T_g^\infty = 170$ °C) at a temperature of 200 °C for 120 h. The complete solvent removal is verified with thermogravimetric analysis (*TGA*). Figure 28 shows the weight loss of samples prepared via *CIPS* at T=80 °C with initially 20 wt % cyclohexane prior to and after the drying procedure. The *TGA* measurement of the sample prior to drying clearly shows a weight loss of 20%, identical with the initial amount of solvent added, up to the onset of decomposition reactions at temperatures above 350 °C. Thus no solvent loss occurs during the preparation of the phase separated samples in the closed tubes. Furthermore it can be concluded, that temperatures above T_g^∞ are required to enable solvent removal. After the thermal drying, no weight loss is detected with *TGA* below the decomposition temperature, and hence the drying is complete. As the drying is performed via a diffusion process, the time for drying depends on the drying temperature, the crosslink density of the network, the molecular structure of the liquid, and the sample dimensions. For samples with co-continuous morphology the drying is simple and rapid since solvent may leave through the channels of the pores.

The solvent removal can also be followed by gravimetric analysis of the weight loss. This method is used to determine the time necessary for complete solvent removal. Shown in Fig. 29 are such drying curves for 5 mm diameter cylindrical samples prepared with 16 wt % and 20 wt % cyclohexane. It is seen that most of the solvent is removed within the first day. However, the samples are held for an additional four days to achieve a porous structure with minimal amount of residual solvent.

SEM including image analysis is performed prior to and after the thermal removal of the liquid phase to investigate the influence of the drying process on the morphology. Typical size distributions of epoxies showing a narrow size distribution prepared with different amounts of hexane and cyclohexane are shown in Fig. 30. These investigations show no significant change in morphology, and thus neither ripening, coarsening, nor collapse of the dispersed domains occur during the drying procedure. Hence the pore size and size distribution is predetermined by the morphology generated during the phase separation process, knowing that the matrix must shrink upon removal of the solvent contained in it after phase separation.

4.3.4
Influence of Reaction Rate

It was concluded in Sect. 4.2 that the phase separation proceeds via a nucleation and growth mechanism. Consequently the domain size depends on the competing effects between the growth rate and the reaction rate to build the highly crosslinked network. Thus the morphology development is closely linked to reaction kinetics. Hence the question arises whether the pore size might be further lowered by increasing the reaction rate. Clearly this might be achieved by increasing the reaction temperature. However it should be kept in mind that a change in reaction temperature automatically effects the onset of phase separation. We were looking to change the reaction rate without affecting the thermodynamics. Therefore we have selected a catalyst which only speeds up the reaction rate and does not change the phase separation gap at a catalyst concentration of 1 wt % [85]. As the morphology should be controlled by changing the curing kinetics, this process is called *kinetically controlled CIPS*.

Isothermal differential scanning calorimetry (DSC) measurements were carried out to investigate the curing kinetics [85]. Conversion vs time curves of DGEBPA-PACP systems prepared with 1 wt % of catalyst and without catalyst at identical curing temperature are overlaid in Fig. 31.

It can be clearly seen that the reaction proceeds considerably faster when 1 wt % catalyst is added to the precursor mixture. The influence of solvent concentra-

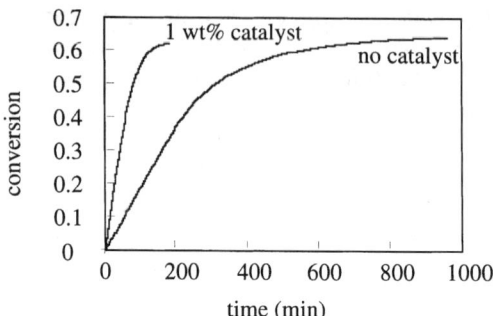

Fig. 31. Influence of catalyst on reaction rate

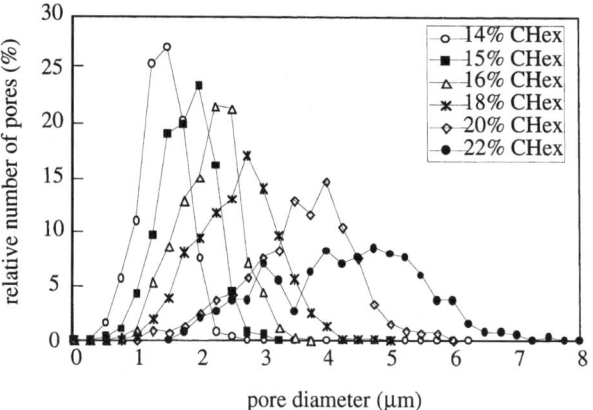

Fig. 32. Pore size distribution of macroporous epoxies prepared via kinetically controlled CIPS with 1 wt % catalyst

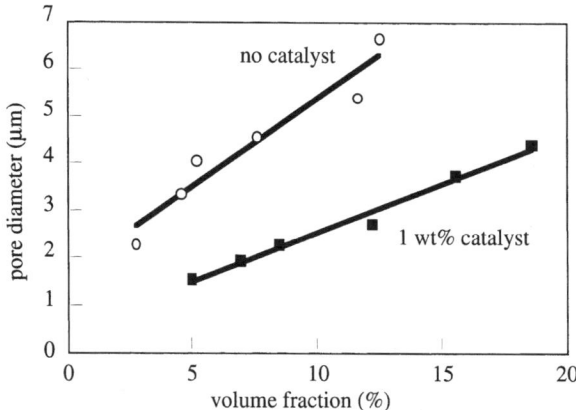

Fig. 33. Influence of catalyst on mean pore diameter

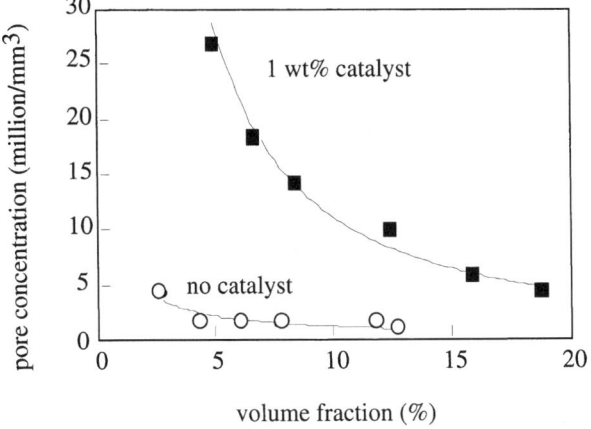

Fig. 34. Influence of catalyst on pore concentration

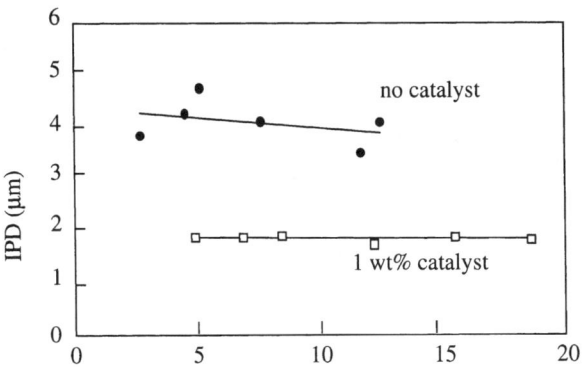

Fig. 35. Influence of catalyst on interparticle distance (IPD)

tion on the curing process has not been tested with isothermal DSC measurements to avoid damage of DSC apparatus. Image analysis on an average of around 400 pores has been performed on systems prepared with 1 wt % catalyst and compared to data obtained on identical systems prepared without catalyst at equal curing temperatures (T=40 °C). The pore size distribution of systems prepared with kinetically controlled CIPS with 1 wt % catalyst and cyclohexane concentrations of 14–22 wt % is shown in Fig. 32. In perfect agreement with the results of Sect. 4.3.1, an increase in mean pore size and broadening of size distribution is observed with increasing cyclohexane concentration.

Plotted in Figs. 33–35 are variations of morphological characteristics such as mean diameter, pore concentration, and IPD as a function of the volume fraction. It is clearly seen that the use of a catalyst allows to greatly change the morphological characteristics. The diameter increases almost linearly with the volume fraction and the use of 1 wt % catalyst leads to a reduction in pore size of more than 50% (Fig. 33). The decrease in pore concentration with increasing volume fraction is a direct consequence of a coalescence mechanism, which is driven by the reduction of surface energy. The pore concentrations of macroporous epoxies prepared with 1 wt % catalyst via kinetically controlled CIPS is substantially higher than in the uncatalyzed sytem (Fig. 34).

Very interestingly, the IPD is found to show practically no or very little dependence on the volume fraction and only depends on the reaction rate (Fig. 35). Here the IPD is calculated from the mean diameter taken from image analysis and the volume fraction, which has been determined directly from the porosity of these macroporous thermosets. Both values could be determined experimentally with high accuracy and no simplifications are needed for the calculations.

4.4
Characteristics of Epoxy Networks Prepared with Hexane and Cyclohexane via CIPS

The influence of the drying procedure on the density, both prior to and after evaporation of the low molecular weight liquid, is plotted in Fig. 36 for samples cured with various amounts of cyclohexane. It can clearly be seen that, after drying, a considerable drop in density results from the formation of a porous morphology at concentrations above ϕ_p. However, practically no decrease in density is measured after the drying of transparent samples. In this case, the low molecular weight liquid is not involved in the formation of separated domains and after drying the fully crosslinked network is achieved by solvent evaporation from the matrix. Therefore the density of transparent samples increases during the drying process and approaches the value of the fully crosslinked network (1.123 g/cm^3). Similar results are also obtained for macroporous epoxies prepared with hexane as shown in Fig. 37.

Prior to drying, the solvent dissolved in the matrix has a plasticizing effect, thus lowering the T_g of the crosslinked product. This is shown in Fig. 38 where

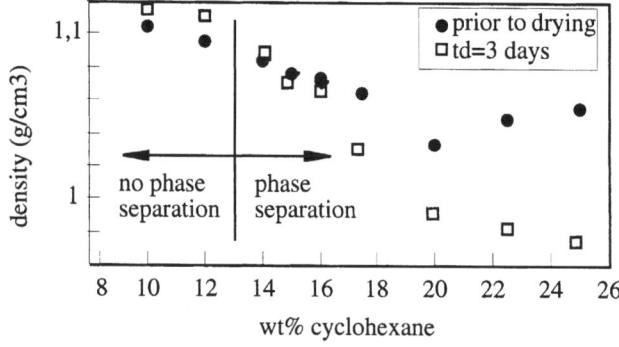

Fig. 36. Density of epoxy networks prepared via CIPS with cyclohexane before and after the drying procedure

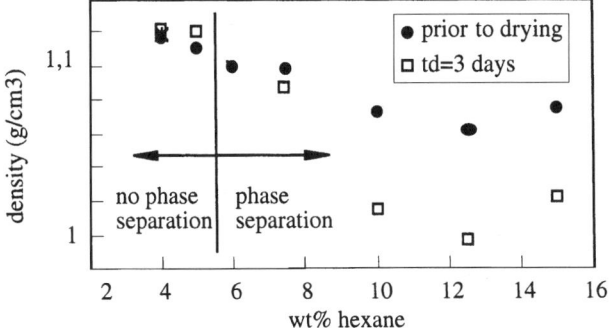

Fig. 37. Density of epoxy networks prepared via *CIPS* with hexane before and after the drying procedure

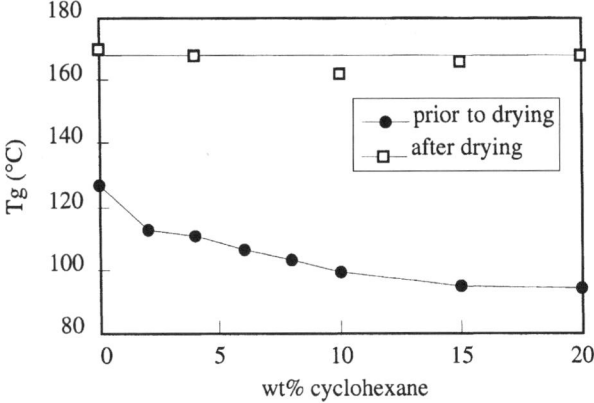

Fig. 38. Glass transition temperature of epoxies prepared with cyclohexane via *CIPS* before and after the drying procedure

the T_g, taken from dynamic mechanical analysis (*DMA*) measurements [49], is plotted against the solvent concentration both prior to and after thermal removal of solvent-modified and macroporous epoxies prepared with cyclohexane. Amazingly, no lowering of final T_g is observed after the drying procedure, even at initial high amounts of cyclohexane. Thus, the final T_g of the matrix is independent of the initial amount of solvent and T_g^∞ of the neat matrix can be reached upon drying. This also shows that solvent removal is completed and an eventual post curing is realized simultaneously with the drying procedure. CIPS therefore allows for the synthesis of macroporous thermosets without lowering the thermal stability of the fully crosslinked network. This is in contrast to rubber-modified epoxies, where the network modifier fraction, which is not involved in the phase separation, remains dissolved in the epoxy matrix after cure and hence leads to a decrease in T_g.

As a conclusion to this section on morphology development it can be said that CIPS allows for the preparation of macroporous epoxies with controlled morphology and closed pores with diameters of 1–10 μm. The pore size distribution is predetermined by the phase separation process and can be altered by carefully adjusting the internal and external reaction parameters to obtain either a narrow size distribution or a bimodal size distribution. The appearance of a bimodal size distribution clearly indicates that the phase separation proceeds via a nucleation and growth mechanism rather than via spinodal decomposition. The nucleation and growth mechanism is favored by low reaction rates, thus enabling for a smooth transition from the stable into the metastable region. The phase separation behavior strongly depends on the chemical nature and amount of solvent, as well as on the curing temperature. The nature of the solvent influences the critical amount for phase separation, ϕ_p. The pore size and volume fraction depend mainly on the difference between the initial amount of solvent and ϕ_p. Pore size and volume fraction increase with the solvent concentration once ϕ_p has been passed. Lowering the curing temperatures enables one to lower ϕ_p for the epoxy-cyclohexane system which is typical for UCST behavior. Morphological characteristics such as pore size, pore concentration, and IPD can be significantly varied simply by using a catalyst which speeds up the curing reaction without affecting the phase separation gap. Thermal removal of the solvent affects neither the size nor distribution and acts as a simultaneous post curing, and hence T_g^∞ is independent of the initial amount of low molecular weight liquid. Macroporous epoxies prepared via CIPS are characterized by considerably lower density without any other alteration of the crosslinked matrix.

In the next section we will examine the influence of phase separation on the toughness of solvent-modified and macroporous epoxies prepared via CIPS.

5
Toughness of Solvent-Modified and Macroporous Epoxies Prepared via CIPS

The enhancement of toughness is of the utmost importance to render thermosetting polymers (which are inherently brittle owing to their high crosslink density) suitable for industrial applications. Since the pioneering work of Sultan and McGarry published in 1973 [107], who discovered that dispersed rubber particles can substantially enhance the toughness of highly crosslinked epoxy networks, an overwhelming number of research projects has been devoted to optimize the morphology with the target to maximize the efficiency of the dispersed phase for toughening. It is well known that the toughenability is largely influenced by the crosslink density of the thermosetting network, and that a lowering of the crosslink density leads to an increase in toughenability [108–112]. On the other hand, a lowering of the crosslink density leads to a reduction in stiffness, strength, and thermal stability of the matrix material [113, 114]. Very often these properties determine the selection of the appropriate precursors to form a highly crosslinked matrix needed to meet the requirements. Consequently it becomes necessary to enhance the toughness for a given matrix material. Any

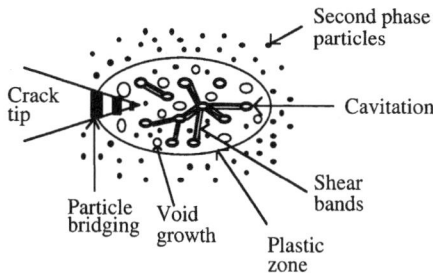

Fig. 39. Micromechanism of toughening encountered in thermoset

strategy to tailor morphology in order to optimize toughness requires a detailed understanding of toughening micromechanisms. Several theories have been proposed and rejected within the last two decades, and the contribution of individual micromechanisms to the overall toughness is still a subject of intensive research and vital discussions.

In the following, we will discuss toughening of thermosets and focus particularly on the role of cavitation. The most widely accepted model of toughening micromechanisms in thermosets goes back to ideas of Garg and Mai [115] as well as Huang and Kinloch [116]. This model is schematically redrawn in Fig. 39 and summarizes the multiple events found ahead of a crack tip in the plastic zone of toughened thermosets. It becomes obvious, that numerous parameters exist which can effect the toughness such as

- nature of particles
- particle dispersion
- particle adhesion
- particle size
- volume fraction
- interparticle distance
- cavitation.

Two different approaches for toughening are most frequently used to achieve the desired two-phase morphology. The first is based on a phase separation process, where a rubbery or thermoplastic phase starts to phase separate during the curing reaction, identical to the CIPS technique, to produce solvent-modified and macroporous thermosets presented in this work. The thermodynamic origins of the phase separation process and the influence of internal and external reaction parameters have been studied extensively for various types of thermoplastic [74, 117] and rubber-modified epoxies [66, 67, 69, 71] as well as for other classes of thermosets such as cyanurates [118] or polyesters [119]. This allows one to control the morphology and to enhance the mechanical properties [68, 120]. However, the generation of a two-phase structure via phase separation usually leads to a reduction in thermal stability owing to the amount of modifier remaining in the matrix. This problem can be overcome by using high T_g-thermoplastics as modifier, which in turn increases the costs [121–125]. One further inconvenient of toughened thermosets which are prepared via phase separation is the fact, that the morphological characteristics such as the particle size and volume fraction cannot be varied independently. Furthermore, this process requires accurate control of modifier concentration and curing temperature to achieve reproducible results.

As an alternative to reaction induced phase separation, preformed particles can be mixed into the precursor mixture. These so-called core-shell particles are prepared via emulsion polymerization [126, 127] and therefore have very narrow size distributions. Consequently it becomes possible to vary the size and volume fraction independently, thus leading to materials with tunable morphologies. However the mixing with preformed particles introduces an additional parameter, which is the particle dispersion, itself strongly depending on the mixing conditions and particle surface chemistry. Generally the surface of these modifiers is chemically designed in order to achieve good adhesion to the matrix and simultaneously avoid particle agglomeration. Compared to to previous strategies to achieve optimum dispersion of core-shell particles, very recent investigations show better toughenability if no surface modification is realized [98, 128–130]. In these cases the higher toughness is attributed to a larger plastic zone size, which is observed in systems clearly showing local agglomerations. This observation is very interesting because these local agglomerations are interconnected to form a network-like structure very similar to interpenetrating polymer networks (IPN). Such IPNs proved to be very effective for toughening PMMA and these systems showed practically no rate dependence of toughness with a small amount of modifier [131]. These recent results demonstrate that much more investigation is needed to understand the influence and importance of particle dispersion for toughening.

After the pioneering work of Sultan and McGarry, it was claimed that the multiple stretching and fracture of rubbery inclusions in front of the crack tip, called rubber bridging or rubber tearing, is the major mechanism responsible for high energy consumption and consequent increase in toughness [132]. However, similar toughening effect can also be achieved with rigid inclusions, such as glassy thermoplastic particles [111, 133–135] or even glass beads [136, 137] and ceramic particles [138], which do not allow for plastic deformation themselves. The desired toughening effect is also found using particles composed of hyperbranched molecules [139, 140] or epoxidized soybean oil [141] phase separating to dispersed domains. Hence the nature of dispersed particles does not play the predominant role for toughening. Recently Huang and Kinloch presented mathematical models to calculate the contribution of individual micromechanisms to the overall toughness and concluded that rubber bridging plays a secondary role for toughening and becomes only important, when the ability of the matrix to undergo plastic deformation is suppressed [116]. Nowadays it is generally accepted that the multiple formation of shear bands is a necessary condition for toughening and shear banding is considered as the most important toughening mechanism. Then the question arises of the origin and the reason for generating shear bands.

The occurrence of shear banding seems to be closely related to cavitation which results either from interfacial debonding or from internal cavitation. The phenomenon of cavitation has been recognized since the original work of Sultan and McGarry. This cavitation was first thought to be unlikely and considerable efforts were undertaken to use end-functionalized rubbers that can react with the thermosetting matrix to give a chemical bonding at the interface, thus avoiding interfacial debonding. However such a treatment to enhance the adhesion also favors the miscibility between the rubbery phase and the matrix. This automatically leads to lower domain sizes and lower volume fractions for identical amounts of non-functionalized and functionalized particles [142]. The most common reactive liquid rubbers used for toughening via reaction induced phase separation are carboxyl terminated (CTBN) [143–146] or amino terminated butadienes (ATBN) [105, 147–149]. Despite the considerable improvement of interfacial adhesion, no significant changes in toughenability are obtained from these approaches [150, 151]. If the interfacial debonding is chemically suppressed owing to surface modification, cavitation can occur internally.

Bucknall relates the cavitation resistance to the modulus and the size of the dispersed phase. These calculations show that the cavitation resistance increases with decreasing particle size [152–154]. Furthermore, it was found that the cavitated particles are aligned in the plastic zone to give a craze-like damage zone. This phenomen has been called croiding [155] in toughened epoxies or dilatational bands [154] in rubber-modified thermoplastics. Experimentally it is found that a critical particle size exists where the cavitation is suppressed. Several research groups observed that no toughening occurs if the second phase particles become smaller than 150–200 nm [154, 156, 157]. This is in agreement with theoretical considerations which predict that this is a lower critical size to induce cavitation [152]. Bucknall claims that the cavitation should occur just prior to yielding to achieve the best toughening effect in thermoplastic polymers and calls this "just in time cavitation" [158, 159].

While the surface modification is not effective to suppress cavitation, Yee and coworkers performed an experiment to suppress the cavitation mechanically in a rubber-modified epoxy network. They applied hydrostatic pressure during mechanical testing of rubber toughened epoxies [160]. At pressures above 30–38 MPa the rubber particles are unable to cavitate and consequently no massive shear yielding is observed, resulting in poor mechanical properties just like with the unmodified matrix. These experiments proved that cavitation is a necessary condition for effective toughening.

In addition to the cavitation process related to the presence of a dispersed phase, the formation of voids in the plastic zone has been observed to occur also in the matrix phase. Kinloch and Huang stated that the plastic void growth succeeding cavitation also contributes to energy absorption and might become as important as the shear banding, especially at fairly elevated temperatures [161].

Several groups tried to unravel whether the cavitation occurs before or after the onset of plastic deformation. Real-time small angle X-ray scattering [162] and light scattering [163, 164] techniques were used to study the deformation of rubber-modified thermoplastics in situ. All these groups conclude that the cavitation and deformation of glassy ligaments between the rubber particles ("micronecking") precedes crazing, which is considered as the most important toughening mechanism in thermoplastic polymers. These studies were realized on rubber toughened thermoplastics and it is believed that the cavitation also precedes the formation of shear bands in toughened thermosets, even though identical in-situ studies on toughened thermosets are still not available.

Cavitation is equivalent to the generation of randomly distributed voids. Indeed such a morphology has been simulated by using either a non-reactive rubber [161] or hollow latex spheres as the dispersed phase [165–167]. In the first study the liquid rubber remained dispersed and only one single composition has been studied. In the second study, hollow particles were used, thus creating an additional interface. Using such pseudo-porous systems, both groups claimed the ability of voids to toughen epoxies in the same manner and to the same magnitude as rubber particles. The same group calculated the stress distributions of spherical inclusions, being either rubber [168, 169] or voids [170, 171], in an isotropic epoxy matrix or in polyamides [172] based on a finite element model. It was concluded that the effect of voids or a rubbery phase is very similar. These inclusions can release the degree of triaxial stresses at the crack tip and lead to the formation of shear bands. Even though this model allowed one to predict the possibility of void toughening, the finite element approach does not allow one to take into account the multiple interactions between voids.

The most widely used model to relate the toughness to morphological characteristics is the interparticle distance (IPD) concept, first presented by Wu [99]. He studied the deformation behavior in rubber toughened nylon which clearly shows a brittle-ductile transition. Together with the results from image analysis, he concludes that a ductile behavior results if the IPD decreases below a critical value. This critical interparticle distance is considered as a material characteristic. This concept also forms the basis of toughening strategies in thermosets, even though no clear experimental evidence of a critical interparticle distance has been reported for rubber toughened thermosets.

To summarize the above observations, it was concluded that the cavitation is at the origin of the shear band formation. Thus the question arises as to whether the toughness can be enhanced by using voids as the dispersed phase. The experimental verification of the above theoretical predictions of void toughening requires a technique which yields porous thermosets having closed pores with sizes and distributions in the µm-range similar to those commonly used for toughening with rubber or thermoplastic particles. The void toughening should lead to materials which would combine low density with high toughness. Such materials would be highly desirable for many applications such as transport industries.

Solvent-modified and macroporous epoxies prepared via the CIPS technique are ideal materials to verify these predictions and to throw some light on the ongoing discussion on the role of the second phase and cavitation for the toughening of thermosets.

5.1
Calculation of Stress Distribution in Macroporous Epoxies

Upon loading a void-containing material, a certain stress distribution in the sample will develop that proceeds and determines the following deformation. Typically the voids (or other dispersed phase) will tend to concentrate stresses to interphases between materials of different modulus. Even though no complete picture exists of what will happen upon deformation, such a stress description may give a better understanding of the relation between stress concentrations in the sample due to the voids and the final fracture behavior.

Fond et al. [84] developed a numerical procedure to simulate a random distribution of voids in a definite volume. These simulations are limited with respect to a minimum distance between the pores equal to their radius. The detailed mathematical procedure to realize this simulation and to calculate the stress distribution by superposition of mechanical fields is described in [173] for rubber toughened systems and in [84] for macroporous epoxies. A typical result for the simulation of a three-dimensional void distribution is shown in Fig. 40, where a cube is subjected to uniaxial tension. The presence of voids induces stress concentrations which interact and it becomes possible to calculate the appearance of plasticity based on a von Mises stress criterion.

The stresses around the particles are determined to find the maximum von Mises equivalent stress. These calculation show that the mean value of the stress concentrations is a few percent greater than the exact value for a single void in an infinite matrix. The effect of mechanical interaction globally increases the

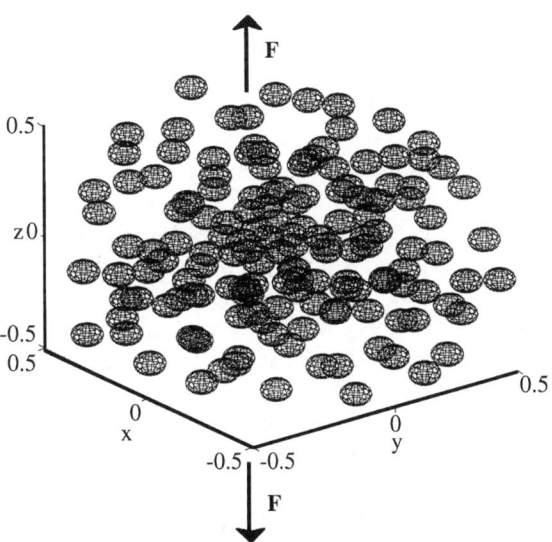

Fig. 40. Distribution of voids resulting from simulation

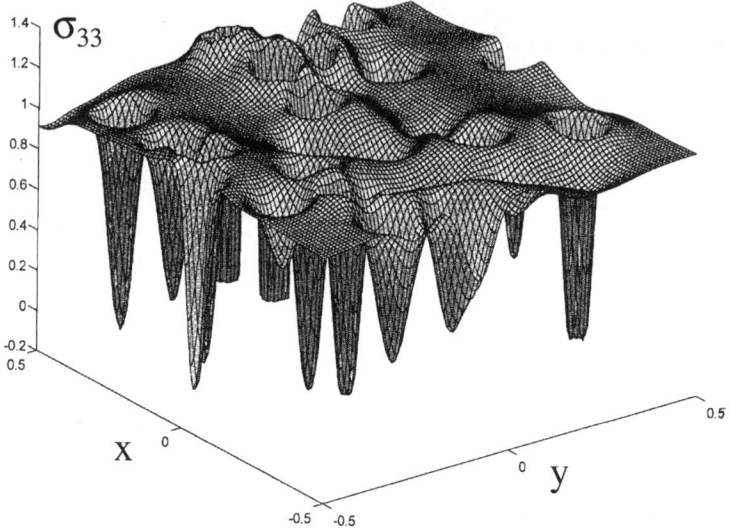

Fig. 41. Simulated stress distribution in the z-direction for a sample containing 138 voids subjected to a deformation of 1%

maximum von Mises equivalent stress and the appearance of plasticity is encountered for a lower external loading. Practically, the yield stress of the material decreases when interaction is present.

A typical example of the stress distribution in the z-direction, σ_{33}, at a constant value of z is shown in Fig. 41 for a sample containing 138 voids subjected to a deformation of 1%. Clearly a large number of points are observed where the stress is completely released. These points correspond to locations where a void can be found either closely below or above the plane at the considered value of z. This picture also demonstrates the importance of the interaction of voids on the stress concentrations. If the distance between two neighboring pores is large, no increase in stress concentration is observed. However, if several voids are located very close to each other, the interaction leads to the buildup of local stresses, which are considerably higher than the imposed loading. Hence these calculations confirm the importance of the distance between the pores in generating local stress concentrations, thus enabling multiple shear band formation.

This observation is also confirmed if one plots the normalized von Mises stresses for the same conditions (Fig. 42). Again, one recognizes a large number of points where the stress is released and, on the other hand, a considerable number of regions where the von Mises stress becomes twice as great as the external loading. Those regions will lead to an effective lowering of the onset of plastic deformation, representing the yield point. The calculations predict the plasticity to appear locally for macroscopic loadings, which are lower than the yield stress of the neat matrix by a factor of about 1.7–2.3. The normalized dis-

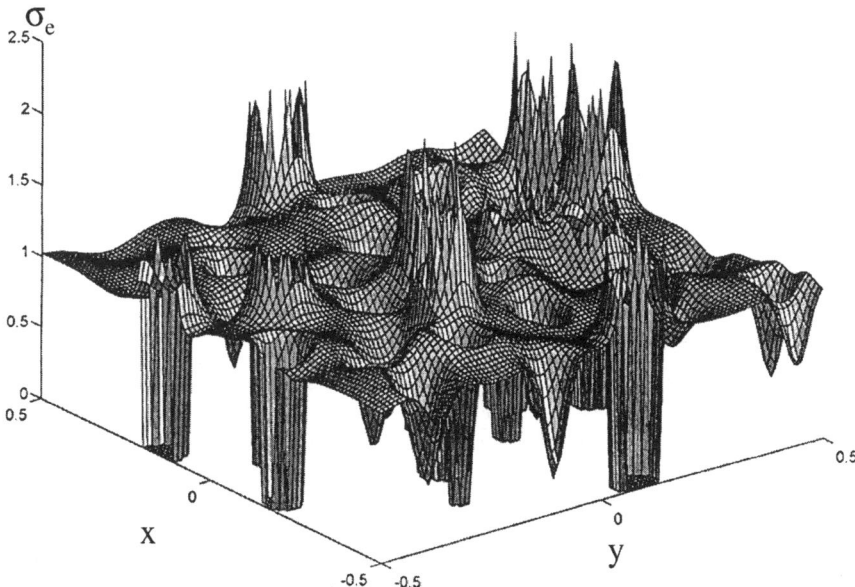

Fig. 42. Von Mises stress distribution for a sample containing 138 voids subjected to a deformation of 1%

Table 4. Characteristics of macroporous epoxies used for compression testing

Wt % cyclohexane	Density after drying (g/cm^3)	Porosity (%)	Pore diameter (µm)	IPD (µm)
15	1.123	–	–	–
16	1.1	2.05	1.3	2.5
18	0.982	12.6	4.4	2.7
20	0.975	13.2	4.9	2.9
22.5	0.935	16.7	5.3	2.4
25	0.94	16.3	6.6	3.1

tribution of stress concentration seems to be insignificantly influenced by the volume fraction of pores in the range 2–10%.

The numerical simulations of the stress distributions are carried out on porous materials submitted to uniaxial loading. In order to check the validity of the numerical simulations, macroporous epoxies are prepared via the CIPS technique. Cyclohexane is selected as the solvent, thus resulting in the formation of a closed porosity, and the statistical distribution of the voids coincides with the random distribution of the model system. The structural characteristics of these materials prepared by curing at T=80 °C are summarized in Table 4.

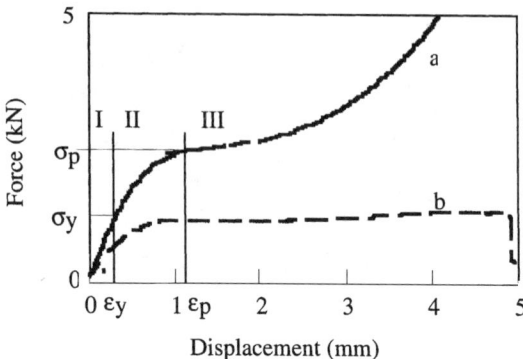

Fig. 43. Load displacement curves under uniaxial compression for the neat epoxy (*line a*) and macroporous epoxy networks prepared via CIPS with 20 wt % cyclohexane (*line b*)

Typical results of stress deformation curves under uniaxial compression of the neat matrix material and a porous epoxy prepared with 20% cyclohexane, thus exhibiting a porosity of around 13%, are plotted in Fig. 43. In these curves one can distinguish mainly three different regions. In the early stage of deformation the material shows an ideally linear behavior. The modulus, given by the slope in this early stage of deformation, decreases as a consequence of the generation of a porous morphology. The onset of plastic deformation leads to a deviation from linear behavior. It is concluded from the simulations that the shear banding starts at a limited number of voids (<5%) which satisfy the yield criterion. As the external load increases further in the second region, the stress concentrations in the material become more important, involving a continuously increasing number of pores to fulfill the yielding criterion. As the stress concentrations exceed the yield stress in the entire material, extensive plastic deformation takes place in region 3.

The calculations are limited to the elastic behavior. Therefore, in the following discussion the onset of plastic deformation will be taken as a comparative value for the yield strength, σ_y, and not the stress at the onset of extensive plastic deformation, σ_p. Figure 44 shows the experimental data of the yield strength, σ_y, as well as σ_p in comparison to the values calculated from the theoretical model. Even though the concentration of solvent has been varied stepwise with steps of 2%, a drastic difference in porosity is obtained for samples prepared with 16 wt % and 18 wt % cyclohexane. Thus, for this particular system, no porous epoxies could be prepared with the CIPS technique with porosities in the range 2–10%. On the other hand, the calculations were limited to 10% porosity. The calculations based on the theoretical model, with the restriction of a separating distance between voids of at least one radius, predict a decrease in yield strength to around 45–55% of the value for the neat matrix, independent of the degree of porosity. The experimental values for the yield strength, taken as the first point of

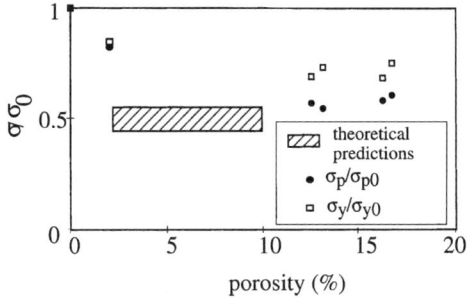

Fig. 44. Comparison of theoretical predictions and experimental results for the lowering of yield strength, σ_y, and stress at the onset of extensive plastic deformation, σ_p, obtained from macroporous epoxies prepared via CIPS related to the values of the neat matrix, σ_{y0} and σ_{p0}

deviation from the linear behavior, as discussed above, are somewhat higher than the predicted values. However, if one compares the strength at the onset of extensive massive plastic deformation, σ_p, to the theoretical predictions, it is confirmed that the theoretical and experimental values are in good agreement.

These theoretical calculations predict that the generation of a porous morphology leads to a decrease in modulus and yield strength, and are in good agreement with experimental data. Furthermore, the calculation of stress distribution, which takes into account the interaction of randomly dispersed voids, predicts the buildup of local stress concentrations which in turn can initiate shear banding.

5.2
Fracture Toughness of Solvent-Modified and Macroporous Epoxies Prepared via CIPS

Cyclohexane is surveyed as selective solvent to give porous morphologies with narrow size distribution and volume fractions up to 20%. In Sect. 4.3.4 it has been shown that the morphology can be varied by using a catalyst. Here, results of single edge notched bending (SENB) tests [85, 87] of solvent-modified and macroporous epoxy networks prepared with cyclohexane using either no or 1 wt % catalyst at identical curing temperatures will be treated. The morphologies of these samples are well characterized by image analysis as presented in the previous sections, thus allowing one to investigate the influence of morphology on the fracture behavior.

In a first testing series, the fracture behavior of the neat, fully crosslinked epoxy network was studied. A fully unstable crack propagation behavior was observed and the critical stress intensity factor, K_{Ic} (0.82 MPa×m$^{1/2}$), and the critical energy release rate, G_{Ic} (0.28 kJ/m^2), were determined [87]. These are typical values for highly crosslinked epoxy networks prepared with *DGEBPA* and aromatic or cycloaliphatic diamines.

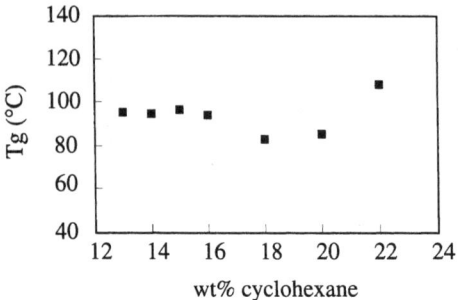

Fig. 45. Glass transition temperature of solvent-modified epoxy networks used for SENB testing and prepared via CIPS with various amounts of cyclohexane

The curing reaction of all series is carried out at 40 °C for 20 h. Even though this period is long enough to ensure phase separation, gelation, and vitrification even for high solvent concentrations, the crosslinking is not completed. Furthermore, a considerable amount of cyclohexane does not participate in the phase separation and remains dissolved in the matrix. Both the plasticization effect of dissolved solvent and the low curing temperature lead to networks with lowered crosslink densities, which cannot be compared to the neat, fully crosslinked matrix.

Two subsequent heat treatments are carried out to process the samples further, thus yielding *solvent-modified* and *macroporous* epoxy networks for the mechanical testing. After the isothermal curing process the samples are heated stepwise with 20 K/h to 180 °C. This process allows for complete reaction at minimum solvent loss. After this heat treatment, the material consists of a highly crosslinked network with dispersed liquid droplets and part of the solvent remaining dissolved in the matrix, thus contributing to plasticization. These materials will be termed *solvent-modified* in the following. *Macroporous* epoxy networks are prepared by holding the samples at 200 °C under vacuum for five days.

Figure 45 shows the glass transition temperature of solvent-modified networks prepared with various amounts of cyclohexane. It is seen, that T_g is independent of, or varies only slightly with, the initial amount of cyclohexane after the heat treatment. The T_g-values of solvent-modified epoxy networks are lower than for the fully crosslinked network, which is a result of the cyclohexane dissolved in the matrix, with a concentration given by the binodal curve and therefore is independent of the initial amount of cyclohexane.

Various types of crack propagation, namely unstable, partially stable, and fully stable can be observed during SENB tests [85, 131]. Solvent-modified epoxy networks prepared via CIPS with 13–16 wt % and 22 wt % cyclohexane show unstable crack propagation. For solvent-concentrations of 18 wt % and 20 wt % cyclohexane, a partially stable crack propagation is observed. The amount of energy consumption upon crack propagation is only 10% and 14% respectively for these two compositions. The change in the fracture behavior indicates that the

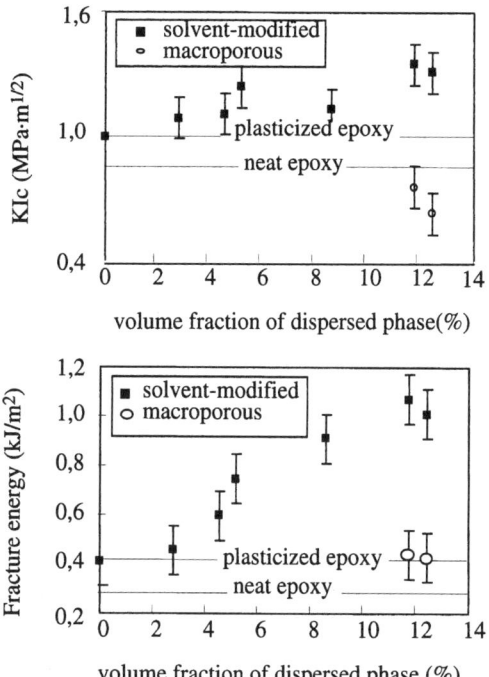

Fig. 46. a Critical stress intensity factor, K_{Ic}, of solvent-modified and macroporous epoxy networks prepared via CIPS with various amounts of cyclohexane. **b** Fracture energy of solvent-modified and macroporous epoxy networks prepared via CIPS with various amounts of cyclohexane

generation of a liquid dispersed phase allows for additional energy absorption upon crack propagation. The values of K_{Ic} and fracture energy, G, are represented in Fig. 46a,b to quantify the increase in toughness. The fracture energy, G, is calculated from the total area under load-displacement curves, thus taking into account the energies for crack initiation and crack propagation.

One might argue that an increase in toughness in solvent-modified networks results from the considerable amount of solvent remaining in the matrix. Therefore solvent-modified networks are prepared with 13 wt % cyclohexane corresponding to the highest solvent concentration which can be added to the precursor mixture without undergoing a phase separation upon curing. This composition serves as reference to take into account the plasticization effect caused by cyclohexane dissolved in the network. The T_g of this material is identical to solvent-modified epoxy networks prepared with higher concentrations and containing up to 12 vol. % dispersed liquid droplets (Fig. 45). Figure 46 clearly reveals the significant increase in toughness which is due to the formation of well dispersed liquid droplets and which cannot be explained only by the plasticization effect.

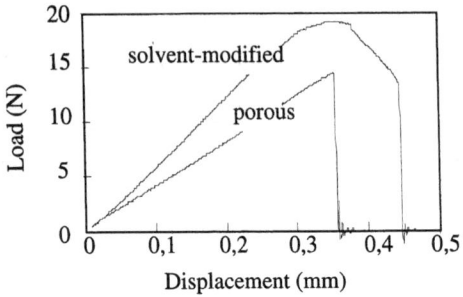

Fig. 47. Load-displacement curves of solvent-modified and macroporous epoxy networks prepared with 20 wt % cyclohexane via CIPS with SENB testing at 1 mm/min

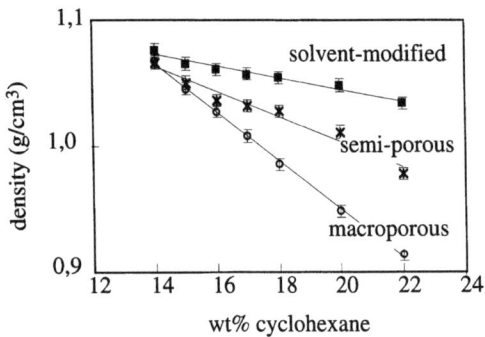

Fig. 48. Density of solvent-modified, semi-porous, and macroporous epoxy networks prepared via kinetically controlled CIPS with 1 wt % catalyst

The toughening efficiency is relatively low for such highly crosslinked epoxy networks. Sue studied the toughness of epoxy networks prepared from *DGEBPA* and a cycloaliphatic diamine without indication of the chemical structure of the curing agent. He determined the critical stress intensity factors with three different experiments and obtains K_{Ic}=0.78 MPa×m$^{1/2}$ for the neat matrix and a value of K_{Ic}=1.07 MPa×m$^{1/2}$ for a system prepared with 10% core shell particles [174]. For comparison, one fully crosslinked epoxy sample modified with 10% core shell particles was prepared with our DGEBPA-PACP system. These experiments gave K_{Ic}=1.43 MPa×m$^{1/2}$ and G=0.628 kJ/m^2. Typical results of load-displacement curves for solvent-modified and macroporous epoxies of identical composition are shown in Fig. 47.

While solvent-modified epoxies show a considerable increase in toughness, a completely brittle behavior is observed after the drying procedure. Only those samples with volume fractions higher than 10% can be tested, because no natural cracks, which are necessary for correct SENB measurements, can be introduced into macroporous epoxies with porosities lower than 10% owing to their

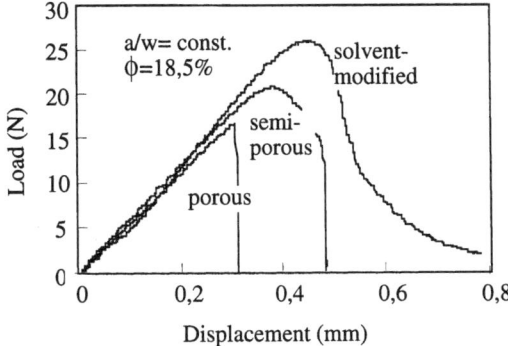

Fig. 49. Load-displacement curves of solvent-modified, semi-porous, and macroporous epoxies prepared with 22 wt % cyclohexane via kinetically controlled CIPS with 1 wt % catalyst

extreme brittleness. The K_{Ic}-values are similar to those of the plasticized epoxy matrix, whereas fracture energy values are even lower. It was not clear if this embrittlement results from the porous morphology or if it is caused by oxidation at the surface during the drying process, which might also drastically effect the macroscopic properties.

In a second series of experiments, similar materials were prepared with 1 wt % catalyst to investigate the influence of morphology on the toughening. In addition to the two heat treatments to generate *solvent-modified* and *macroporous* epoxies as presented before, a third heat treatment was carried out to give a *semi-porous* morphology. A brief heating above T_g and under vacuum results in partial solvent removal. The differences in the three heat treatments is clearly revealed with density measurements as shown in Fig. 48.

As for the samples prepared without catalyst, the ability for energy absorption after crack propagation decreases strongly as the solvent is removed. This is reflected in Fig. 49, where the load displacement curves of solvent-modified, semi-porous, and macroporous epoxies prepared with initially 22 wt % cyclohexane (ϕ=18.5%) are shown. The crack length is the same in all three cases. Therefore the decrease in maximum load is directly proportional to the decrease in K_{Ic}. It is also clearly seen that the fracture behavior changes drastically and that the surface under the load-displacement curve, which is used to calculate the fracture energy is significantly lowered.

The quantitative results of fracture energy, which are calculated from the total area under load-displacement curves, are presented in Fig. 50. It becomes obvious, that a brittle-tough transition exists at a volume fraction of around 10%. This brittle-tough transition is observed for solvent-modified as well as semi-porous epoxies. A brittle behavior is observed in macroporous epoxies after complete solvent removal, thus giving low fracture energies similar to the neat epoxy for each porosity.

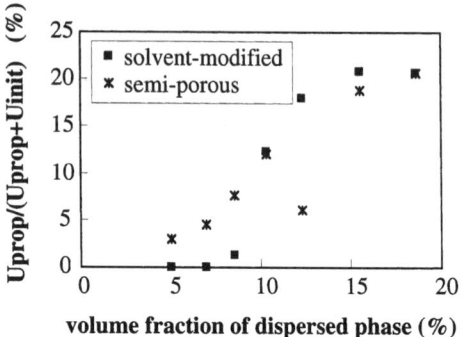

Fig. 50. Contribution of energy for crack propagation to the total fracture energy for solvent-modified epoxies prepared via CIPS with 1 wt % catalyst porous, and macroporous epoxies prepared via kinetically controlled CIPS with 1 wt % catalyst

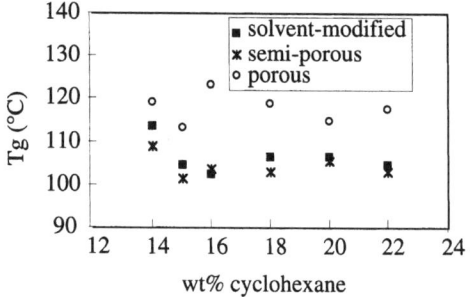

Fig. 51. Glass transition temperature of solvent-modified, semi-porous and macroporous epoxies prepared via kinetically controlled CIPS with 1 wt% catalyst

It might be argued that the brittle tough transition and the embrittlement upon drying is related to variations in the extent of reaction and hence to T_g. However, no significant changes in T_g as a function of solvent concentration were found (Fig. 51).

The brittle-ductile transition which is observed above clearly results from a change in the morphology. However, it is not clear whether this is a result of decreasing the particle size and inter-pore distance (IPD) or increasing concentration. It has earlier been shown that the IPD remains constant for this system prepared with 1 wt % catalyst (Fig. 35). Therefore the IPD-concept introduced by Wu [99] cannot be the single morphological parameter which determines the toughness. Moreover the toughness is the result of synergistic effects, which are not yet fully understood.

To complete the analysis of the fracture toughness the values of K_{Ic} and G_{Ic} calculated from the maximum load and the energy for crack initiation are summarized in Fig. 52. It becomes obvious that the brittle-ductile transition, which

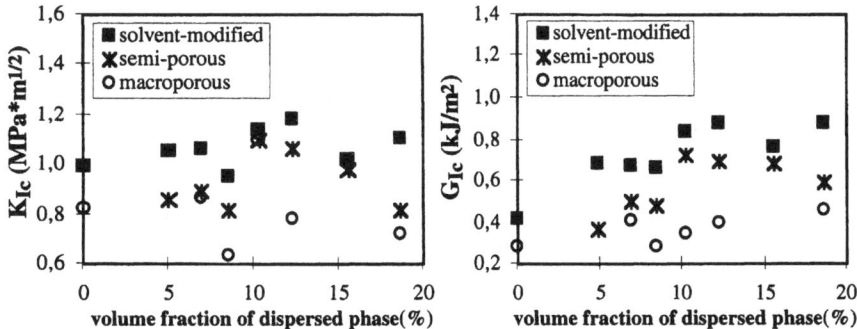

Fig. 52. Critical stress intensity factor, K_{Ic}, and critical stress energy release rate, G_{Ic}, of solvent-modified, semi-porous, and macroporous epoxies prepared via kinetically controlled CIPS with 1 wt % catalyst calculated from SENB tests

can be shown from Fig. 50 is also observed from the G_{Ic}-values. However this transition can hardly be shown from the K_{Ic}-values because the experimental error is nearly identical to the increase in toughness which is typical for the toughening of very highly crosslinked systems. As mentioned previously, the toughenability depends strongly on the crosslink density. Therefore precursor monomers yielding lower crosslink densities should be selected to demonstrate better the ability of dispersed liquids or voids to increase the toughness. Changing the precursor monomers automatically influences the miscibility and phase separation behavior and therefore requires one to select convenient solvents and evaluate optimum reaction conditions to prepare solvent-modified and macroporous thermosets with controlled morphologies via CIPS.

As a *summary of the findings* the synthesis of solvent-modified epoxies via CIPS can lead to a substantial increase in fracture energy of up to 400% even though a very highly crosslinked matrix material is used in the present study. The macroscopic fracture behavior changes from completely brittle to a partially stable or even fully stable crack propagation behavior with increasing volume fraction of the dispersed phase. This improvement in toughness is only partly due to the plasticization effect, which is caused by the solvent portion that remains dissolved in the crosslinked matrix and is not involved in the phase separation. Moreover, it results from the formation of randomly distributed liquid droplets. It is thought that the dispersed, liquid phase which has a very low cavitation resistance, drastically affects the stress distribution upon loading, thus initiating shear banding, which is responsible for the increase in toughness. It remains questionable whether the embrittlement which is observed upon the thermal removal of the liquid phase might be due to a skin effect resulting from oxidation at the surface or result from the formation of a truly porous morphology. We intend to overcome the thermal ageing problem by using alternative methods such as supercritical extraction to generate the pores [22].

Two systems have been studied which both display narrow size distributions but significant differences in size, concentration, and IPD. The toughness is the result of synergistic effects of morphological parameters and clearly the two series of morphologies and measurements presented here are not sufficient to draw general conclusions for the complex relationships between the morphology and toughening. However, these results confirm much of the observations described by other research groups and are combined to yield a better understanding on toughening. Our results clearly show that the nature of the dispersed phase, which in our case is liquid droplets, is not responsible for the improvement in toughness. Furthermore, it can be concluded that the IPD distance cannot be the single morphological parameter to explain the improvement in toughness. For solvent-modified epoxies prepared via CIPS by using 1 wt % catalyst, a brittle-tough transition is found, whereas the IPD, which is determined from image analysis and density measurements with high accuracy, is found to be constant for the brittle and the tough systems and the IPD does not depend on the volume fraction. However the fracture behavior might be influenced by the size or concentration of the dispersed phase, which cannot be varied independently for two-phase morphologies prepared via a phase separation process. A continuous increase in fracture energy is observed in systems prepared without catalyst, whereas those systems prepared by using 1 wt % catalyst, which have lower domain sizes and higher concentrations, show a brittle-tough transition.

One morphological parameter that cannot be quantified and only qualitatively described might be the key to explain the above findings – particle dispersion. The brittle-tough transition is observed at around 10 vol. %. At higher concentrations the image analysis reveals a considerable number of particles with spacings lower than one radius. The number and concentration of such local agglomerations, consisting of well dispersed single domains which are neither chemically nor physically linked, increase automatically with increasing volume fraction until percolation occurs. The stress distributions in the vicinity of such agglomerations cannot be calculated with the actual model, where the simulation of a random distribution is limited to a minimum spacing equal to the diameter of the domains [84, 173]. Such local agglomerations might drastically influence the stress distribution and initiate multiple shear banding, thus allowing for increased toughness. This explanation agrees very well with the recent observations of Pearson et al. [128, 129] and Sue et al. [98, 130], who all observe higher toughness in systems clearly showing such agglomerations. This might also explain the results of Béguelin [131] on the toughness of PMMA modified with polyurethane to give an interpenetrating network. Additionally the percolation model proposed by Margolina and Wu must be cited [175], where it is argued that the brittle-tough transition results from the percolation of matrix ligaments with distances lower than a critical value. This model coincides with the present explanation and also accounts for the alignment of cavitated particles, which has been reported by Sue [155] and Lazzeri and Bucknall [154]. Even though Wu's percolation model [175] is a refinement of his IPD concept and seems to be appropriate to give a reasonable explanation of the phenomena observed in toughened thermoplastic and thermosetting polymers, this model has not gained as much attention as the IPD concept [99].

The above hypothesis, that the toughening efficiency is enhanced by the formation of local agglomerations, requires intensive theoretical calculations of stress distributions as well as the preparation and characterization of morphologies with a controlled level of dispersion and local agglomeration respectively.

6
Macroporous Cyanurate Networks

In the previous sections it has been shown that the CIPS technique allows one to prepare solvent-modified and macroporous epoxy networks with controlled morphologies. However, it is possible to extend the CIPS technique to other classes of thermosetting polymers, thus demonstrating that the strategy of CIPS is general and can be applied to nearly any type of crosslinking chemistry involving the formation of a highly crosslinked network starting from low molecular weight precursor monomers. It is a matter of enough change in free energy of the system either by entropy or by enthalpy changes. Here we aim to show potential of the CIPS technique for lowering the dielectric constant of thermosetting cyanurate esters by the formation of a porous structure.

Cyanate ester resins are an emerging class of high performance thermosetting polymers which offer a substantial potential for applications in microelectronics based on their unique combination of desirable properties such as low dielectric constant, low water absorption, good adhesion to metals, high thermal stability, and initially low monomer viscosity [176, 177]. Due to these favorable properties, cyanate ester resins are used in microelectronics as insulators for high performance applications in printed wiring boards [178] or multichip modules [179]. Further improvements in printed circuit technology require materials exhibiting a lower dielectric constant since the propagation delay of electronic signals varies with the inverse of the square root of dielectric constant, which then allows for denser wiring at acceptable cross-talk noises [180, 181]. Hedrick and coworkers have demonstrated that lowering the dielectric constant can be successfully achieved by the generation of a porous morphology [27, 28, 31–33, 35] where the material is substituted by air, having a dielectric constant of 1.

As a consequence of high crosslink density, unmodified cyanurates are inherently brittle. Similar to the toughening of epoxies described previously, cyanate ester resins can also be toughened by second phase particles [182, 183]. For cyanurates, thermoplastic particles are mostly used as toughener in order to maintain the high thermal stability of the matrix and due to the enhanced thermal stability of the modifier required for high temperature cure [182]. Toughened cyanate ester resins have entered applications as high performance matrix materials for aircraft composites [184], especially in radome applications [185]. However, the incorporation of second phase particles is often at the expense of the desired dielectric properties. The synthesis of macroporous cyanurates with tunable morphologies via CIPS therefore seems a promising route to combine lowering of dielectric constant with improved toughness.

6.1
Materials Selection and Chemistry of Cyanate Ester Resins

Bisphenol-E cyanate ester, *BPECN*, is selected as the cyanate ester precursor because it provides high thermal stability starting from monomers with low viscosity (Table 5). Due to the high degree of symmetry, cyanurates exhibit a low dipole moment and consequently a low dielectric constant. The cyclotrimerization results from the reaction of two cyanate groups with imidocarbonate [176, 177]. The favorable intermediate product, imidocarbonate, is easily obtained during the reaction of the precursor monomer and an active hydrogen source, such as an aryl phenol. As the low dielectric constant results from the symmetrical arrangement of the ring structure, it is recommended to use the parent aryl phenol as the H-donor. Therefore we have chosen bisphenol-E (*BPE*) as the hydrogen source. It is well known that the final extent of conversion increases with the amount of active hydrogen source [186]. For our experiments 1% of *BPE* was necessary to achieve sufficient conversion to induce phase separation.

The reaction rate of cyanate ester resins can be increased by using catalysts such as carboxylate salts or chelates of transition metal ions. The role of transition metal ions in the polymerization reaction consists of facilitating the cyclization reaction of three cyanate monomer functionalities by the formation of

Table 5. Molecular structure and properties of bisphenol-E cyanate ester resin (*BPECN*)

Product name	Molecular structure	T_g °C	Viscosity at 25 °C cP	Water absorption %	ε (1 MHz)
Bisphenol-E cyanate ester (AroCy L10)	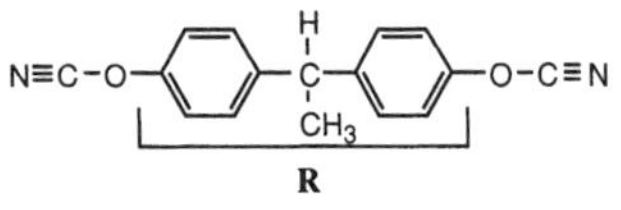 N≡C−O−⟨⟩−CH(CH₃)−⟨⟩−O−C≡N R	258	90–140	2.4	2.98

$$\text{Ar}-\text{O}-\text{C}\equiv\text{N} \ +\ \text{H}_2\text{O}\ \longrightarrow\ \text{Ar}-\text{O}-\underset{\text{OH}}{\text{C}}=\text{NH}\ \longrightarrow\ \text{Ar}-\text{O}-\underset{\text{O}}{\overset{\|}{\text{C}}}-\text{NH}_2$$

(cyanate ester)   (carbamate)

$$\text{Ar}-\text{O}-\underset{\text{O}}{\overset{\|}{\text{C}}}-\text{NH}_2\ \longrightarrow\ \text{R}-\text{NH}_2\ +\ \text{CO}_2$$

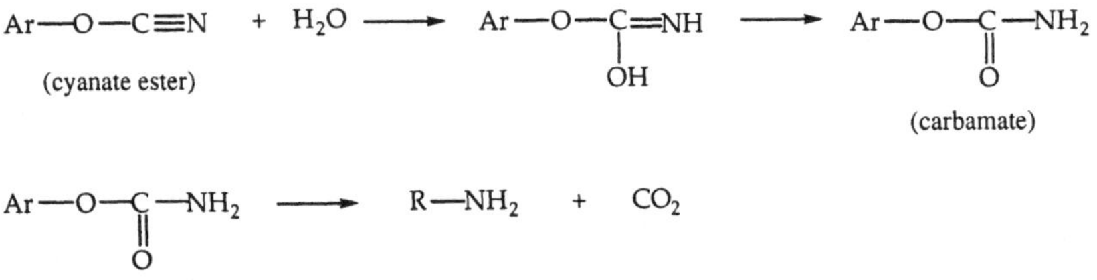

Scheme 2

co-ordination complexes [187, 188]. The reactivity of different catalysts towards the curing of *BPECN* was investigated in [187, 188]. Based on this knowledge, cobaltacetylacetonate is chosen as the catalyst because it combines exceptional long pot life with fairly high reactivity.

Cyanate ester monomers must be stored under dry conditions, because water can react with cyanate ester resins and deliver carbamates as undesired side products (Scheme 2) [186]. Even when the carbamates are only formed in the presence of a catalyst, this catalytic effect can be caused by traces remaining from the monomer synthesis. The formation of carbamates is critical, as they can decompose to amines and CO_2. While the amine easily reacts with another cyanate ester, the CO_2 can act as a blowing agent and hence leads to uncontrolled porosity during the processing.

Experimentally a cyanate ester precursor mixture consisting of *BPEC*, 1 wt % *BPE*, and 100 ppm cobaltacetylacetonate was prepared and subsequently mixed with the cyclohexane phase separating solvent [86]. Essentially the same procedure as for the epoxy is used for sample preparation with the difference that the curing was done at 80 °C and post drying at 240 °C.

6.2
Morphology, Thermal and Dielectric Properties

The cyclotrimerization is usually performed at temperatures above 150 °C without any catalyst required. Preliminary experiments showed that such high curing temperatures lead to samples showing very large domains of some millimeters when neat *BPECN* is cured in the presence of cyclohexane. These inhomogeneities are largely caused by exceeding the boiling point of cyclohexane (78 °C). In order to prepare homogeneous samples, it was necessary to lower the curing temperature. We found experimentally that the curing of *BPECN* can be performed at temperatures as low as $T=80$ °C within 24 h by adding 1 wt % of *BPE* as the hydrogen donor and 100 ppm of cobaltacetyl-acetonate as catalyst [86].

When *BPECN* is used as the precursor for the synthesis of macroporous cyanurates, phase separation resulting in the formation of opaque samples is achieved by using at least 16 wt % cyclohexane as the dispersed phase at a curing temperature of 80 °C. It was observed that phase separation started after approximately 18–20 h, whereas gelation took place around 1 h after the start of the phase separation. The maximum amount of cyclohexane which could be dissolved in *BPECN* by slight stirring at $T=60$ °C for around 2 h, was 20 wt %. Samples at concentrations of 15 wt % cyclohexane or lower remained transparent and no separated domains were found with SEM even at highest magnification. Hence this system exhibits a very narrow phase separation gap with 16–20 wt % cyclohexane. The influence of concentration on the domain size was visible from SEM-micrographs of macroporous cyanurate networks prepared via CIPS with 16, 17.5, and 20 wt % cyclohexane [86]. It was found that the domain sizes and volume fraction increase with increasing amount of cyclohexane and that the

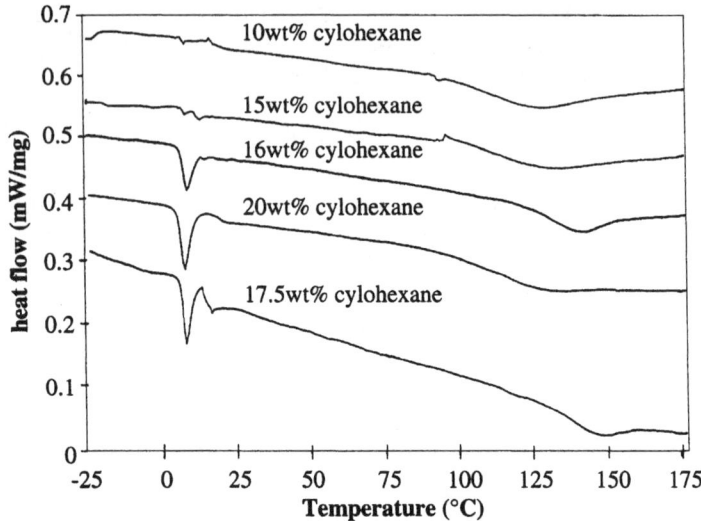

Fig. 53. DSC scans of cyclohexane-modified cyanurates after curing at $T=80$ °C showing melting peaks of cyclohexane

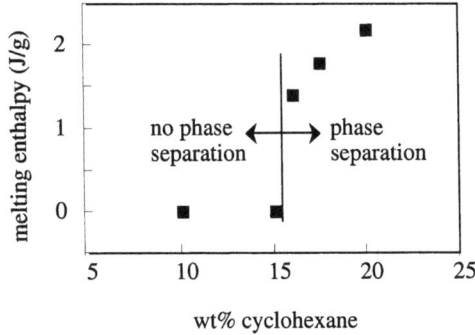

Fig. 54. Melting enthalpies of cyclohexane detected in cyanurates cured with various amounts of cyclohexane at $T=80$ °C

pores were spherical with sizes of around 5–20 µm. The pore sizes are considerably larger in macroporous cyanurates than in epoxies. We believe that this is due to lower reaction rates in cyclohexane-modified cyanurates.

The increase of pore size with increasing amount of solvent can also be monitored with dynamic DSC-measurements. An endothermic peak at $T=7$ °C, corresponding to the melting point of crystalline cyclohexane, is observed in the opaque samples after the phase separation resulting from the formation of dispersed cyclohexane droplets (Fig. 53).

The melting enthalpy related to the sample mass, represented by the area of this melting peak, increases nearly linearly with increasing amount of solvent once phase separation has occurred (Fig. 54). The appearance of such a melting peak has also been observed for cyclohexane-modified epoxies prepared with more than 20 wt % cyclohexane, having domain sizes larger than around 5 µm and a volume fraction of higher than around 13%.

It is not clear whether the melting peak results from the appearance of domains larger than a critical size, or if this endothermic peak is caused by the formation of interconnected pores. A more reasonable argument derives from the differences in the phase separation behavior of the two systems. During the phase separation process a polymer-rich and a solvent-rich phase are generated with compositions which are ideally given by the binodal line representing thermodynamic equilibrium. Usually the continuous as well as the dispersed phase still contain a certain amount of the minor phase. The low molecular weight liquid remaining in the polymer-rich phase leads to a plasticization, thus lowering the T_g of the material. On the other hand the liquid droplets always contain some polymeric moieties. In the case of cyclohexane-modified epoxies, the formation of a crystalline cyclohexane phase might be suppressed due to the existence of a considerable amount of epoxy precursors remaining inside the liquid droplets. However, the equilibrium compositions are different in cyclohexane-modified cyanurates and might be very close to the pure phases. Indeed we surmise that the large change in entropy, which results from the transition of small, liquid precursor monomers into a rigid network structure, is responsible for pushing out close to the entire amount of solvent. Consequently the dispersed droplets, consisting of pure cyclohexane, can crystallize upon cooling and therefore form the endothermic melting peak upon reheating during DSC-measurements.

The chemical conversion of cyanate esters to form the triazine ring, resulting in a three-dimensional network structure, can be monitored with FTIR spectroscopy since both precursors and crosslinked resin absorb strongly at wave numbers with little interference from other peaks (Fig. 55). The stretching vibrations of the cyanate group create a significant doublet at 2270 and 2240 cm^{-1} [177]. This group is detected in the monomer precursor as well as in cyclohexane-modified cyanurates after phase separation, but prior to the drying procedure. After drying, the doublet of the cyanate stretching vibration has almost completely vanished, thus indicating that a simultaneous post curing has been effectively realized. The triazine ring stretching band at 1365 cm^{-1} is observed after the phase separation. Its intensity is further increased after the drying procedure owing to the post curing. The large absorption band at 1562 cm^{-1} is associated with the phenol oxygen-triazine ring stretching. These two bands can also be observed in the precursor monomer, thus indicating the existence of a small amount of oligomers even in the precursor mixture.

Density measurements were carried out on a gradient column and show a considerable decrease in density for macroporous cyanurates (Fig. 56).

The density drop between the transparent sample with initially 15 wt % cyclohexane, and the opaque sample with 16 wt % cyclohexane, is around 17%. We

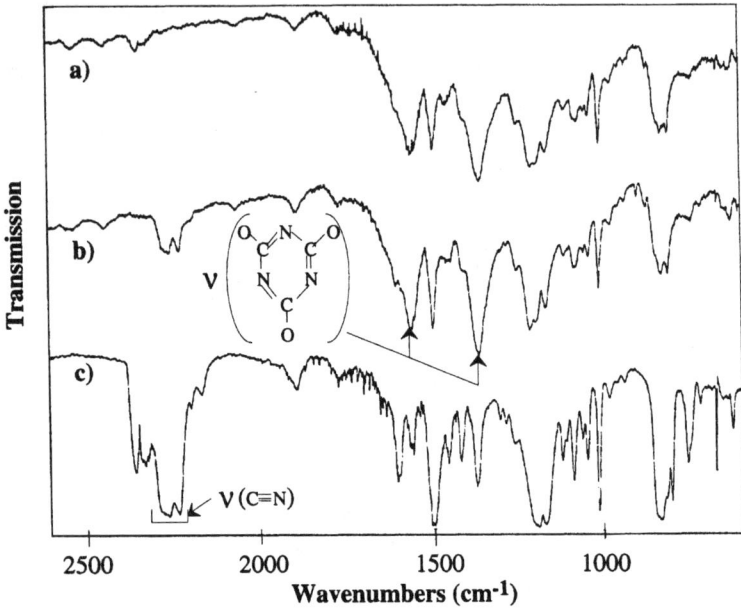

Fig. 55. FTIR spectra of cyanurates: uncured *BPECN* (AroCy-L10 (*line a*); cyclohexane modified cyanurate prepared via CIPS with 20 wt % cyclohexane after curing at $T=80$ °C (*line b*); macroporous cyanurate prepared via CIPS with 20 wt % cyclohexane after drying (*line c*)

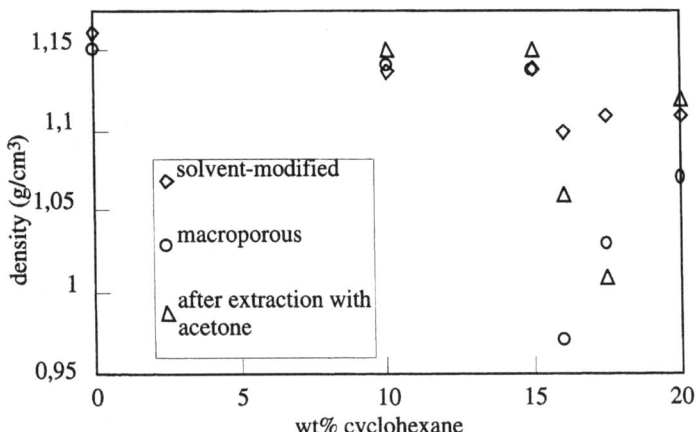

Fig. 56. Density of solvent-modified and macroporous cyanurates prepared with various amounts of cyclohexane via CIPS

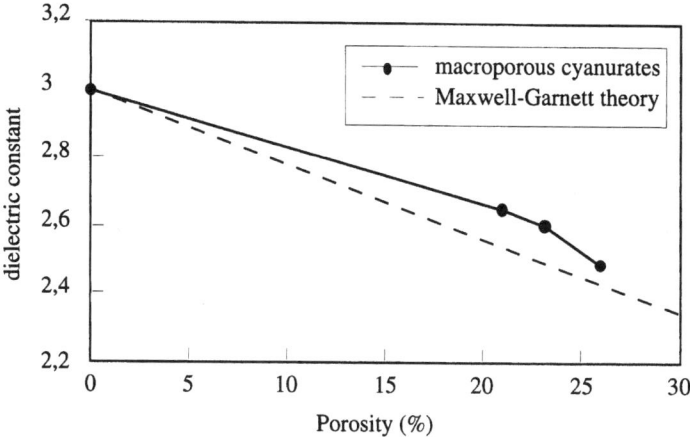

Fig. 57. Dielectric constant of macroporous cyanurates prepared via CIPS

surmise that once the phase separation starts, nearly all of the cyclohexane is involved in the phase separation owing to the substantial changes in entropy in cyanurates as mentioned above. This agrees well with the appearance of melting peaks resulting from crystalline cyclohexane droplets, which are observed with dynamic DSC measurements. The subsequent increase in density with a higher amount of solvent can be explained by interconnected and open pores, allowing the solvent mixture in the density column to penetrate into the samples. Such interconnected pores are visible from the corresponding SEM micrographs [86]. According to this observation, the volume fraction of pores cannot be determined exactly with density measurements and is therefore calculated based on the assumption that all the cyclohexane is involved in the phase separation to yield pure cyclohexane droplets which become voids after the drying procedure.

The dielectric constant of the pure cyanurate network under dry nitrogen atmosphere at 20 °C is 3.0 (at 1 MHz). For the macroporous cyanurate networks, the dielectric constant decreases with the porosity as shown in Fig. 57, where the solid and dotted lines represent experimental dielectric results together with the prediction of the dielectric constant from Maxwell-Garnett theory (MGT) [189]. The small discrepancies between experimental results and MGT might be due to the error in estimated porosities, which are calculated from the density of the matrix material and cyclohexane assuming that the entire amount of cyclohexane is involved in the phase separation. It is supposed that a small level of miscibility after phase separation would result in closer agreement of dielectric constants measured and predicted. Dielectric constant values as low as 2.5 are measured for macroporous cyanurates prepared with 20 wt % cyclohexane.

With electrical applications in mind, the chemically induced phase separation technique has been extended to prepare macroporous cyanurate networks with pore sizes in the micrometer range, thus demonstrating the general applicability of the technique to any type of thermosets. Closed cell morphologies are obtained with 16 wt % cyclohexane, while higher solvent concentration leads to interconnected pores. These materials are characterized by a low dielectric constant without influencing the thermal stability of the matrix material.

7
Conclusions

General strategies to prepare solvent-modified and macroporous thermosets via chemically induced phase separation have been reviewed. Phase separation resulting in the generation of two-phase morphology leads to desirable properties such as low density, low dielectric constant, high thermal stability, and enhanced toughness. To benefit from these properties and allow for industrial applications it is necessary to control the size and distribution of the secondary phase, be it either liquid droplets or simply voids. A key to control the morphology and design new materials is a basic understanding of the thermodynamic origins of phase separation. A first criterion for solvent selection can be realized while following the solubility parameter approach to express changes in the free enthalpy of mixing as a function of conversion. These thermodynamic considerations enable one to calculate and reconstruct phase diagrams which in turn form the basis of various strategies to prepare materials having closed or interconnected pores with sizes in the low µm-range. A gradient oven has been presented which allows one to screen rapidly a wide temperature and composition range to identify systems which undergo the desired phase separation. The influence of internal and external reaction parameters on the morphology has been reported and it was concluded that low reaction rates favor a nucleation and growth mechanism leading to spherical domains with narrow or alternatively bimodal size distributions. Experimental studies were limited to thermal curing systems and considerable size reduction was achieved while using a catalyst which increases the reaction rate.

Solvent-modified thermosets display enhanced toughness due to the incorporation of a second phase material. A brittle-tough transition has been observed which cannot be attributed to changes in the interparticle distance. The chemically induced phase separation technique offers new routes and strategies to prepare such materials and enter new areas of applications. Hence, engineered porosity is demonstrated as a research concept developed into a toolbox for material scientists.

Further reduction of pore sizes might be achieved for UV-curable systems, thus resulting in nanometer sized pores. Such a scenario is highly desirable to prepare materials characterized by low dielectric constant useful for microelectronic applications. Nanoporous thermosets are also interesting for the preparation of microdevices. Open porosities and thermosets serving as membranes can be prepared by careful adjustment of solvent concentration and reaction kinetics, allowing for high reaction rates, thus leading to a chemical quench ending in the spinodal decomposition area of the phase diagram.

Acknowledgements. The authors gratefully acknowledge the financial support (No. 20-45417.95) from the Swiss National Science Foundation and the valuable comments on the manuscript by Prof. Vipin Kumar.

References

1. Schaefer DW (1994) MRS Bulletin XIX(4):14
2. Khemani KC (eds) (1997) Polymeric foams: science and technology. American Chemical Society, Washington, DC
3. Klempner DC, Frisch KC (eds) (1991) Handbook of polymeric foams and technology. Hanser, München
4. Shutov FA (1991) Blowing agents for polymer foams. In: Klempner DC, Frisch KC (eds) Handbook of polymeric foams and technology. Hanser, München, chap 17
5. Bailey FE (1991) Flexible polyurethane foams. In: Klempner DC, Frisch KC (eds) Handbook of polymeric foams and technology. Hanser, München, chap 2
6. Neff R, Adedeji A, Macosko CW (1996) In: Fradet A (ed) Polycondensation '96, Paris, p 17
7. Kumar V (1993) Cellular Polymers 12:207
8. Kumar V, Suh NP (1990) Polym Eng Sci 30:1323
9. Ramesh NS, Rasmussen DH, Campbell GA (1994) Polym Eng Sci 34:1685
10. Goel SK, Beckman EJ (1993) Polymer 34:1410
11. Beckman EJ (1995) Polymer nanostructures via critical fluid processing. In: Vincenzini P (ed) Adv Sci Technol New Horizons for Materials, Techna Srl, p 151
12. Goel SK, Beckman EJ (1994) Polym Eng Sci 34:1137
13. Guan Z, DeSimone JM (1994) Macromolecules 27:5527
14. Shaffer KA, DeSimone JM (1995) Trends in Polymer Science 3:146
15. Tomalia DA, Fréchet JMJ, Moore JS (1996) In: MRS Spring Meeting Symposium Y, San Francisco
16. Ruckenstein E (1997) Adv Polym Sci 127:1
17. Ruckenstein E, Park JS (1992) Polymer 33:405
18. Ruckenstein E, Sun F (1993) J Membr Sci 81:191
19. Sun F, Ruckenstein E (1993) J Membr Sci 85:59
20. Even WRJ, Gregory DP (1994) MRS Bulletin XIX(4):29
21. Svec F, Fréchet JMJ (1996) Science 273:205
22. Della Martina A, Hilborn JG, Kiefer J, Hedrick JL, Srinivasan S, Miller RD (1997) ACS Symp Ser 669:8
23. Qutubuddin S, Lin CS, Tajuddin Y (1994) Polymer 35:4606
24. Chieng TH, Gan LM, Chew CH, Ng SC (1995) Polymer 36:1941
25. Gan LM, Chew CH, Chieng TH (1994) Macromol Symp 84:237
26. Gan LM, Chieng TH, Chew CH, Ng SC (1994) Langmuir 10:4022
27. Hedrick JL, Russell TP, Labadie J, Lucas M, Swanson S (1995) Polymer 36(14):2685
28. Hedrick JL, Pietro RD, Charlier Y, Jérôme R (1995) High Perform Polym 7:133
29. Hedrick JL, Hawker CJ, DiPietro R, Jérôme R, Charlier R (1995) Polymer 36:4855
30. Hedrick JL, Hawker CJ, DiPietro R, Jérôme R, Charlier R (1995) Polymer 36:4855
31. Hedrick JL, Brown HR, Volksen W, Sanchez M, Plummer CJG, Hilborn JG (1997) Polymer 38:605
32. Hedrick JL, Charlier Y, Russell TP (1994) MRS Symp Proc 323:277
33. Hedrick JL, DiPietro R, Plummer CJG, Hilborn J, Jérôme R (1996) Polymer 37:5229
34. Hedrick JL, Hilborn JG, Palmer TD, Labadie JW (1990) J Polym Sci Part A: Polym
35. Hedrick JL, Labadie J, Russell TP, Wakharkar V, Hofer D (1992) MRS Symp Proc 274:37
36. Plummer CJG, Hilborn JG, Hedrick JL (1995) Polymer 36:2845
37. Cha HJ, Hedrick JL, Yoon DY (1996) Appl Phys Lett 68:1930

38. Hedrick JC, Shaw JM, Buchwalter SL, Gelorme JD, Kang SK, Kosbar LL, Lewis DA, Purushotham S, Saraf R, Viehbeck A (1996) Polym. Prepr. 37(1):172
39. Aubert JH, Clough RL (1985) Polymer 26:2047
40. Lal J, Bansil R (1991) Macromolecules 24:290
41. Torkelson JM, Song SW (1993) Polymer Prepr 34(2):496
42. Torkelson JM, Song SW (1994) Macromolecules 27:6389
43. Loeb S, Sourirajan S (1962) Adv Chem Ser 38:117
44. Haggin J (1990) C&EN Chicago October:22
45. Michaels AS (1989) Chemtech 3:162
46. Strathmann H, Kock K (1977) Desalination 21:241
47. Young TH, Chen LW (1995) Desalination 103:233
48. Fréchet JM (1993) Makromol Chem Macromol Symp 70/71:289
49. Kiefer J, Hilborn JG, Hedrick JL (1996) Polymer 37:5715
50. Kiefer J, Hilborn JG, Månson JAE, Leterrier Y, Hedrick JL (1996) Macromolecules 29:4158
51. Champetier G, Buvet R, Néel J, Sigwalt P (eds) (1972) Chimie macromoléculaire. Hermann, Paris
52. de Gennes P-G (ed) (1979) Scaling concepts in polymer physics. Cornell University Press, Ithaca
53. Flory JP (ed) (1953) Principles of polymer chemistry. Cornell University Press, Ithaca
54. Olabisi O, Robeson LM, Shaw MT (eds) (1979) Polymer-polymer miscibility. Academic Press, New York
55. Flory JP (1942) J Chem Phys 10:51
56. Flory PJ (1941) J Chem Phys 9:660
57. Huggins ML (1941) J Chem Phys 9:660
58. Huggins ML (1942) J Am Chem Soc 64:1712
59. Russell TP, Hjelm RP Jr., Seeger PA (1990) Macromolecules 23:890
60. Kumar S, Taylor J, Debenedetti P, Graessley W (1994) Polym Mat Sci Eng 71:358
61. Nakamoto C, Kitada T, Kato E (1996) Polymer Gels & Networks 4:17
62. Petri HM, Schuld N, Wolf BA (1995) Macromolecules 28:4975
63. Balsara NP (1996) Thermodynamics of polymer blends. In: Mark JE (ed) Physical properties of polymers handbook. AIP Press, New York, p 257
64. Einaga Y (1994) Prog Polym Sci 19:1
65. Utracki L (ed) (1990) Polymer blends. Hanser, München
66. Riccardi CC, Borrajo J, Williams RJJ (1994) Polymer 35:5541
67. Vazquez A, Rojas AJ, Adabbo HE, Borrajo J, Williams RJJ (1987) Polymer 28:1156
68. Williams RJJ, Rozenberg BA, Pascault JP (1997) Adv Polym Sci 128:95
69. Williams RJJ, Borrajo J, Adabbo HE, Rojas AJ (1984) Adv Chem Ser 208:195
70. Macosko CW, Miller DR (1976) Macromolecules 9:199
71. Vérchère D, Sautereau H, Pascault JP, Riccardi CC, Moschiar SM, Williams RJJ (1991) J Appl Polym Sci 42:717
72. Moschiar SM, Riccardi CC, Williams RJJ, Vérchère D, Sautereau H, Pascault JP (1991) J Appl Polym Sci 42:717
73. Borrajo J, Riccardi CC, Williams RJJ, Cao ZQ, Pascault JP (1995) Polymer 36:3541
74. Riccardi CC, Borrado J, Williams RJJ, Girard-Reydet E, Sautereau H, Pascault JP (1996) J Appl Polym Sci 34:349
75. Al-Saigh ZY (1997) Trends Polym Sci 5:97
76. Munk P, Hattam P, Du Q (1990) Makromol Chem Macromol Symp 48:205
77. Londono JD, Wignall GD (1997) Macromolecules 30:3821
78. Koningsveld R, Dusek K (1995) Macromolecules 28:1103
79. Hildebrand JH (ed) (1950) The solubility of non-electrolytes. Reinhold, New York
80. Van Krevelen DW, Hoftyzer PJ (eds) (1976) Properties of polymers. Elsevier, Amsterdam
81. David DJ, Sincock TF (1992) Polymer 33:4505

82. Ellis B (ed) (1993) Chemistry and technology of epoxy resins. Black Academic & Professional, London
83. May CA (ed) (1988) Epoxy resins chemistry and technology. Marcel Dekker, New York
84. Fond C, Kiefer J, Mendels D, Ferrer JB, Kausch HH, Hilborn JG (1998) J Mat Sci (33:3975)
85. Kiefer J (1997) Macroporous thermosets via chemically induced phase separation. Thesis N°1725 at Swiss Federal Institute of Technology, Lausanne
86. Kiefer J, Hilborn JG, Hedrick JL, Cha HH, Yoon DY, Hedrick JC (1997) Macromolecules 30:8546
87. Kiefer J, Kausch HH, Hilborn JG (1997) Polym Bull 38:477
88. Kiefer J, Porouchani R, Hilborn JG (1996) In: Fradet A (ed) Polycondensation, Paris, p 586
89. Kiefer J, Porouchani R, Mendels D, Ferrer JB, Fond C, Hedrick JL, Kausch HH, Hilborn JG (1996) MRS Symp Proc 431:527
90. Meijer HEH, Venderbosch RW, Goossens JGP, Lemstra PJ (1996) In: Eur Symp Polym Blends, Maastricht, p 525
91. Pascault JP (1995) Macromol Symp 93:43
92. Basil R, Liao G (1997) Trends Polym Sci 5:146
93. Inoue T (1995) Prog Polym Sci 20:119
94. Sue HJ, Yee AF (1988) Polymer 29:1619
95. Kim SC, Ko MB, Jo WH (1995) Polymer 36:2189
96. Kinloch AJ, Hunston DL (1986) J Mat Sci Lett 5:1207
97. Rozenberg BA, Sgigalov GM (1996) Macromol Symp 102:329
98. Sue HJ, Garcia-Meitin EI (1996) Fracture behavior of rubber-modified high-performance epoxies. In: Arends CB (ed) Polymer toughening. Marcel Dekker, New York, p 131
99. Wu S (1988) J Appl Polym Sci 35:549
100. Dompas D, Groeninckx G, Isogawa M, Hasegawa T, Kadokura M (1995) Polymer 35:4760
101. Wagner R, Kampmann R (1991) Phase transformations in materials. In: Haasen P (ed) Materials science and technology. VCH, Weinheim, p 243
102. Char K, Cha BJ, Kim JJ, Kim SS, Kim CK (1995) J Membr Sci 108:219
103. Vérchère D, Sautereau H, Pascault JP, Riccardi CC, Moschiar SM, Williams RJJ (1991) J Appl Polym Sci 42:701
104. Chan LC, Gillham JK, Kinloch AJ, Shaw SJ (1984) Adv Chem Ser 208:235
105. Kim DS, Kim SC (1990) Polym Adv Techn 1:211
106. Manzione LT, Gillham JK, McPherson CA (1981) J Appl Polym Sci 26:899
107. Sultan JN, McGarry FJ (1973) Polym Eng Sci 13:29
108. Lu AF, Cantwell WJ, Kausch HH (1995) Macromol Symp 93:317
109. Lu AF, Kausch HH, Cantwell WJ, Fischer M (1996) J Mat Sci Lett 15:1018
110. Lu AF, Plummer CJG, Cantwell WJ, Kausch HH (1996) Polym Bull 37:399
111. Pearson RA (1993) Adv Chem Ser 233:405
112. Pearson RA, Yee AF (1983) Polym Mat Sci Eng 49:316
113. Fischer M (1992) Adv Polym Sci 100:313
114. LeMay JD, Kelley FN (1986) Adv Polym Sci 78:115
115. Garg AC, Mai YW (1988) Comp Sci Technol 31:179
116. Huang Y, Kinloch AJ (1992) J Mat Sci 27:2763
117. Oyanguren PA, Frontini PM, Williams RJJ, Girard-Reydet E, Pascault JP (1996) Polymer 37:3079
118. Williams RJJ, Pascault JP (1995) Polymer 36:3541
119. Suspene L, Yand YS, Pascault JP (1993) Adv Chem Ser 233:163
120. Vérchère D, Pascault JP, Sautereau H, Moschiar SM, Riccardi CC, Williams RJJ (1991) J Appl Polym Sci 43:293
121. Hedrick JC, McGrath, J.E. (1993) Adv. Chem. Ser 233:293
122. Hedrick JL, Hilborn JG, Prime RB, Labadie JW, Dawson DJ, Russell TP, Warkhaker V (1994) Polymer 35:291

123. Hedrick JL, Yilgor I, Jurek M, Hedrick JC, Wilkes GL, McGrath JE (1991) Polymer 32:2020
124. Kinloch AJ, Yuen ML, Jenkins SD (1994) J Mat Sci 29:3781
125. Kishi H, Shi YB, Huang J, Yee AF (1997) J Mat Sci 32:761
126. Antonietti M, Basten R, Lohmann S (1995) Macromol Chem Phys 196:441
127. Swarup S, Schoff CK (1993) Prog Org Coatings 23:1
128. Bagheri R, Pearson RA (1996) J Mat Sci 31:3945
129. Qian JY, Pearson RA, Dimonie VL, Shaffer OL, El-Aasser MS (1997) Polymer 38:21
130. Sue HJ, Garcia-Meitin EI, Pickelman DM, Yang PC (1993) Adv Chem Ser 233:259
131. Béguelin P (1996) Approche expérimentale du comportement mécanique des polymères en sollicitation rapide. Thesis N°1572 at Swiss Federal Institute of Technology, Lausanne
132. Kunz-Douglass S, Beaumont PWR, Ashby MF (1980) J Mat Sci 15:1109
133. Bucknall CB, Partridge IK (1983) Polymer 24:639
134. Hedrick JL, Yilgor I, Wilkes GL, McGrath JE (1985) Polym Bull 13:201
135. Kim SC, Brown HR (1987) J Mat Sci 22:2589
136. Spanoudakis J, Young RJ (1984) J Mat Sci 19:473
137. Spanoudakis J, Young RJ (1984) J Mat Sci 19:487
138. Moloney AC, Kausch HH, Kaiser T, Beer HR (1987) J Mat Sci 22:381
139. Boogh L, Pettersson B, Japon S, Månson JAE (1995) Proc. ICCM 10:VI389
140. Boogh L, Pettersson B, Månson JAE (1997) In: Pick R (ed) International Conference on Surfaces and Interfaces in Polymers and Composites, Lausanne, p 30
141. Frischinger I, Dirlikov S (1993) Adv Chem Ser 233:451
142. Sohn JE, Emerson JA, Koberstein JT (1989) J Appl Polym Sci 37:2627
143. Azimi HR, Pearson RA, Hertzberg RW (1996) J Mat Sci 31:3777
144. Daghyani HR, Ye L, Mai YW, Wu J (1994) J Mat Sci Lett 13:1330
145. Manzione LT, Gillham JK (1981) J Appl Polym Sci 26:889
146. Yanamaka K, Inoue T (1990) J Mat Sci 25:241
147. Hwang JF, Manson JA, Hertzberg RW, Miller GA, Sperling LH (1989) Polym Eng Sci 29:1466
148. Kim DS, Kim SG (1994) Polym Eng Sci 34:1598
149. Yamanaka K, Takagi Y, Inoue T (1989) Polymer 30:1839
150. Chen TK, Jan, YH. (1991) J Mat Sci 26:5848
151. Huang Y, Kinloch AJ, Bertsch RJ, Siebert AR (1993) Adv Chem Ser 233:189
152. Bucknall CB, Karpodinis A, Zhang XC (1994) J Mat Sci 29:3377
153. Bucknall CB, Soares VLP, Yang HH, Zhang XC (1996) Macromol Symp 101:265
154. Lazzeri A, Bucknall, CB (1993) J Mat Sci 28:6799
155. Sue HJ (1992) J Mat Sci 27:3098
156. Dompas D, Groeninckx G, Isogawa M, Hasegawa T, Kadokura M (1994) Polymer 35:4750
157. Kim DS, Cho K, Park CE (1996) Polym Eng Sci 36:755
158. Bucknall CB (1997) In: Pick R (ed) Surfaces and interfaces in polymers and composites. Lausanne, p 91
159. Bucknall CB, Ayre D (1997) In: Institute of Materials (ed) 10th Int Conf on Deformation, Yield and Fracture of Polymers, Cambridge, p 179
160. Li D, Yee AF, Chen IW, Chang SC, Takahashi K (1994) J Mat Sci 29:2205
161. Huang Y, Kinloch, AJ (1992) J Mat Sci Lett 11:484
162. Bubeck RA, Buckley DJ, Kramer EJ, Brown HR (1991) J Mat Sci 26:6249
163. Schirrer R, Fond C (1995) Rev Met CIT 92:1027
164. He C, Donald AM, Butler MF (1998) Macromolecules 31:158
165. Bagheri R, Pearson RA (1993) 25th International SAMPE Technical Conference, p 25
166. Bagheri R, Pearson RA (1995) Polymer 36:4883
167. Bagheri R, Pearson RA (1996) Polymer 37:4529
168. Guild FJ, Young RJ (1989) J Mat Sci 24:2454

169. Huang Y, Kinloch AJ (1992) J Mat Sci 27:2753
170. Guild FJ, Kinloch AJ (1995) J Mat Sci 30:1689
171. Huang Y, Kinloch AJ (1992) J Mat Sci 27:2763
172. Fukui T, Kikiuchi Y, Inoue T (1991) Polymer 32:2367
173. Fond C, Lobbrecht A, Schirrer R (1996) Int J Fract 77:141
174. Sue HJ (1991) Polym Eng Sci 31:270
175. Margolina A, Wu S (1988) Polymer 29:2170
176. Hamerton I (ed) (1994) Chemistry and technology of cyanate ester resins. Chapman & Hall, London
177. Shimp DA, Fang T (1995) Prog Polym Sci 20:61
178. Paulus JR (1989) Circuit World 15:19
179. Shimp DA (1994) Polym Mat Sci Eng 71:561
180. Balde J, Messner G (1987) Circuit World 14:1
181. Tummala RR, Rymaszewski EJ (eds) (1989) Microelectronics packaging handbook. Van Nostrand Reinhold, New York
182. Hedrick JC, Goro JT, Viehbeck A (1994) International SAMPE Electronics Conference 7:171
183. Pascault JP (1994) ACS Polym Mat Sci Eng 71:559
184. Shimp DA, Wentworth JE (1992) SAMPE Symp 37:293
185. Speak SC, Sitt H, Fuse RH (1991) SAMPE Symp 36:336
186. Pascault JP, Galy J, Méchin F (1994) Additives and modifiers for cyanate ester resins. In: Hamerton I (ed) (1994) Chemistry and technology of cyanate ester resins. Chapman & Hall, London, p 112
187. Hay JN (1994) Processing and cure schedules for cyanate ester resins. In: Hamerton I (ed) (1994) Chemistry and technology of cyanate ester resins. Chapman & Hall, London, p 151
188. Shimp DA (1989) Int SAMPE Symp 32:1063
189. Maxwell-Garnett JC (1905) Phil Trans R Soc 205:237

Received: August 1998

Author Index Volumes 101–147

Author Index Volumes 1–100 see Volume 100

de, Abajo, J. and *de la Campa, J.G.*: Processable Aromatic Polyimides. Vol. 140, pp. 23-60.
Adolf, D. B. see Ediger, M. D.: Vol. 116, pp. 73-110.
Aharoni, S. M. and *Edwards, S. F.*: Rigid Polymer Networks. Vol. 118, pp. 1-231.
Améduri, B., Boutevin, B. and *Gramain, P.*: Synthesis of Block Copolymers by Radical Polymerization and Telomerization. Vol. 127, pp. 87-142.
Améduri, B. and *Boutevin, B.*: Synthesis and Properties of Fluorinated Telechelic Monodispersed Compounds. Vol. 102, pp. 133-170.
Amselem, S. see Domb, A. J.: Vol. 107, pp. 93-142.
Andrady, A. L.: Wavelenght Sensitivity in Polymer Photodegradation. Vol. 128, pp. 47-94.
Andreis, M. and *Koenig, J. L.*: Application of Nitrogen-15 NMR to Polymers. Vol. 124, pp. 191-238.
Angiolini, L. see Carlini, C.: Vol. 123, pp. 127-214.
Anseth, K. S., Newman, S. M. and *Bowman, C. N.*: Polymeric Dental Composites: Properties and Reaction Behavior of Multimethacrylate Dental Restorations. Vol. 122, pp. 177-218.
Armitage, B. A. see O'Brien, D. F.: Vol. 126, pp. 53-58.
Arndt, M. see Kaminski, W.: Vol. 127, pp. 143-187.
Arnold Jr., F. E. and *Arnold, F. E.*: Rigid-Rod Polymers and Molecular Composites. Vol. 117, pp. 257-296.
Arshady, R.: Polymer Synthesis via Activated Esters: A New Dimension of Creativity in Macromolecular Chemistry. Vol. 111, pp. 1-42.

Bahar, I., Erman, B. and *Monnerie, L.*: Effect of Molecular Structure on Local Chain Dynamics: Analytical Approaches and Computational Methods. Vol. 116, pp. 145-206.
Ballauff, M. see Dingenouts, N.: Vol. 144, pp. 1-48.
Baltá-Calleja, F. J., González Arche, A., Ezquerra, T. A., Santa Cruz, C., Batallón, F., Frick, B. and *López Cabarcos, E.*: Structure and Properties of Ferroelectric Copolymers of Poly(vinylidene) Fluoride. Vol. 108, pp. 1-48.
Barshtein, G. R. and *Sabsai, O. Y.*: Compositions with Mineralorganic Fillers. Vol. 101, pp. 1-28.
Batallán, F. see Baltá-Calleja, F. J.: Vol. 108, pp. 1-48.
Batog, A. E., Pet'ko, I. P., Penczek, P.: Aliphatic-Cycloaliphatic Epoxy Compounds and Polymers. Vol. 144, pp. 49-114.
Barton, J. see Hunkeler, D.: Vol. 112, pp. 115-134.
Bell, C. L. and *Peppas, N. A.*: Biomedical Membranes from Hydrogels and Interpolymer Complexes. Vol. 122, pp. 125-176.
Bellon-Maurel, A. see Calmon-Decriaud, A.: Vol. 135, pp. 207-226.
Bennett, D. E. see O'Brien, D. F.: Vol. 126, pp. 53-84.
Berry, G.C.: Static and Dynamic Light Scattering on Moderately Concentraded Solutions: Isotropic Solutions of Flexible and Rodlike Chains and Nematic Solutions of Rodlike Chains. Vol. 114, pp. 233-290.
Bershtein, V. A. and *Ryzhov, V. A.*: Far Infrared Spectroscopy of Polymers. Vol. 114, pp. 43-122.
Bigg, D. M.: Thermal Conductivity of Heterophase Polymer Compositions. Vol. 119, pp. 1-30.

Binder, K.: Phase Transitions in Polymer Blends and Block Copolymer Melts: Some Recent Developments. Vol. 112, pp. 115-134.
Binder, K.: Phase Transitions of Polymer Blends and Block Copolymer Melts in Thin Films. Vol. 138, pp. 1-90.
Bird, R. B. see *Curtiss, C. F.*: Vol. 125, pp. 1-102.
Biswas, M. and *Mukherjee, A.*: Synthesis and Evaluation of Metal-Containing Polymers. Vol. 115, pp. 89-124.
Bolze, J. see *Dingenouts, N.*: Vol. 144, pp. 1-48.
Boutevin, B. and *Robin, J. J.*: Synthesis and Properties of Fluorinated Diols. Vol. 102. pp. 105-132.
Boutevin, B. see Amédouri, B.: Vol. 102, pp. 133-170.
Boutevin, B. see Améduri, B.: Vol. 127, pp. 87-142.
Bowman, C. N. see Anseth, K. S.: Vol. 122, pp. 177-218.
Boyd, R. H.: Prediction of Polymer Crystal Structures and Properties. Vol. 116, pp. 1-26.
Briber, R. M. see Hedrick, J. L.: Vol. 141, pp. 1-44.
Bronnikov, S. V., Vettegren, V. I. and *Frenkel, S. Y.*: Kinetics of Deformation and Relaxation in Highly Oriented Polymers. Vol. 125, pp. 103-146.
Bruza, K. J. see Kirchhoff, R. A.: Vol. 117, pp. 1-66.
Burban, J. H. see Cussler, E. L.: Vol. 110, pp. 67-80.
Burchard, W.: Solution Properties of Branched Macromolecules. Vol. 143, pp. 113-194.

Calmon-Decriaud, A. Bellon-Maurel, V., Silvestre, F.: Standard Methods for Testing the Aerobic Biodegradation of Polymeric Materials. Vol 135, pp. 207-226.
Cameron, N. R. and *Sherrington, D. C.*: High Internal Phase Emulsions (HIPEs)-Structure, Properties and Use in Polymer Preparation. Vol. 126, pp. 163-214.
de la Campa, J. G. see *de Abajo, , J.*: Vol. 140, pp. 23-60.
Candau, F. see Hunkeler, D.: Vol. 112, pp. 115-134.
Canelas, D. A. and *DeSimone, J. M.*: Polymerizations in Liquid and Supercritical Carbon Dioxide. Vol. 133, pp. 103-140.
Capek, I.: Kinetics of the Free-Radical Emulsion Polymerization of Vinyl Chloride. Vol. 120, pp. 135-206.
Capek, I.: Radical Polymerization of Polyoxyethylene Macromonomers in Disperse Systems. Vol. 145, pp. 1-56.
Capek, I.: Radical Polymerization of Polyoxyethylene Macromonomers in Disperse Systems. Vol. 146, pp. 1-56.
Carlini, C. and *Angiolini, L.*: Polymers as Free Radical Photoinitiators. Vol. 123, pp. 127-214.
Carter, K. R. see Hedrick, J. L.: Vol. 141, pp. 1-44.
Casas-Vazquez, J. see Jou, D.: Vol. 120, pp. 207-266.
Chandrasekhar, V.: Polymer Solid Electrolytes: Synthesis and Structure. Vol 135, pp. 139-206
Charleux, B., Faust R.: Synthesis of Branched Polymers by Cationic Polymerization. Vol. 142, pp. 1-70.
Chen, P. see Jaffe, M.: Vol. 117, pp. 297-328.
Choe, E.-W. see Jaffe, M.: Vol. 117, pp. 297-328.
Chow, T. S.: Glassy State Relaxation and Deformation in Polymers. Vol. 103, pp. 149-190.
Chung, T.-S. see Jaffe, M.: Vol. 117, pp. 297-328.
Comanita, B. see Roovers, J.: Vol. 142, pp. 179-228.
Connell, J. W. see Hergenrother, P. M.: Vol. 117, pp. 67-110.
Criado-Sancho, M. see Jou, D.: Vol. 120, pp. 207-266.
Curro, J.G. see Schweizer, K.S.: Vol. 116, pp. 319-378.
Curtiss, C. F. and *Bird, R. B.*: Statistical Mechanics of Transport Phenomena: Polymeric Liquid Mixtures. Vol. 125, pp. 1-102.
Cussler, E. L., Wang, K. L. and *Burban, J. H.*: Hydrogels as Separation Agents. Vol. 110, pp. 67-80.

DeSimone, J. M. see Canelas D. A.: Vol. 133, pp. 103-140.
DiMari, S. see Prokop, A.: Vol. 136, pp. 1-52.

Dimonie, M. V. see Hunkeler, D.: Vol. 112, pp. 115-134.
Dingenouts, N., Bolze, J., Pötschke, D., Ballauf, M.: Analysis of Polymer Latexes by Small-Angle X-Ray Scattering. Vol. 144, pp. 1-48
Dodd, L. R. and *Theodorou, D. N.:* Atomistic Monte Carlo Simulation and Continuum Mean Field Theory of the Structure and Equation of State Properties of Alkane and Polymer Melts. Vol. 116, pp. 249-282.
Doelker, E.: Cellulose Derivatives. Vol. 107, pp. 199-266.
Dolden, J. G.: Calculation of a Mesogenic Index with Emphasis Upon LC-Polyimides. Vol. 141, pp. 189-245.
Domb, A. J., Amselem, S., Shah, J. and *Maniar, M.:* Polyanhydrides: Synthesis and Characterization. Vol.107, pp. 93-142.
Dubois, P. see Mecerreyes, D.: Vol. 147, pp. 1-60.
Dubrovskii, S. A. see Kazanskii, K. S.: Vol. 104, pp. 97-134.
Dunkin, I. R. see Steinke, J.: Vol. 123, pp. 81-126.
Dunson, D. L. see McGrath, J. E.: Vol. 140, pp. 61-106.

Economy, J. and *Goranov, K.:* Thermotropic Liquid Crystalline Polymers for High Performance Applications. Vol. 117, pp. 221-256.
Ediger, M. D. and *Adolf, D. B.:* Brownian Dynamics Simulations of Local Polymer Dynamics. Vol. 116, pp. 73-110.
Edwards, S. F. see Aharoni, S. M.: Vol. 118, pp. 1-231.
Endo, T. see Yagci, Y.: Vol. 127, pp. 59-86.
Erman, B. see Bahar, I.: Vol. 116, pp. 145-206.
Ewen, B, Richter, D.: Neutron Spin Echo Investigations on the Segmental Dynamics of Polymers in Melts, Networks and Solutions. Vol. 134, pp. 1-130.
Ezquerra, T. A. see Baltá-Calleja, F. J.: Vol. 108, pp. 1-48.

Faust, R. see Charleux, B: Vol. 142, pp. 1-70.
Fekete, E see Pukánszky, B: Vol. 139, pp. 109-154.
Fendler, J.H.: Membrane-Mimetic Approach to Advanced Materials. Vol. 113, pp. 1-209.
Fetters, L. J. see Xu, Z.: Vol. 120, pp. 1-50.
Förster, S. and *Schmidt, M.:* Polyelectrolytes in Solution. Vol. 120, pp. 51-134.
Freire, J. J.: Conformational Properties of Branched Polymers: Theory and Simulations. Vol. 143, pp. 35-112.
Frenkel, S. Y. see Bronnikov, S. V.: Vol. 125, pp. 103-146.
Frick, B. see Baltá-Calleja, F. J.: Vol. 108, pp. 1-48.
Fridman, M. L.: see Terent´eva, J. P.: Vol. 101, pp. 29-64.
Funke, W.: Microgels-Intramolecularly Crosslinked Macromolecules with a Globular Structure. Vol. 136, pp. 137-232.

Galina, H.: Mean-Field Kinetic Modeling of Polymerization: The Smoluchowski Coagulation Equation. Vol. 137, pp. 135-172.
Ganesh, K. see Kishore, K.: Vol. 121, pp. 81-122.
Gaw, K. O. and *Kakimoto, M.:* Polyimide-Epoxy Composites. Vol. 140, pp. 107-136.
Geckeler, K. E. see Rivas, B.: Vol. 102, pp. 171-188.
Geckeler, K. E.: Soluble Polymer Supports for Liquid-Phase Synthesis. Vol. 121, pp. 31-80.
Gehrke, S. H.: Synthesis, Equilibrium Swelling, Kinetics Permeability and Applications of Environmentally Responsive Gels. Vol. 110, pp. 81-144.
de Gennes, P.-G.: Flexible Polymers in Nanopores. Vol. 138, pp. 91-106.
Giannelis, E.P., Krishnamoorti, R., Manias, E.: Polymer-Silicate Nanocomposites: Model Systems for Confined Polymers and Polymer Brushes. Vol. 138, pp. 107-148.
Godovsky, D. Y.: Electron Behavior and Magnetic Properties Polymer-Nanocomposites. Vol. 119, pp. 79-122.
González Arche, A. see Baltá-Calleja, F. J.: Vol. 108, pp. 1-48.
Goranov, K. see Economy, J.: Vol. 117, pp. 221-256.

Gramain, P. see Améduri, B.: Vol. 127, pp. 87-142.
Grest, G.S.: Normal and Shear Forces Between Polymer Brushes. Vol. 138, pp. 149-184
Grosberg, A. and Nechaev, S.: Polymer Topology. Vol. 106, pp. 1-30.
Grubbs, R., Risse, W. and *Novac, B.*: The Development of Well-defined Catalysts for Ring-Opening Olefin Metathesis. Vol. 102, pp. 47-72.
van Gunsteren, W. F. see Gusev, A. A.: Vol. 116, pp. 207-248.
Gusev, A. A., Müller-Plathe, F., van Gunsteren, W. F. and *Suter, U. W.*: Dynamics of Small Molecules in Bulk Polymers. Vol. 116, pp. 207-248.
Guillot, J. see Hunkeler, D.: Vol. 112, pp. 115-134.
Guyot, A. and *Tauer, K.*: Reactive Surfactants in Emulsion Polymerization. Vol. 111, pp. 43-66.

Hadjichristidis, N., Pispas, S., Pitsikalis, M., Iatrou, H., Vlahos, C.: Asymmetric Star Polymers Synthesis and Properties. Vol. 142, pp. 71-128.
Hadjichristidis, N. see Xu, Z.: Vol. 120, pp. 1-50.
Hadjichristidis, N. see Pitsikalis, M.: Vol. 135, pp. 1-138.
Hall, H. K. see Penelle, J.: Vol. 102, pp. 73-104.
Hammouda, B.: SANS from Homogeneous Polymer Mixtures: A Unified Overview. Vol. 106, pp. 87-134.
Harada, A.: Design and Construction of Supramolecular Architectures Consisting of Cyclodextrins and Polymers. Vol. 133, pp. 141-192.
Haralson, M. A. see Prokop, A.: Vol. 136, pp. 1-52.
Hawker, C. J. Dentritic and Hyperbranched Macromolecules – Precisely Controlled Macromolecular Architectures. Vol. 147, pp. 113-160
Hawker, C. J. see Hedrick, J. L.: Vol. 141, pp. 1-44.
Hedrick, J. L., Carter, K. R., Labadie, J. W., Miller, R. D., Volksen, W., Hawker, C. J., Yoon, D. Y., Russell, T. P., McGrath, J. E., Briber, R. M.: Nanoporous Polyimides. Vol. 141, pp. 1-44.
Hedrick, J. L., Labadie, J. W., Volksen, W. and *Hilborn, J. G.*: Nanoscopically Engineered Polyimides. Vol. 147, pp. 61-112.
Hedrick, J. L. see Hergenrother, P. M.: Vol. 117, pp. 67-110.
Hedrick, J. L. see Kiefer, J.: Vol. 147, pp. 161-247.
Hedrick, J.L. see McGrath, J. E.: Vol. 140, pp. 61-106.
Heller, J.: Poly (Ortho Esters). Vol. 107, pp. 41-92.
Hemielec, A. A. see Hunkeler, D.: Vol. 112, pp. 115-134.
Hergenrother, P. M., Connell, J. W., Labadie, J. W. and *Hedrick, J. L.*: Poly(arylene ether)s Containing Heterocyclic Units. Vol. 117, pp. 67-110.
Hernández-Barajas, J. see Wandrey, C.: Vol. 145, pp. 123-182.
Hervet, H. see Léger, L.: Vol. 138, pp. 185-226.
Hilborn, J. G. see Hedrick, J. L.: Vol. 147, pp. 61-112.
Hilborn, J. G. see Kiefer, J.: Vol. 147, pp. 161-247.
Hiramatsu, N. see Matsushige, M.: Vol. 125, pp. 147-186.
Hirasa, O. see Suzuki, M.: Vol. 110, pp. 241-262.
Hirotsu, S.: Coexistence of Phases and the Nature of First-Order Transition in Poly-N-isopropylacrylamide Gels. Vol. 110, pp. 1-26.
Hornsby, P.: Rheology, Compoundind and Processing of Filled Thermoplastics. Vol. 139, pp. 155-216.
Hult, A., Johansson, M., Malmström, E.: Hyperbranched Polymers. Vol. 143, pp. 1-34.
Hunkeler, D., Candau, F., Pichot, C., Hemielec, A. E., Xie, T. Y., Barton, J., Vaskova, V., Guillot, J., Dimonie, M. V., Reichert, K. H.: Heterophase Polymerization: A Physical and Kinetic Comparision and Categorization. Vol. 112, pp. 115-134.
Hunkeler, D. see Prokop, A.: Vol. 136, pp. 1-52; 53-74.
Hunkeler, D see Wandrey, C.: Vol. 145, pp. 123-182.

Iatrou, H. see Hadjichristidis, N.: Vol. 142, pp. 71-128
Ichikawa, T. see Yoshida, H.: Vol. 105, pp. 3-36.
Ihara, E. see Yasuda, H.: Vol. 133, pp. 53-102.
Ikada, Y. see Uyama, Y.: Vol. 137, pp. 1-40.

Ilavsky, M.: Effect on Phase Transition on Swelling and Mechanical Behavior of Synthetic Hydrogels. Vol. 109, pp. 173-206.
Imai, Y.: Rapid Synthesis of Polyimides from Nylon-Salt Monomers. Vol. 140, pp. 1-23.
Inomata, H. see Saito, S.: Vol. 106, pp. 207-232.
Inoue, S. see Sugimoto, H.: Vol. 146, pp. 39-120.
Irie, M.: Stimuli-Responsive Poly(N-isopropylacrylamide), Photo- and Chemical-Induced Phase Transitions. Vol. 110, pp. 49-66.
Ise, N. see Matsuoka, H.: Vol. 114, pp. 187-232.
Ito, K., Kawaguchi, S,:Poly(macronomers), Homo- and Copolymerization. Vol. 142, pp. 129-178.
Ivanov, A. E. see Zubov, V. P.: Vol. 104, pp. 135-176.

Jacob, S. and Kennedy, J.: Synthesis, Characterization and Properties of OCTA-ARM Polyisobutylene-Based Star Polymers. Vol. 146, pp. 1-38.
Jaffe, M., Chen, P., Choe, E.-W., Chung, T.-S. and *Makhija, S.*: High Performance Polymer Blends. Vol. 117, pp. 297-328.
Jancar, J.: Structure-Property Relationships in Thermoplastic Matrices. Vol. 139, pp. 1-66.
Jerôme, R.: see Mecerreyes, D.: Vol. 147, pp. 1-60.
Jiang, M., Li, M., Xiang, M. and Zhou, H.: Interpolymer Complexation and Miscibility and Enhancement by Hydrogen Bonding. Vol. 146, pp. 121-194.
Johansson, M. see Hult, A.: Vol. 143, pp. 1-34.
Joos-Müller, B. see Funke, W.: Vol. 136, pp. 137-232.
Jou, D., Casas-Vazquez, J. and *Criado-Sancho, M.*: Thermodynamics of Polymer Solutions under Flow: Phase Separation and Polymer Degradation. Vol. 120, pp. 207-266.

Kaetsu, I.: Radiation Synthesis of Polymeric Materials for Biomedical and Biochemical Applications. Vol. 105, pp. 81-98.
Kakimoto, M. see Gaw, K. O.: Vol. 140, pp. 107-136.
Kaminski, W. and *Arndt, M.*: Metallocenes for Polymer Catalysis. Vol. 127, pp. 143-187.
Kammer, H. W., Kressler, H. and *Kummerloewe, C.*: Phase Behavior of Polymer Blends - Effects of Thermodynamics and Rheology. Vol. 106, pp. 31-86.
Kandyrin, L. B. and *Kuleznev, V. N.*: The Dependence of Viscosity on the Composition of Concentrated Dispersions and the Free Volume Concept of Disperse Systems. Vol. 103, pp. 103-148.
Kaneko, M. see Ramaraj, R.: Vol. 123, pp. 215-242.
Kang, E. T., Neoh, K. G. and *Tan, K. L.*: X-Ray Photoelectron Spectroscopic Studies of Electroactive Polymers. Vol. 106, pp. 135-190.
Kato, K. see Uyama, Y.: Vol. 137, pp. 1-40.
Kawaguchi, S. see Ito, K.: Vol. 142, p 129-178.
Kazanskii, K. S. and *Dubrovskii, S. A.*: Chemistry and Physics of „Agricultural" Hydrogels. Vol. 104, pp. 97-134.
Kennedy, J. P. see Jacob, S.: Vol. 146, pp. 1-38.
Kennedy, J. P. see Majoros, I.: Vol. 112, pp. 1-113.
Khokhlov, A., Starodybtzev, S. and *Vasilevskaya, V.*: Conformational Transitions of Polymer Gels: Theory and Experiment. Vol. 109, pp. 121-172.
Kiefer, J., Hedrick J. L. and *Hiborn, J. G.*: Macroporous Thermosets by Chemically Induced Phase Separation. Vol. 147, pp. 161-247.
Kilian, H. G. and *Pieper, T.*: Packing of Chain Segments. A Method for Describing X-Ray Patterns of Crystalline, Liquid Crystalline and Non-Crystalline Polymers. Vol. 108, pp. 49-90.
Kishore, K. and *Ganesh, K.*: Polymers Containing Disulfide, Tetrasulfide, Diselenide and Ditelluride Linkages in the Main Chain. Vol. 121, pp. 81-122.
Kitamaru, R.: Phase Structure of Polyethylene and Other Crystalline Polymers by Solid-State ^{13}C/MNR. Vol. 137, pp 41-102.
Klier, J. see Scranton, A. B.: Vol. 122, pp. 1-54.
Kobayashi, S., Shoda, S. and *Uyama, H.*: Enzymatic Polymerization and Oligomerization. Vol. 121, pp. 1-30.
Koenig, J. L. see Andreis, M.: Vol. 124, pp. 191-238.

Kokufuta, E.: Novel Applications for Stimulus-Sensitive Polymer Gels in the Preparation of Functional Immobilized Biocatalysts. Vol. 110, pp. 157-178.
Konno, M. see Saito, S.: Vol. 109, pp. 207-232.
Kopecek, J. see Putnam, D.: Vol. 122, pp. 55-124.
Koßmehl, G. see Schopf, G.: Vol. 129, pp. 1-145.
Kressler, J. see Kammer, H. W.: Vol. 106, pp. 31-86.
Kricheldorf, H. R.: Liquid-Cristalline Polyimides. Vol. 141, pp. 83-188.
Krishnamoorti, R. see Giannelis, E.P.: Vol. 138, pp. 107-148.
Kirchhoff, R. A. and *Bruza, K. J.*: Polymers from Benzocyclobutenes. Vol. 117, pp. 1-66.
Kuchanov, S. I.: Modern Aspects of Quantitative Theory of Free-Radical Copolymerization. Vol. 103, pp. 1-102.
Kudaibergennow, S.E.: Recent Advances in Studying of Synthetic Polyampholytes in Solutions. Vol. 144, pp. 115-198.
Kuleznev, V. N. see Kandyrin, L. B.: Vol. 103, pp. 103-148.
Kulichkhin, S. G. see Malkin, A. Y.: Vol. 101, pp. 217-258.
Kummerloewe, C. see Kammer, H. W.: Vol. 106, pp. 31-86.
Kuznetsova, N. P. see Samsonov, G. V.: Vol. 104, pp. 1-50.Labadie, J. W. see Hergenrother, P. M.: Vol. 117, pp. 67-110.

Labadie, J. W. see Hedrick, J. L.: Vol. 141, pp. 1-44.
Labadie, J. W. see Hedrick, J. L.: Vol. 147, pp. 61-112.
Lamparski, H. G. see O´Brien, D. F.: Vol. 126, pp. 53-84.
Laschewsky, A.: Molecular Concepts, Self-Organisation and Properties of Polysoaps. Vol. 124, pp. 1-86.
Laso, M. see Leontidis, E.: Vol. 116, pp. 283-318.
Lazár, M. and *RychlΩ, R.*: Oxidation of Hydrocarbon Polymers. Vol. 102, pp. 189-222.
Lechowicz, J. see Galina, H.: Vol. 137, pp. 135-172.
Léger, L., Raphaël, E., Hervet, H.: Surface-Anchored Polymer Chains: Their Role in Adhesion and Friction. Vol. 138, pp. 185-226.
Lenz, R. W.: Biodegradable Polymers. Vol. 107, pp. 1-40.
Leontidis, E., de Pablo, J. J., Laso, M. and *Suter, U. W.*: A Critical Evaluation of Novel Algorithms for the Off-Lattice Monte Carlo Simulation of Condensed Polymer Phases. Vol. 116, pp. 283-318.
Lesec, J. see Viovy, J.-L.: Vol. 114, pp. 1-42.
Li, M. see Jiang, M.: Vol. 146, pp. 121-194.
Liang, G. L. see Sumpter, B. G.: Vol. 116, pp. 27-72.
Lienert, K.-W.: Poly(ester-imide)s for Industrial Use. Vol. 141, pp. 45-82.
Lin, J. and *Sherrington, D. C.*: Recent Developments in the Synthesis, Thermostability and Liquid Crystal Properties of Aromatic Polyamides. Vol. 111, pp. 177-220.
López Cabarcos, E. see Baltá-Calleja, F. J.: Vol. 108, pp. 1-48.

Majoros, I., Nagy, A. and *Kennedy, J. P.*: Conventional and Living Carbocationic Polymerizations United. I. A Comprehensive Model and New Diagnostic Method to Probe the Mechanism of Homopolymerizations. Vol. 112, pp. 1-113.
Makhija, S. see Jaffe, M.: Vol. 117, pp. 297-328.
Malmström, E. see Hult, A.: Vol. 143, pp. 1-34.
Malkin, A. Y. and *Kulichkhin, S. G.*: Rheokinetics of Curing. Vol. 101, pp. 217-258.
Maniar, M. see Domb, A. J.: Vol. 107, pp. 93-142.
Manias, E., see Giannelis, E.P.: Vol. 138, pp. 107-148.
Mashima, K., Nakayama, Y. and *Nakamura, A.*: Recent Trends in Polymerization of a-Olefins Catalyzed by Organometallic Complexes of Early Transition Metals. Vol. 133, pp. 1-52.
Matsumoto, A.: Free-Radical Crosslinking Polymerization and Copolymerization of Multivinyl Compounds. Vol. 123, pp. 41-80.
Matsumoto, A. see Otsu, T.: Vol. 136, pp. 75-138.
Matsuoka, H. and *Ise, N.*: Small-Angle and Ultra-Small Angle Scattering Study of the Ordered Structure in Polyelectrolyte Solutions and Colloidal Dispersions. Vol. 114, pp. 187-232.

Matsushige, K., Hiramatsu, N. and *Okabe, H.:* Ultrasonic Spectroscopy for Polymeric Materials. Vol. 125, pp. 147-186.
Mattice, W. L. see Rehahn, M.: Vol. 131/132, pp. 1-475.
Mays, W. see Xu, Z.: Vol. 120, pp. 1-50.
Mays, J. W. see Pitsikalis, M.: Vol.135, pp. 1-138.
McGrath, J. E. see Hedrick, J. L.: Vol. 141, pp. 1-44.
McGrath, J. E., Dunson, D. L., Hedrick, J. L.: Synthesis and Characterization of Segmented Polyimide-Polyorganosiloxane Copolymers. Vol. 140, pp. 61-106.
McLeish, T.C. B., Milner, S. T.: Entangled Dynamics and Melt Flow of Branched Polymers. Vol. 143, pp. 195-256.
Mecerreyes, D., Dubois, P. and *Jerôme, R.:* Novel Macromolecular Architectures Based on Aliphatic Polyesters: Relevance of the „Coordination-Insertion" Ring-Opening Polymerization. Vol. 147, pp. 1 -60.
Mecham, S. J. see McGrath, J. E.: Vol. 140, pp. 61-106.
Mikos, A. G. see Thomson, R. C.: Vol. 122, pp. 245-274.
Milner, S. T. see McLeish, T. C. B.: Vol. 143, pp. 195-256.
Mison, P. and Sillion, B.: Thermosetting Oligomers Containing Maleimides and Nadiimides End-Groups. Vol. 140, pp. 137-180.
Miyasaka, K.: PVA-Iodine Complexes: Formation, Structure and Properties. Vol. 108. pp. 91-130.
Miller, R. D. see Hedrick, J. L.: Vol. 141, pp. 1-44.
Monnerie, L. see Bahar, I.: Vol. 116, pp. 145-206.
Morishima, Y.: Photoinduced Electron Transfer in Amphiphilic Polyelectrolyte Systems. Vol. 104, pp. 51-96.
Mours, M. see Winter, H. H.: Vol. 134, pp. 165-234.
Müllen, K. see Scherf, U.: Vol. 123, pp. 1-40.
Müller-Plathe, F. see Gusev, A. A.: Vol. 116, pp. 207-248.
Mukherjee, A. see Biswas, M.: Vol. 115, pp. 89-124.
Mylnikov, V.: Photoconducting Polymers. Vol. 115, pp. 1-88.

Nagy, A. see Majoros, I.: Vol. 112, pp. 1-11.
Nakamura, A. see Mashima, K.: Vol. 133, pp. 1-52.
Nakayama, Y. see Mashima, K.: Vol. 133, pp. 1-52.
Narasinham, B., Peppas, N. A.: The Physics of Polymer Dissolution: Modeling Approaches and Experimental Behavior. Vol. 128, pp. 157-208.
Nechaev, S. see Grosberg, A.: Vol. 106, pp. 1-30.
Neoh, K. G. see Kang, E. T.: Vol. 106, pp. 135-190.
Newman, S. M. see Anseth, K. S.: Vol. 122, pp. 177-218.
Nijenhuis, K. te: Thermoreversible Networks. Vol. 130, pp. 1-252.
Noid, D. W. see Sumpter, B. G.: Vol. 116, pp. 27-72.
Novac, B. see Grubbs, R.: Vol. 102, pp. 47-72.
Novikov, V. V. see Privalko, V. P.: Vol. 119, pp. 31-78.

O'Brien, D. F., Armitage, B. A., Bennett, D. E. and *Lamparski, H. G.:* Polymerization and Domain Formation in Lipid Assemblies. Vol. 126, pp. 53-84.
Ogasawara, M.: Application of Pulse Radiolysis to the Study of Polymers and Polymerizations. Vol.105, pp. 37-80.
Okabe, H. see Matsushige, K.: Vol. 125, pp. 147-186.
Okada, M.: Ring-Opening Polymerization of Bicyclic and Spiro Compounds. Reactivities and Polymerization Mechanisms. Vol. 102, pp. 1-46.
Okano, T.: Molecular Design of Temperature-Responsive Polymers as Intelligent Materials. Vol. 110, pp. 179-198.
Okay, O. see Funke, W.: Vol. 136, pp. 137-232.
Onuki, A.: Theory of Phase Transition in Polymer Gels. Vol. 109, pp. 63-120.
Osad'ko, I.S.: Selective Spectroscopy of Chromophore Doped Polymers and Glasses. Vol. 114, pp. 123-186.

Otsu, T., Matsumoto, A.: Controlled Synthesis of Polymers Using the Iniferter Technique: Developments in Living Radical Polymerization. Vol. 136, pp. 75-138.

de Pablo, J. J. see Leontidis, E.: Vol. 116, pp. 283-318.
Padias, A. B. see Penelle, J.: Vol. 102, pp. 73-104.
Pascault, J.-P. see Williams, R. J. J.: Vol. 128, pp. 95-156.
Pasch, H.: Analysis of Complex Polymers by Interaction Chromatography. Vol. 128, pp. 1-46.
Penczek, P. see Batog, A. E.: Vol. 144, pp. 49-114.
Penelle, J., Hall, H. K., Padias, A. B. and *Tanaka, H.*: Captodative Olefins in Polymer Chemistry. Vol. 102, pp. 73-104.
Peppas, N. A. see Bell, C. L.: Vol. 122, pp. 125-176.
Peppas, N. A. see Narasimhan, B.: Vol. 128, pp. 157-208.
Pet'ko, I. P. see Batog, A. E.: Vol. 144, pp. 49-114.
Pichot, C. see Hunkeler, D.: Vol. 112, pp. 115-134.
Pieper, T. see Kilian, H. G.: Vol. 108, pp. 49-90.
Pispas, S. see Pitsikalis, M.: Vol. 135, pp. 1-138.
Pispas, S. see Hadjichristidis: Vol. 142, pp. 71-128.
Pitsikalis, M., Pispas, S., Mays, J. W., Hadjichristidis, N.: Nonlinear Block Copolymer Architectures. Vol. 135, pp. 1-138.
Pitsikalis, M. see Hadjichristidis: Vol. 142, pp. 71-128.
Pötschke, D. see Dingenouts, N.: Vol 144, pp. 1-48.
Pospíšil, J.: Functionalized Oligomers and Polymers as Stabilizers for Conventional Polymers. Vol. 101, pp. 65-168.
Pospíšil, J.: Aromatic and Heterocyclic Amines in Polymer Stabilization. Vol. 124, pp. 87-190.
Powers, A. C. see Prokop, A.: Vol. 136, pp. 53-74.
Priddy, D. B.: Recent Advances in Styrene Polymerization. Vol. 111, pp. 67-114.
Priddy, D. B.: Thermal Discoloration Chemistry of Styrene-co-Acrylonitrile. Vol. 121, pp. 123-154.
Privalko, V. P. and *Novikov, V. V.*: Model Treatments of the Heat Conductivity of Heterogeneous Polymers. Vol. 119, pp 31-78.
Prokop, A., Hunkeler, D., Powers, A. C., Whitesell, R. R., Wang, T. G.: Water Soluble Polymers for Immunoisolation II: Evaluation of Multicomponent Microencapsulation Systems. Vol. 136, pp. 53-74.
Prokop, A., Hunkeler, D., DiMari, S., Haralson, M. A., Wang, T. G.: Water Soluble Polymers for Immunoisolation I: Complex Coacervation and Cytotoxicity. Vol. 136, pp. 1-52.
Pukánszky, B. and *Fekete, E.*: Adhesion and Surface Modification. Vol. 139, pp. 109-154.
Putnam, D. and *Kopecek, J.*: Polymer Conjugates with Anticancer Acitivity. Vol. 122, pp. 55-124.

Ramaraj, R. and *Kaneko, M.*: Metal Complex in Polymer Membrane as a Model for Photosynthetic Oxygen Evolving Center. Vol. 123, pp. 215-242.
Rangarajan, B. see Scranton, A. B.: Vol. 122, pp. 1-54.
Raphaël, E. see Léger, L.: Vol. 138, pp. 185-226.
Reddinger, J. L. and *Reynolds, J. R.*: Molecular Engineering of π-Conjugated Polymers. Vol. 145, pp. 57-122.
Reichert, K. H. see Hunkeler, D.: Vol. 112, pp. 115-134.
Rehahn, M., Mattice, W. L., Suter, U. W.: Rotational Isomeric State Models in Macromolecular Systems. Vol. 131/132, pp. 1-475.
Reynolds, J.R. see Reddinger, J. L.: Vol. 145, pp. 57-122.
Richter, D. see Ewen, B.: Vol. 134, pp.1-130.
Risse, W. see Grubbs, R.: Vol. 102, pp. 47-72.
Rivas, B. L. and *Geckeler, K. E.*: Synthesis and Metal Complexation of Poly(ethyleneimine) and Derivatives. Vol. 102, pp. 171-188.
Robin, J. J. see Boutevin, B.: Vol. 102, pp. 105-132.
Roe, R.-J.: MD Simulation Study of Glass Transition and Short Time Dynamics in Polymer Liquids. Vol. 116, pp. 111-114.
Roovers, J., Comanita, B.: Dendrimers and Dendrimer-Polymer Hybrids. Vol. 142, pp 179-228.

Rothon, R. N.: Mineral Fillers in Thermoplastics: Filler Manufacture and Characterisation. Vol. 139, pp. 67-108.
Rozenberg, B. A. see Williams, R. J. J.: Vol. 128, pp. 95-156.
Ruckenstein, E.: Concentrated Emulsion Polymerization. Vol. 127, pp. 1-58.
Rusanov, A. L.: Novel Bis (Naphtalic Anhydrides) and Their Polyheteroarylenes with Improved Processability. Vol. 111, pp. 115-176.
Russel, T. P. see Hedrick, J. L.: Vol. 141, pp. 1-44.
Rychlý, J. see Lazár, M.: Vol. 102, pp. 189-222.
Ryzhov, V. A. see Bershtein, V. A.: Vol. 114, pp. 43-122.

Sabsai, O. Y. see Barshtein, G. R.: Vol. 101, pp. 1-28.
Saburov, V. V. see Zubov, V. P.: Vol. 104, pp. 135-176.
Saito, S., Konno, M. and *Inomata, H.*: Volume Phase Transition of N-Alkylacrylamide Gels. Vol. 109, pp. 207-232.
Samsonov, G. V. and *Kuznetsova, N. P.*: Crosslinked Polyelectrolytes in Biology. Vol. 104, pp. 1-50.
Santa Cruz, C. see Baltá-Calleja, F. J.: Vol. 108, pp. 1-48.
Sato, T. and *Teramoto, A.*: Concentrated Solutions of Liquid-Christalline Polymers. Vol. 126, pp. 85-162.
Scherf, U. and *Müllen, K.*: The Synthesis of Ladder Polymers. Vol. 123, pp. 1-40.
Schmidt, M. see Förster, S.: Vol. 120, pp. 51-134.
Schopf, G. and *Koßmehl, G.*: Polythiophenes - Electrically Conductive Polymers. Vol. 129, pp. 1-145.
Schweizer, K. S.: Prism Theory of the Structure, Thermodynamics, and Phase Transitions of Polymer Liquids and Alloys. Vol. 116, pp. 319-378.
Scranton, A. B., Rangarajan, B. and *Klier, J.*: Biomedical Applications of Polyelectrolytes. Vol. 122, pp. 1-54.
Sefton, M. V. and *Stevenson, W. T. K.*: Microencapsulation of Live Animal Cells Using Polycrylates. Vol. 107, pp. 143-198.
Shamanin, V. V.: Bases of the Axiomatic Theory of Addition Polymerization. Vol. 112, pp. 135-180.
Sherrington, D. C. see Cameron, N. R., Vol. 126, pp. 163-214.
Sherrington, D. C. see Lin, J.: Vol. 111, pp. 177-220.
Sherrington, D. C. see Steinke, J.: Vol. 123, pp. 81-126.
Shibayama, M. see Tanaka, T.: Vol. 109, pp. 1-62.
Shiga, T.: Deformation and Viscoelastic Behavior of Polymer Gels in Electric Fields. Vol. 134, pp. 131-164.
Shoda, S. see Kobayashi, S.: Vol. 121, pp. 1-30.
Siegel, R. A.: Hydrophobic Weak Polyelectrolyte Gels: Studies of Swelling Equilibria and Kinetics. Vol. 109, pp. 233-268.
Silvestre, F. see Calmon-Decriaud, A.: Vol. 207, pp. 207-226.
Sillion, B. see Mison, P.: Vol. 140, pp. 137-180.
Singh, R. P. see Sivaram, S.: Vol. 101, pp. 169-216.
Sivaram, S. and *Singh, R. P.*: Degradation and Stabilization of Ethylene-Propylene Copolymers and Their Blends: A Critical Review. Vol. 101, pp. 169-216.
Starodybtzev, S. see Khokhlov, A.: Vol. 109, pp. 121-172.
Steinke, J., Sherrington, D. C. and *Dunkin, I. R.*: Imprinting of Synthetic Polymers Using Molecular Templates. Vol. 123, pp. 81-126.
Stenzenberger, H. D.: Addition Polyimides. Vol. 117, pp. 165-220.
Stevenson, W. T. K. see Sefton, M. V.: Vol. 107, pp. 143-198.
Sumpter, B. G., Noid, D. W., Liang, G. L. and *Wunderlich, B.*: Atomistic Dynamics of Macromolecular Crystals. Vol. 116, pp. 27-72.
Sugimoto, H. and *Inoue, S.*: Polymerization by Metalloporphyrin and Related Complexes. Vol. 146, pp. 39-120
Suter, U. W. see Gusev, A. A.: Vol. 116, pp. 207-248.
Suter, U. W. see Leontidis, E.: Vol. 116, pp. 283-318.
Suter, U. W. see Rehahn, M.: Vol. 131/132, pp. 1-475.

Suzuki, A.: Phase Transition in Gels of Sub-Millimeter Size Induced by Interaction with Stimuli. Vol. 110, pp. 199-240.
Suzuki, A. and *Hirasa, O.*: An Approach to Artifical Muscle by Polymer Gels due to Micro-Phase Separation. Vol. 110, pp. 241-262.

Tagawa, S.: Radiation Effects on Ion Beams on Polymers. Vol. 105, pp. 99-116.
Tan, K. L. see Kang, E. T.: Vol. 106, pp. 135-190.
Tanaka, T. see Penelle, J.: Vol. 102, pp. 73-104.
Tanaka, H. and *Shibayama, M.*: Phase Transition and Related Phenomena of Polymer Gels. Vol. 109, pp. 1-62.
Tauer, K. see Guyot, A.: Vol. 111, pp. 43-66.
Teramoto, A. see Sato, T.: Vol. 126, pp. 85-162.
Terent´eva, J. P. and *Fridman, M. L.*: Compositions Based on Aminoresins. Vol. 101, pp. 29-64.
Theodorou, D. N. see Dodd, L. R.: Vol. 116, pp. 249-282.
Thomson, R. C., Wake, M. C., Yaszemski, M. J. and *Mikos, A. G.*: Biodegradable Polymer Scaffolds to Regenerate Organs. Vol. 122, pp. 245-274.
Tokita, M.: Friction Between Polymer Networks of Gels and Solvent. Vol. 110, pp. 27-48.
Tsuruta, T.: Contemporary Topics in Polymeric Materials for Biomedical Applications. Vol. 126, pp. 1-52.

Uyama, H. see Kobayashi, S.: Vol. 121, pp. 1-30.
Uyama, Y: Surface Modification of Polymers by Grafting. Vol. 137, pp. 1-40.

Vasilevskaya, V. see Khokhlov, A.: Vol. 109, pp. 121-172.
Vaskova, V. see Hunkeler, D.: Vol.:112, pp. 115-134.
Verdugo, P.: Polymer Gel Phase Transition in Condensation-Decondensation of Secretory Products. Vol. 110, pp. 145-156.
Vettegren, V. I.: see Bronnikov, S. V.: Vol. 125, pp. 103-146.
Viovy, J.-L. and *Lesec, J.*: Separation of Macromolecules in Gels: Permeation Chromatography and Electrophoresis. Vol. 114, pp. 1-42.
Vlahos, C. see Hadjichristidis, N.: Vol. 142, pp. 71-128.
Volksen, W.: Condensation Polyimides: Synthesis, Solution Behavior, and Imidization Characteristics. Vol. 117, pp. 111-164.
Volksen, W. see Hedrick, J. L.: Vol. 141, pp. 1-44.
Volksen, W. see Hedrick, J. L.: Vol. 147, pp. 61-112.

Wake, M. C. see Thomson, R. C.: Vol. 122, pp. 245-274.
Wandrey C., Hernández-Barajas, J. and *Hunkeler, D.*: Diallyldimethylammonium Chloride and its Polymers. Vol. 145, pp. 123-182.
Wang, K. L. see Cussler, E. L.: Vol. 110, pp. 67-80.
Wang, S.-Q.: Molecular Transitions and Dynamics at Polymer/Wall Interfaces: Origins of Flow Instabilities and Wall Slip. Vol. 138, pp. 227-276.
Wang, T. G. see Prokop, A.: Vol. 136, pp.1-52; 53-74.
Whitesell, R. R. see Prokop, A.: Vol. 136, pp. 53-74.
Williams, R. J. J., Rozenberg, B. A., Pascault, J.-P.: Reaction Induced Phase Separation in Modified Thermosetting Polymers. Vol. 128, pp. 95-156.
Winter, H. H., Mours, M.: Rheology of Polymers Near Liquid-Solid Transitions. Vol. 134, pp. 165-234.
Wu, C.: Laser Light Scattering Characterization of Special Intractable Macromolecules in Solution. Vol 137, pp. 103-134.
Wunderlich, B. see Sumpter, B. G.: Vol. 116, pp. 27-72.

Xiang, M. see Jiang, M.: Vol. 146, pp. 121-194.
Xie, T. Y. see Hunkeler, D.: Vol. 112, pp. 115-134.
Xu, Z., Hadjichristidis, N., Fetters, L. J. and *Mays, J. W.*: Structure/Chain-Flexibility Relationships of Polymers. Vol. 120, pp. 1-50.

Yagci, Y. and *Endo, T.*: N-Benzyl and N-Alkoxy Pyridium Salts as Thermal and Photochemical Initiators for Cationic Polymerization. Vol. 127, pp. 59-86.
Yannas, I. V.: Tissue Regeneration Templates Based on Collagen-Glycosaminoglycan Copolymers. Vol. 122, pp. 219-244.
Yamaoka, H.: Polymer Materials for Fusion Reactors. Vol. 105, pp. 117-144.
Yasuda, H. and *Ihara, E.*: Rare Earth Metal-Initiated Living Polymerizations of Polar and Nonpolar Monomers. Vol. 133, pp. 53-102.
Yaszemski, M. J. see Thomson, R. C.: Vol. 122, pp. 245-274.
Yoon, D. Y. see Hedrick, J. L.: Vol. 141, pp. 1-44.
Yoshida, H. and *Ichikawa, T.*: Electron Spin Studies of Free Radicals in Irradiated Polymers. Vol. 105, pp. 3-36.

Zhou, H. see Jiang, M.: Vol. 146, pp. 121-194.
Zubov, V. P., Ivanov, A. E. and *Saburov, V. V.*: Polymer-Coated Adsorbents for the Separation of Biopolymers and Particles. Vol. 104, pp. 135-176.

Subject Index

ABA triblock copolymers 46
Adhesion 61
Adipic anhydride 8
Agglomeration, local 235
Alkanes, aliphatic/cycloaliphatic 185
Alkyl ester-perfluoroether 84
Aluminum alkoxides 6
Aluminum isopropoxide 8
Amic ester-aryl ether 73, 74
3-Aminophenol 69
Anionic polymerization 27
Arborols 116, 118
Aryl ether phenylquinoxaline 78
Aryl fluorides 69
Arylacetylene dendrimers 131

Back-biting reactions 6
Back-end-of-the-line wiring 62
Benzoxazole 65
BEOL 62
Bimetallic (Zn,AP)-oxo alkoxides 7
Binodal line 172, 174
Biodegradable polymers 3
Bioglass, degradable 52
2,2'-Bis(4-amino-cyclohexyl)propane (PACP) 169
Bisphenol-E cyanate ester (BPECN) 236
Block copolymers 10, 61
Blowing agents 165
Bonding 178
BPDA/PDA polyimide 64
BPECN 236
Branching, degree 153
Buckminsterfullerene 122, 139
Butyrolactone 4

Caprolactone 4
Carborane nucleus 140
N-Carboxy anhydride 24
Cascade molecules 115, 116
Catalytic systems 152
Cationic polymerization 30
Cavitation resistance 221

Ceramers 51
Chemically induced phase separation (CIPS) 164, 169
Chips 62
Chiral dendrimers 131
Chlorofluorocarbons 165
CIPS 164, 169
–, kinetically controlled 213
Cobaltacetylacetonate 237
Continuous process 24
Controlled polymerization 6
Convergent growth approach 119, 124
Conversion, extent of 176
Coordination chemistry 136
Coordination-insertion mechanism 8
Crack initiation 229
Crack propagation 228
Critical concentration 182
Critical energy release rate 227
Critical stress intensity factor 227
Curing temperature 209
Cyanate ester resins 235
Cyanate esters 239
Cyanurate networks, macroporous 235
Cyclic ester, functional 19
Cyclohexane 206
N-Cyclohexylpyrrolidone 72
Cyclotrimerization 236

Degradation 154
Dendrimers 41
–, arylacetylene 131
–, chiral 131
–, poly(phenylene) 131
–, polyether 120
–, redox active 151
–, starburst 115, 116
–, transition-metal containing 135
Dendritic-linear block copolymers 142
Density 216
DGEBPA 169
Dielectric constant 61, 63, 235
Diffusion, up-hill 172

Diglycidyl ether of bisphenol-A (DGEBPA) 169
Dimethyl acetamide (DMAC) 64
1,5-Dioxepan-2-one 4
Dispersions, colloidal 54
Dispersive forces 178
Distribution, bimodal 189
Divergent growth approach 115, 124
DMAC 64
Drug delivery systems 53
Drying procedure 211, 216
DSM research 118
Dual living polymerization 33

Elastomer, thermoplastic 21, 47
Elastomeric properties 46
Emulsion, concentrated 166
End-functionalization 13
End-reactive polymers 13
Energy release rate, critical 227
Enzymatic synthesis 4
Epoxies, macroporous 218
Epoxy networks, cyclohexane-modified 206
Esters, cyclic 3

3FDA/PMDA 98
Flexus stress analyzer 106
Flory, P.J. 153
Flory-Huggins theory 172
Fluorescence behavior 138
1,6-(4-Fluorophenyl)perfluorohexane 83
Foams, emulsion-derived 166
-, microcellular thermoplastic 165
-, polymeric 164
-, siloxane 166
Focal point 114, 119, 137
Fractionation, selective 12
Fracture toughness 227
Free energy 171
Free enthalpy of mixing 177
Freedom, conformational 187
FTIR spectroscopy 239

Gas nucleation 165
Gibbs equation 171
Glass transition temperature 4, 150, 157
Glycolides 4
Gradient column 239
Gradient oven 183
Graft copolymers 29
Gravimetric analysis 213
Group contributions 177
Growth mechanism 174, 203
Growth rate 204

Hybrid materials 50
Hydrogen bonding 178
Hypercores 122

Image analysis 197
Imide copolymers, adhesion characteristics 82
Imide-aryl ether 77, 81
Imide-aryl ether benzoxazole 65
Imide-aryl ether phenylquinoxaline 65
Imide-dimethyldiphenylsiloxane 105
Imide-perfluoroalkyl ether 83
Imide-perfluoroether 85
Imide-siloxane 104
Imidocarbonate 236
Initiators, difunctional 15
-, macromolecular 33
Insulators, polymeric 62
Interaction parameter 174
Interparticle distance 215
Interpore distance 232
Isobutylene polymers 30

Kharasch addition 153
Kinetic studies 16

L-peel test 82
Lactides 4
Lactone copolymerization 11
Lanthanide alkoxide 8
Lattice model 172
LCST 174
Lewis bases 19
Living polymerization 6
Living-dormant polymerization 17
Load displacement curves 226
Lower critical solution temperature (LCST) 174
Luminescence 134

Macromonomers 35, 142
Macroporous materials 164, 169
MALDI mass spectrometry 126
Maxwell-Garnett theory 85, 86, 241
MCM 62
Melting temperature 4
Membranes, polymeric 167
Mesoporous materials 164
Metal alkoxides 6
Metastable region 174
N-Methyl-2-pyrrolidone 64
-Methylstyrene coblock 98
MGT 85, 86, 241
Micelles 149, 157
Microelectronic packaging 47

Subject Index

Microemulsion polymerization 166
Micromechanisms, toughening 218
Microphase separations 49
Microporous materials 164
Microspheres 54
Morphologies, tailored 182
Morphology, semi-porous 231
Morphology development 181
Multiblock copolymers 34
Multichip modules 62

Nanocarriers 54
Nanofoams 95, 97
Nanoparticles 54
Network swelling rate 54
Networks, amphiphilic 52
NMP 64
NMR spectroscopy 127
Nucleation mechanism 174, 203
Nucleation rate 204

Open porosity 182
Organic-inorganic nanocomposites 50
Organolanthanide complexes 45
Oxidation 231
Oxide, nanoporous inorganic 52

PACP 169
Particles, dispersed 220
PCL, dendri-graft 41
PEEK 70
PEO 22
Perfluoroaryl ether 84
PHA 4
Phase diagrams 173, 186
Phase inversion 167, 182
Phase separation 183
-, reaction induced 181
-, thermally induced 167
Phase separation gap 193
Phenylquinoxaline 65
Ping-opening polymerization 3
Plasticization 229, 233
PMDA/ODA 63, 68
PMDA/PDA polyimide 64
Polar bonding 178
Poly(amic alkyl ester) 61, 71
Poly(amic ethyl ester) 72
Poly(amidoamines) PAMAM 116, 118
Poly(aryl ether) homopolymers 72
Poly(aryl ether ether ketone) (PEEK) 70
Poly(aryl ether ketones) 65
Poly(aryl ether phenylquinoxaline) 68
Poly(aryl ether sulfones) 65
Poly(aryl ethers), heterocycle-containing 68

Poly(chloroethylvinylether) 30
Poly(di)lactose 4
Poly(ester-alt-ether), alternating 12
Poly(ester urethane) 39
Poly(ethylene oxide) 22
Poly(hydroxy alkanoic acids) 4
Poly(3-hydroxybutyrate) 4
Poly(lysine) 116
Poly(methyl methacrylate) 99
Poly(-methylstyrene) 89, 90
Poly(methylene-1,3-cyclopentane) 44
Poly(phenylene) dendrimers 131
Poly(phenylquinoxaline) (PPQ) 67
Poly(propylene oxide) 89, 99
Polycondensation 3, 39
Polyesters, aliphatic 3
-, crosslinked aliphatic 27
-, hyperbranched aliphatic 41
-, star-shaped 16
Polyether dendrimers 120
Polyether dendrons 41
Polyglutamic acid 24
Polyimide, auto-adhesion 67
Polyimide copolymers 71
-, synthesis/characterization 91
Polyimide dielectric constant 83
Polyimide foams 95, 97
Polyimide nanofoams 47, 85-87
Polyimide-perfluoroalkyl ether 83
Polyimide-polyester 41
Polylactone 4
Polymacromonomers 37
Polymer decomposition temperature 76
Polymer-rich phase 172, 239
Polymerization mechanism 5
Polynorbornene 39
Polystyrene 89
Polyurethane foams, flexible 165
Polyurethane networks 39
Pore concentration 200
Pore diameter 197
Pore distance 197, 232
Pores, interconnected 182
Porosity, narrow-sized 189
-, open 182
Porous structure 47
Porphyrins 122
PPO 89, 99
PPQ 67
Propiolactone 4
PS beads, porous 168
Pseudoliving polymerization 17
Pulse propagation 62
Pyralolactone 4
Pyromellitic dianhydride (PMDA) 63, 66

Quench, chemical 181

Radial block copolymers 44
Radical polymerization 32
Random copolymers 10
Reaction, extent of 176
Reaction kinetics 213
Redox active dendrimers 151
Ring opening polymerization 155
ROMP 36
ROP 3
-, anionic 24
-, heterogeneous catalytic coordination 23

Self adhesion 63
Self assembly 140, 166
Self condensing procedures 155
Semiconductor device performance 62
Shape transition 137
Shear bands 220
Sialinic acid 151
Siloxane foams 166
Size distributions 205
Size exclusion chromatography 126
Sol-gel process 50
Solubility limit 206
Solubility parameter 177
Solubilization 149
Solvatochromic probe 137
Solvent-modified systems 170
Solvent-rich phase 239
Solvents 18
Spinodal decomposition 203
Spinodal line 174
Stable region 175
Stannous octoate 6
Starburst dendrimers 115, 116

Stress concentration 223
Stress distribution 223
-, simulated 224
Stress intensity factor, critical 227
Stresses, von Mises 224
Surface, fractured 197
Surface groups 133
Surfactants 24
System cycle time 62

Telechelic macromolecules 13
Temperature quench 167
Thermal decomposition 47
Thermal expansion coefficients 77, 103
Thermal stress 61, 63
Thermally induced phase separation 167, 175
Thermolysis 47
Thermoplastic polymers, porous 175
Thermosets, porous 164
TIPS 167, 175
para-Toluene sulfonic acid hydrate (PTS) 75
Toughness 218
Transesterification reactions 5
Transfer agent 18
Transition-metal containing dendrimers 135
Triblock copolyesters 13
Triblock copolymers, amphiphilic 27

UCST 174
Upper critical solution temperature (UCST) 174

Valerolactone 4
Viscosity, intrinsic 157
Void toughness 222
Yield point 224

Springer and the environment

At Springer we firmly believe that an international science publisher has a special obligation to the environment, and our corporate policies consistently reflect this conviction.

We also expect our business partners – paper mills, printers, packaging manufacturers, etc. – to commit themselves to using materials and production processes that do not harm the environment. The paper in this book is made from low- or no-chlorine pulp and is acid free, in conformance with international standards for paper permanency.

Printing: Saladruck, Berlin
Binding: Buchbinderei Lüderitz & Bauer, Berlin

T
green